European Forest Policy and Governance

This book provides a state-of-the-art overview covering distinct and relevant aspects of forest policy processes in Europe, presenting a fresh perspective on different analytical approaches, theories, and frameworks.

Set against the background of a changing world, driven by significant social, environmental, and economic developments, in Europe and elsewhere, there is a growing need for an improved understanding of forest governance and how to analyse the forest policymaking processes. This book introduces the reader to some of the key issues typically encountered in reviewing proposed as well as established forest policies, focusing on five socially relevant topics for the forest-based sector today, namely:

- European forest governance under a green new deal
- Systemic changes and the circular (bio-)economy
- Social changes connected with forest ownership and forest actors
- Nature conservation and the pursuit of multifunctional forests
- Living with forest fires and climate change.

In so doing, this book presents a set of timely and rich case studies relevant to the study of forest governance. In the final chapter, it puts forward an innovative and systematic method for selecting the most appropriate analytical tool that accounts for the constraints and objectives involved in monitoring forest policy. The book is accompanied by chapter-level exercises and online Support Material which details the various approaches, theories, and frameworks discussed in the book, providing direct links back to individual chapters, discussion points, and a step-by-step guide for how each method can be applied.

This book will be an essential read for forestry students and scholars, and professionals and policymakers working on forest policy and forest management.

Filip Aggestam is a Guest Researcher at the European Forest Institute's Forest Policy Research Network (EFI-FPRN) and works as a Scientific and Technical Officer for the Convention on Wetlands in Gland, Switzerland. Previously, he has worked as a Technical Expert for the United Nations Environment Programme (UNEP) and as a Senior Researcher for the Institute of Forest, Environmental and Natural Resource Policy at the University of Natural Resources and Life Sciences (BOKU) in Vienna, Austria.

European Forest Policy and Governance

An Integrative Analytical Framework

Filip Aggestam

Routledge
Taylor & Francis Group
LONDON AND NEW YORK

earthscan
from Routledge

Designed cover image: Getty images /Reikkinen

First published 2025
by Routledge
4 Park Square, Milton Park, Abingdon, Oxon OX14 4RN

and by Routledge
605 Third Avenue, New York, NY 10158

Routledge is an imprint of the Taylor & Francis Group, an informa business

British Library Cataloguing-in-Publication Data
A catalogue record for this book is available from the British Library

Library of Congress Cataloging-in-Publication Data
Names: Aggestam, Filip, author.
Title: European forest policy and governance : an integrative analytical framework / Filip Aggestam
Description: New York, NY : Routledge, 2025 | Includes bibliographical references and index
Identifiers: LCCN 2024020802 (print) | LCCN 2024020803 (ebook) | ISBN 9781032057477 (hardback) | ISBN 9781032057460 (paperback) | ISBN 9781003198987 (ebook)
Subjects: LCSH: Forest policy—Europe. | Forest management—Europe. | Forests and forestry—Europe.
Classification: LCC SD599 ,A44 2024 (print) | LCC SD599 (ebook) | DDC 333.75094—dc23/eng/20240715
LC record available at https://lccn.loc.gov/2024020802
LC ebook record available at https://lccn.loc.gov/2024020803

ISBN: 9781032057477 (hbk)
ISBN: 9781032057460 (pbk)
ISBN: 9781003198987 (ebk)

DOI: 10.4324/9781003198987

Typeset in Times New Roman
by Apex CoVantage, LLC

Access the Support Material: www.routledge.com/9781032057460

To Frida, Felix, and Lorena

Contents

Lists of boxes, tables, and figures

Boxes

Tables

Figures

Additional boxes, tables, and figures are included in the online Support Material at www.routledge.com/9781032057460

Abbreviations

9R	Nine R's (refuse, rethink, reduce, reuse, repair, refurbish, remanufacture, repurpose, recycle and recover)
ACER	Agency for the Cooperation of Energy Regulators
ACF	Advocacy Coalition Framework
CAP	Common Agricultural Policy
CBD	Convention on Biological Diversity
CEI-BOIS	European Confederation of Woodworking Industries
CEPF	Confederation of European Forest Owners
CEPI	Confederation of European Paper Industries
CGRI	Center for Governmental Research and Innovation
CICES	Common International Classification of Ecosystem Services
CITES	Convention on International Trade in Endangered Species of Wild Fauna and Flora
CO_2	Carbon dioxide
COFFI	Committee on Forests and the Forest Industry
COGECA	General Confederation of Agricultural Cooperatives
COPA	Committee of Professional Agricultural Organisations
COST	European Cooperation in Science and Technology
CPA	Critical policy analysis
CSD	Commission on Sustainable Development
DG	Directorate-General
EAFRD	European Agricultural Fund for Rural Development
EAGF	European Agricultural Guarantee Fund
EC	European Commission
ECE	United Nations Economic Commission for Europe
EEA	European Environment Agency
EFC	European Forestry Commission
EFFIS	European Forest Fire Information System
EFI	European Forest Institute
ELO	European Landowners' Organization
EMF	Ellen MacArthur Foundation
EOS	European Organisation of the Sawmill Industry
EP	European Parliament
ERCC	Emergency Response Coordination Centre
EU	European Union
EUROSTAT	Statistical Office of the European Union

EUSTAFOR	European State Forest Association
EUTR	EU Timber Regulation
FAO	United Nations Food and Agriculture Organization
FECOF	Federation of European Communal Forest Owners
FES	Forest ecosystem services
FLEGT	Forest Law Enforcement, Governance and Trade
FSC	Forest Stewardship Council
GDP	Gross domestic product
GHG	Greenhouse gas
GPS	Global Positioning System
ha	Hectares
HWP	Harvested wood products
IAF	International Arrangement on Forests
ICPDR	International Commission for the Protection of the Danube River
IEEP	Institute for European Environmental Policy
IFF	Intergovernmental Forum on Forests
IIASA	International Institute for Applied Systems Analysis
ILUC	Indirect land use change
IMF	International Monetary Fund
INFORMAR	Integrated Forest Management Learning Architecture
IPBES	Intergovernmental Science-Policy Platform on Biodiversity and Ecosystem Services
IPF	Intergovernmental Panel on Forests
ITTA	International Tropical Timber Agreement
IUCN	International Union for Conservation of Nature
IUFRO	International Union of Forest Research Organisations
JRC	Joint Research Centre
LBA	Legally binding agreement
LIFE	Financial Instrument for the Environment
LULUCF	Land Use, Land Use Change, and Forestry
MAES	Mapping and Assessment of Ecosystems and their Services
MCPFE	Ministerial Conference on the Protection of Forests in Europe
NACE	Nomenclature of Economic Activities
NFI	National Forest Inventory
NWFP	Non-wood forest products
OECD	Organisation for Economic Co-operation and Development
PEBLDS	Pan-European Biological and Landscape Diversity Strategy
PEFC	Program for the Endorsement of Forest Certification
PES	Payment for ecosystem services
PESTEL	Political, Economic, Social, Technological, Environmental, and Legal (analysis)
REACH	Registration, Evaluation, Authorisation and Restriction of Chemicals
REDD+	Reducing Emissions from Deforestation and Forest Degradation
SDG	Sustainable Development Goals
SEEA	System of Environmental-Economic Accounting
SFC	Standing Forestry Committee
SFM	Sustainable Forest Management
SLU	Swedish University of Agricultural Sciences
STEEP	Social, Technological, Economical, Environmental, and Political (analysis)

SWD	Staff working document
SWOT	Strengths, weaknesses, opportunities, threats (analysis)
TEEB	The Economics of Ecosystems and Biodiversity
TFEU	Treaty on the Functioning of the European Union
UEF	Union of European Foresters
UN	United Nations
UNCCD	United Nations Convention to Combat Desertification
UNCED	United Nations Conference on Environment and Development
UNDP	United Nations Development Programme
UNEP	United Nations Environment Programme
UNFCCC	United Nations Framework Convention on Climate Change
UNFF	United Nations Forum on Forests
UNSPF	UN Strategic Plan for Forests
USSE	Union of Forestry Societies in Southern Europe
VCA	Value chain analysis
WB	World Bank
WPF	Working Party on Forestry
WWF	World Wildlife Fund

Foreword

Helga Pülzl

This book on European Forest Governance arrives at a critical moment when the world faces urgent climate change, biodiversity loss, and sustainability challenges. Forests are a crucial part of the European land area, covering 35 per cent (Forest Europe, 2020). They are important in carbon storage, forming climate sinks, and acting as biodiversity and water reservoirs. Forests also provide clean air and renewable materials, making management and governance essential. Forests offer recreational spaces for citizens, keep cities cooler, and enhance our well-being considerably. Forests also face increasing threats due to disturbances that have risen significantly in the past 50 years. According to Patacca et al. (2023), these disturbances include wind, forest fires, bark beetles, pests, diseases, and other disturbances caused by snow and ice. All of these increase the risk of maintaining forests' ability to store carbon and provide sinks (Verkerk et al. 2022), maintain and enhance forest biodiversity (Muys et al. 2022), and supply ecosystem service in Europe (Lecina-Diaz et al. 2024).

Who decides what and how? Which interests prevail, and how do discourses, values, beliefs, and power play out in forest governance? These are only some of the questions that become relevant to answer when wanting to get a grip on how forest governance works in Europe. For example, some scholars still struggle with the question of whether the European Union has the competencies to regulate forests. Even though forests are not referred to in the EU's treaties and the Union has no exclusive or shared competence to legislate, many legal and non-legal forest-related policy instruments have been passed. They are linked to policy domains of environment, climate, agriculture, trade, civil protection and forest genetics. Only in the past four years have 28 new legislative acts and about 20 non-legislative instruments directly relevant to forests or the wood-processing industries entered into force, while 20 legal proposals still need to be debated. Forests remain high on the political agenda, and the complexity of studying how agendas are set, who becomes involved or not involved, and how decisions are taken increases as demands increase.

Implementing those decisions in Europe becomes the second part of the puzzle, and here, forest owners play a crucial role. On the one hand, they risk losing their investments due to forest disturbances. On the other hand, they must react to climate change and biodiversity loss by adapting climate and biodiversity smart management forms for their forests. Who owns the forests in Europe is not easy to answer as not only are forests unequally distributed across countries, but also private and public ownership, as well as the size of ownership, varies a lot.

A comprehensive book that tackles those questions on European forest governance was missing up to this point. Who could be better suited than the brilliant Filip Aggestam to write such a book that provides deep insights into how forests are governed in Europe and delve into which policy theories help to analyse what we see to make sense of forest governance? This book provides a timely resource in our collective quest to understand more about European decision-making on forests. This resource will not only be useful for researchers and students as a teaching aid, but it may also be relevant to EU Member State representatives, stakeholders, and NGOs that wish to learn more about European forest governance.

Helga Pülzl is Assistant Director for Policy Support at the European Forest Institute (Finland) and Senior *Scientist at the University of Natural Resources and Life Sciences, Vienna, Department of Economics and Social Sciences, Institute of Forest, Environmental and Natural Resource Policy, Vienna (Austria).*

References

Forest Europe (2020) *State of Europe's Forests 2020*. Bratislava: Ministerial Conference on the Protection of Forests in Europe.

Lecina-Diaz, J., et al. Ecosystem services at risk from disturbance in Europe's forests. *Glob. Chang. Biol.* 2024; 30e17242. https://doi.org/10.1111/gcb.17242.

Muys, B., Angelstam, P., Bauhus, J., Bouriaud, L., Jactel, H., Kraigher, H., Müller, J., Pettorelli, N., Pötzelsberger, E., Primmer, E., Svoboda, M., Thorsen, B.J., Van Meerbeek, K. 2022. *Forest Biodiversity in Europe*. From Science to Policy 13. European Forest Institute. https://doi.org/10.36333/fs13.

Patacca, M., et al. Significant increase in natural disturbance impacts on European forests since 1950. *Glob. Chang. Biol.* 2022; 29: 1359–1376. https://doi.org/10.1111/gcb.16531.

Verkerk, P.J., Delacote, P., Hurmekoski, E., Kunttu, J., Matthews, R., Mäkipää, R., Mosley, F., Perugini, L., Reyer, C. P. O., Roe, S., Trømborg, E. 2022. *Forest-Based Climate Change Mitigation and Adaptation in Europe*. From Science to Policy 14. European Forest Institute. https://doi.org/10.36333/fs14.

1 Introduction

1. Overview

Let us start with a fundamental question. What exactly does 'forest policy' mean, not only within the context of this book but also from a broader perspective, whether we are interested in forests at the national, regional, or international level. Naturally, you may think that forest policy is only about forests (or trees) or perhaps wood-based products; however, it can be challenging to delimit what we mean when referring to forest policy. This requires context. For instance, from a forestry perspective, forest policy could consist of the laws and regulations that guide the protection and management of forests, including the rules that forest owners must follow when managing and using their forests. In comparison, from a forest-based industry[1] perspective, forest policy could encompass laws and regulations related to energy efficiency, sustainable finance, bioplastics, air quality and trade. In addition, if we think about forest policy from a vertical perspective (local to international), each level can have its own legal and institutional competencies on forests (Box 1.1). Forest policy in Europe is, for this reason, very diverse. For example, for the European Union (EU), forests per se remain outside the realm of the EU's (exclusive and shared) competencies and the treaty on the functioning of the EU. This means the EU can only implement non-legally binding (or soft) policy instruments relating to forests. The same commonly applies to global or pan-European voluntary agreements on forests, such as the United Nations Non-legally Binding Instrument on All Types of Forests (or the 'Forest Instrument') or the more recent 'Declaration on Forests and Land Use' from the 2021 United Nations Climate Change Conference (COP26).[2]

Box 1.1 Forest policy and governance in Sweden, the United Kingdom, Greece, and Romania

Sweden

Sweden is a highly forested country, where around 69 per cent of the country is forest land. It has one of the highest rates of forest land per capita globally. It furthermore has a large and export-oriented forest-based sector that contributes significantly to the Swedish economy. In Sweden, forest policy and actions, such as financial support, advice, information, and education, are implemented by Skogstyrelsen (Swedish Forest Agency). The agency is the leading authority for forestry and associated environmental and conservation issues. The Swedish parliament passed the first forestry act (Skogsvårdslag) in 1903, which, among other things, contained

DOI: 10.4324/9781003198987-1

provisions that required landowners to replant forests after felling (SLU, 2016). The current forest regime has, however, evolved significantly. The most recent revision of the forestry act (SFS 1993:553) occurred in 1993. This revision represented a significant policy shift whereby the Swedish parliament relaxed national forestry regulations and established environmental quality objectives to maintain high wood production (Ekelund and Hamilton, 2001). It can be noted that the timber market in Sweden is less regulated than in most other countries. It is also interesting to note that any forest-related measure or support provided to forest owners in Sweden cannot affect the enterprise's income. This has, for example, implications for implementing forestry measures under the EU's Common Agricultural Policy (CAP).[3]

United Kingdom

The United Kingdom (UK) remains one of the least forested countries in Europe. Approximately 13 per cent of the UK is covered in woodland compared to an average forest cover in Europe of around 40 per cent. There is such a high variability across countries in Europe. For example, Sweden, Finland, Spain, France, Germany, and Poland account for two-thirds of the EU's total forest area. The same variation in forest cover is, to some extent, also the case within the UK, where Scotland is covered by 18 per cent woodland. Historically, forest policy in the UK was introduced during the First World War when woodland stock declined significantly, and emergency felling controls were introduced. This resulted in, among other things, the formation of the Forestry Commission and the Forestry Act in 1919. Today, forest policy and action in the UK is largely devolved to the respective country administrations in England, Scotland, Wales, and Northern Ireland. However, some areas, such as international forestry policy and plant health, are the responsibility of the UK Government (Forestry Commission, 2017). For this reason, it is the responsibility of the government administration in each of the UK countries to set forest policy. Each country has its institutional setup responsible for implementing, regulating, and monitoring forest policy. For instance, in England, the Department for Environment, Food and Rural Affairs (DEFRA) is responsible for setting policies and providing incentives. In Scotland, it is the Scottish Government (Humphreys, 2004).

Greece

Most of the forest land in Greece is publicly owned (around 77 per cent), and forest land represents approximately 26,2 per cent of the country's total area. In contrast to Sweden and the UK, most Greek forests are naturally regenerated, and the role of forests is largely environmental (e.g., non-productive forests). For example, while forest management in Greece primarily focuses on timber production, historically, it is based on the trade-offs between timber and other ecosystem functions (e.g., soil protection and watershed management) (Humphreys, 2004). Moreover, as part of the Mediterranean basin, Greece is among the most vulnerable countries due to increasing temperatures and reduced precipitation (e.g., due to climate change). The country's Forest Code (Law 86/1969), which constitutes the main body of forest-relevant legislation, regulates the protection and management of public and private forests, property rights, and taxation. (Spanos et al., 2018). The main body for managing public and private forests is the Greek Forest Service, through its Forest District Offices. The Forest Service is, in turn, an integral part of the Ministry of Agriculture (now under the Ministry of Environment and Climatic Change), whereas the Ministry of Environment, Planning and Public Works oversees forest policy for protected areas. Decisions are generally made centrally, and management practices tend to include strict protection.

Romania

In 2020, Romania's forests covered approximately 6.6 million ha, constituting 28 per cent of the country's total land area. After the fall of communism, land reforms and restitution processes beginning in 1991 resulted in the state retaining ownership of about 64.7 per cent of these forests, while the remaining 35.3 per cent transitioned to private ownership (Giurcă and Dima, 2022). The Romanian Ministry of Environment, Waters and Forests oversees forest governance, with operational management delegated to entities like the National Forest Administration (Romsilva). Romsilva is responsible for managing around 3 million hectares of this forest land (EUSTAFOR).[4] Private owners and local municipalities manage their forest lands, often through private forest districts. Regardless of the forest's size or ownership type, forest management practices must comply with national standards and technical guidelines. This includes developing forest management plans, which are valid for ten years and provide detailed management guidelines for each forest stand. Specialised forest management planning companies develop these plans and require approval from the national forest authority. Once approved, implementing these management plans is compulsory, ensuring a standardised approach to forest management across Romania.

Box 1.1 provides examples of how forest policy and governance vary across European countries. They illustrate that what we consider as forest policy generally depends on the natural and historical context. For example, the current forest ownership structure in Romania is largely due to the restitution process, which gave people the right to claim back property that had been nationalised by the regime (Paul Dragos and Adina, 2003). In contrast, Greece underwent significant land reforms after World War II and the civil war. These reforms reinforced state ownership of forests, partly as a measure to prevent large-scale privatisation and the fragmentation of forest lands (Humphreys, 2004). Considering this background, this book applies a rather broad definition of forest policy. More specifically, it is seen as a much wider policy domain than what is commonly applied nationally (where competencies and roles are often strictly applied). It refers to any policy instrument that could be seen as affecting forests (directly or indirectly) as a natural resource. This could, for example, concern **forest-focused policy** that addresses forests directly (e.g., affecting forest management, biodiversity conservation, or renewable energy production) or **forest-related policy** that addresses the forest-based sector more indirectly (e.g., regulations covering the use of chemical substances or legal frameworks for treating and managing waste). Consequently, forest policy and governance comprise a wide spectrum of instruments, actors, and interests interacting and utilising the same natural resource. This is also why international forest policy is so interesting. It is complex and varied, covering multiple policy domains. Yet, it is fundamentally about trees and all that we can use trees towards, whether to build houses, produce energy, make paper, plastics, and textiles, recreation, or simply for their natural beauty and biodiversity.

2. What is forest policy and governance?

The preceding section has described a basic definition of forest policy and governance; however, this definition also implies that forest policy could mean many things. For example, when you read about forest policy in a European context, you will commonly come across different terms, such as 'forest-focused', 'forest-related' or 'forest-relevant' policy. These terms are not consistently applied and can mean different things. Therefore, it is important to further elaborate on what forest policy and governance mean in this book. This is mainly so that there is a shared understanding of what the respective terms mean. It should also be noted that this is not meant to be a detailed description; it is primarily meant to provide a common vernacular between the topics that will be addressed.

When talking about **forest-focused policy**, this principally means the legal or regulatory frameworks,[5] including subsidies and financial instruments,[6] that refer to forests directly, whether binding or voluntary. Legally binding examples would be the Forestry Act in the UK or the EU Timber Regulation (EUTR),[7] while examples of soft (or non-legally binding) instruments would include the EU Forest Strategy[8] or the Declaration on Forests and Land Use.[9] These instruments explicitly address forests regardless of their legal status or classification (e.g., no direct mandate for addressing forests). **Forest-related policy** is applied differently and depends on the context and what part of the forest-based sector you are investigating. For example, from an industry perspective, any regulation that affects industrial operations would be of direct relevance, such as the EU's Registration, Evaluation, Authorisation and Restriction of Chemicals (REACH Regulation),[10] which provides a comprehensive framework for chemical manufacture and use in Europe or the EU's air quality directives. The critical distinction here is that these regulations do not refer to forests directly in the legal text, even though they directly impact the sector. The label 'forest-related' is commonly applied to these policy instruments, meaning legal or regulatory frameworks, including soft instruments, that affect the forest-based sector but do not explicitly focus on forests.

Forest governance is, within this context, even broader, referring to the norms, processes, instruments, people, institutions, and organisations that interact with forests or wood-based products, including how they make decisions about the management, use, and conservation of forest resources at multiple scales, from local to global. It can cover many things, such as public and private actors, including formal and informal institutions, small and medium-sized enterprises (SMEs) and large enterprises, civil society organisations, and traditional culture. It also encompasses forest-focused and forest-related policies that determine how forests should be governed, including relevant national and international regulatory frameworks. This means that forest governance systems include legal and regulatory frameworks and the structures and interactions of public and private actors, including government, that influence forest management and use.

> Chapter 2 includes additional information on the terms and concepts utilised throughout this book.

3. History of forest policy developments in the EU and internationally

The historical evolution of European policy and legal frameworks on forests can be traced back several hundred years to when national and industry-related interests characterised forest governance, focusing mainly on preventing deforestation and ensuring sustainable timber supply. These early rules and regulations were commonly developed to ensure timber supply to Europe's mining and shipping industries as a crucial raw material. For example, maintaining a steady wood supply for shipbuilding for coastal countries, such as Britain, France, and Sweden, was critical. From a forest management perspective, the scientific approach to forestry arguably began with Hans Carl von Carlowitz, who published his book *Silvicultura oeconomica* in 1713 (Carlowitz, 2013). This work advocated for the conservation, growing, and sustainable use (or continuous supply) of timber and is widely credited for introducing the concept of 'sustained yield' in forestry management.[11] von Carlowitz's contributions can be used as a conceptual starting point for developing sustainable forest management (SFM) as a concept which has spread from Central Europe to the rest of the world. Today, several EU and international instruments, agreements, and processes shape forest management and use, considering not only timber production but a wide range of ecosystem services that serve both socio-economic and environmental functions (Lazdinis et al., 2009, Edwards and Kleinschmit, 2013, Aggestam and Pülzl, 2018, Pülzl et al., 2018). The following sections will briefly outline the historical developments of EU and international forest policy.

3.1. International forest policy

International efforts to address forest use started in the 1980s, primarily linked to a growing global concern over tropical forest deforestation due to logging, mining, development, and the conversion of forests to agricultural land. For example, although deforestation has slowed down substantially since 1990 (FAO, 2020), it is estimated that 73 per cent of tropical deforestation is driven by agriculture (Hosonuma et al., 2012, Ritchie, 2021).[12] At the time, attention was also focused on the loss of boreal and temperate forests due to poor forest management and air pollution (e.g., forest dieback in Central Europe due to acid rain). However, despite this window for policy change, attempts to create a comprehensive and legally binding forests-focused international policy framework (or forest regime)[13] have, to date, not been successful (Arts and Buizer, 2009, Glück et al., 2010, Arts et al., 2013). There is, as such, no global convention on forests. Instead, what has emerged from this period are several international governance mechanisms, including legally and non-legally binding policy instruments, that either push for SFM, address timber trade, or try to facilitate forest protection (McDermott et al., 2007, Rayner et al., 2010, Edwards and Kleinschmit, 2013, Pülzl and Mayer, 2015, Cashore et al., 2016, Kleinschmit et al., 2018).

3.1.1. The Rio Conventions

It can be noted that the 1992 Earth Summit was a significant turning point in international environmental policy as it signified the launch of three global conventions (the Rio Conventions) at the United Nations Conference on Environment and Development (UNCED). While no agreement was reached on forests, the United Nations Convention on Biological Diversity (CBD), the United Nations Framework Convention on Climate Change (UNFCCC), and the United Nations Convention to Combat Desertification in Those Countries Experiencing Serious Drought and/or Desertification, Particularly in Africa (UNCCD) came out of the Earth Summit (Box 1.2). Moreover, the Conferences of the Parties (COP) to these conventions have, over time, increasingly addressed forest-relevant issues within the context of their respective mandates.

Box 1.2 The Rio Conventions

Convention on Biological Diversity

The Convention on Biological Diversity (CBD)[14] opened for signatures at the Earth Summit and entered into force in 1993. It is a legally binding treaty under the UN, and it is the only global framework addressing forest biological diversity and genetic resources, conservation, and sustainable use. The relevance of the CBD cannot be overestimated. The most important for forests was the 2020 Aichi Biodiversity Strategic Goals and Targets[15] that are included in the Strategic Plan for Biodiversity 2011–2020. These targets and goals have recently been replaced by the Kunming-Montreal Global Biodiversity Framework,[16] which contains 4 goals and 23 targets to halt and reverse biodiversity loss by 2030. National biodiversity strategies, plans, or programs and the expanded work program on forest biological diversity are the main policy tools for global forest biodiversity conservation.

United Nations Framework Convention on Climate Change

The United Nations Framework Convention on Climate Change (UNFCCC),[17] which also opened for signature at the Earth Summit and entered into force in 1994. It is the first global environmental treaty addressing climate change and setting targets for reducing greenhouse gas (GHG) emissions. Relating to forests, the Kyoto Protocol and Paris Agreement set out a legally binding obligation for accounting, monitoring, and reporting of GHG emissions/sinks from the Land Use, Land Use Change, and Forestry (LULUCF) sector.[18]

United Nations Convention to Combat Desertification in Those Countries Experiencing Serious Drought and/or Desertification, Particularly in Africa

The United Nations Convention to Combat Desertification in Those Countries Experiencing Serious Drought and/or Desertification, Particularly in Africa (UNCCD)[19] stems from a recommendation in Agenda 21,[20] which also came out of the Earth Summit. The Convention was adopted in Paris in 1994 and entered into force in 1996. It is the only legally binding framework that addresses the problem of desertification, linking the environment and development to sustainable land management. It sets out a commitment to achieve Land Degradation Neutrality (LDN) to restore the productivity of degraded land and reduce the impacts of drought.

3.1.2. United Nations Forum on Forests

After the Earth Summit and the breakdown in negotiations on a global forest convention, the Intergovernmental Panel on Forests (IPF) was established under the United Nations Commission on Sustainable Development (CSD) in 1995. The IPF was later replaced by the Intergovernmental Forum on Forests (IFF) in 1997 and, more recently, by the United Nations Forum on Forests (UNFF)[21] in 2000. The UNFF represents the main intergovernmental fora[22] for international forest policy at present, governments negotiating under the UNFF still need to find consensus around a legally binding convention on forests. Nevertheless, the UN Strategic Plan for Forests (UNSPF) came out of a UNFF session in 2017, covering the 2020–2030 period. The plan features a set of six Global Forest Goals and 26 targets to be reached by 2030 (such as increasing forest area by 3 per cent worldwide), which are voluntary.

As part of the International Arrangement on Forests (IAF),[23] which also came out of the Earth Summit, countries have adopted two soft policy documents that concern forests. These are the '*Non-Legally Binding Authoritative Statement of Principles for a Global Consensus on the Management, Conservation and Sustainable Development of All Types of Forests*' (or the Forest Principles) and Chapter 11 '*Combating Deforestation*' in Agenda 21. In addition, under the UNFF, countries have also adopted the '*UN Non-Legally Binding Instrument on All Types of Forests*' – commonly known as the Forest Instrument – which is the first time countries have agreed on a common definition for SFM.[24] Moreover, following the Climate Summit held in New York in 2014, the voluntary and non-legally binding New York Declaration on Forests (hereinafter referred to as the New York Declaration) was released as a pledge to halve the rate of deforestation by 2020 and to end it by 2030.[25]

3.1.3. Declaration on Forests and Land Use

More recently, countries have signed a Declaration on Forests and Land Use,[26] (hereinafter referred to as the Glasgow Declaration) under the umbrella of the UNFCCC and the COP26 in Glasgow in 2021. Like the New York Declaration, it aims to end and reverse deforestation by 2030. It aims towards a sustainable land-use transition. However, in contrast to other occasions, substantial progress has also been made concerning the mobilisation of finance for forests. As part of issuing a Global Forest Finance Pledge,[27] several countries committed to providing financial support to restore degraded land, tackle wildfires, and advance the rights of indigenous peoples and local communities. Major financial companies have likewise made additional commitments to end investments in deforestation-related activities. While these political and private sector commitments are encouraging, it should be noted that earlier efforts have yet to be successful. For example, the assessment of the New York Declaration, which set out similarly ambitious targets, argued that it has failed to deliver on crucial pledges (UNDP, 2019). The assessment also argues that forest finance is only a tiny part of what is needed to restore and save forests.

3.1.4. Other international instruments and frameworks

Several other international instruments, processes, commitments, and goals can be seen as forest-relevant; examples include the 2030 Agenda for Sustainable Development, the Forest Law Enforcement, Governance and Trade (FLEGT), Reducing emissions from deforestation and forest degradation (REDD+), the Bonn Challenge, and supply chain certification (Box 1.3).

Box 1.3 Forest-relevant policy instruments

2030 Agenda for Sustainable Development

Agenda 2030, adopted in 2015, consists of a Declaration, 17 Sustainable Development Goals (SDGs) and 169 associated targets.[28] The SDGs were developed in the Post-2015 Development Agenda as the future global development framework succeeding the Millennium Development

Goals. The SDGs are inherently integrative, focusing on multiple cross-cutting issues like climate, energy, gender, and education. For example, SDG 15 'Life on land', states the need to 'protect, restore and promote sustainable use of terrestrial ecosystems, sustainably manage forests, combat desertification, and halt and reverse land degradation and halt biodiversity loss'. This goal recognises the synergistic interactions between forest, water and biodiversity and calls for integrative and collective action.

Forest Law Enforcement, Governance and Trade

The EU and its member states adopted the EU's Forest Law Enforcement, Governance and Trade (FLEGT) Action Plan in 2003, supported by the EU's FLEGT and Timber Regulation (EC, 2003, Council Regulation, 2173/2005, Regulation, 995/2010). While this is an EU initiative, the FLEGT Regulation has international implications. It empowers the European Commission to negotiate bilateral trade deals (voluntary partnership agreements (VPAs)) with timber-exporting countries. Under a VPA, partner countries agree to export only legal timber products to the EU. The EU FLEGT Facility[29] supports the implementation of the EU FLEGT Action Plan, focusing on VPAs. The FLEGT aims to control trade in illegally logged wood and wood products.

Reducing Emissions from Deforestation and Forest Degradation

Reducing Emissions from Deforestation and Forest Degradation (REDD+) is a mechanism developed by Parties to the UNFCCC as part of Article 5 of the Paris Agreement.[30] It creates a financial value for the carbon stored in forests by incentivising developing countries to reduce emissions from forested lands and invest in low-carbon paths to sustainable development. This can take the form of direct payments or in exchange for carbon credits, which represent reductions in GHG emissions to compensate for emissions made elsewhere. The UN-REDD Programme[31] supports nationally led REDD+ processes while, at the EU level, there is also the EU REDD Facility,[32] which supports developing countries to improve land-use governance.

Bonn Challenge

The Bonn Challenge[33] is a transnational collaborative partnership launched by the German Federal Ministry of Environment Protection and the International Union for the Conservation of Nature (IUCN) in 2011. It is based on an informal declaration of intent that set out to restore 160 million hectares (ha) of the world's degraded landscapes and forestlands by 2020[34] and 350 million ha by 2030. The Bonn Challenge is a practical approach for realising international commitments, such as Aichi Target 15, the UNFCCC REDD+ goal, the Rio+ 20 land degradation neutrality goal, the New York Declaration on Forests and goal 15 of the SDGs.

Forest supply chain sustainability certification

There have likewise been initiatives driven mainly by the private sector. One prominent example concerns forest certification. Third-party certification of SFM practices is provided by organisations such as the Forest Stewardship Council (FSC)[35] and the Program for the

Endorsement of Forest Certification (PEFC).[36] FSC is an international non-profit organisation established in 1993 that promotes responsible management of forests. It is an example of a market-based certification program used as a global environmental policy. In contrast, PEFC has also emerged as an industry-led SFM certification scheme run by state forest enterprises, domestic forest industries, and multinational companies. The majority of the world's certified forest area is located in North America and Europe's boreal and temperate forests.

There are also legally binding agreements that relate directly to the forest value chain and the international trade of forest products, such as the Convention on International Trade in Endangered Species of Wild Fauna and Flora (CITES) from 1973,[37] the International Tropical Timber Agreement (ITTA) from 1983,[38] and the World Trade Agreement (WTA) from 1994. Other examples include the Convention on Wetlands from 1971, the Århus Convention on Access to Justice in environmental matters from 1998 and the UN Declaration on the Rights of Indigenous Peoples from 2007.

More governing instruments could be listed; however, this introduction showcases that the number and complexity of forest-relevant international instruments have evolved. In the absence of a legally binding forest convention or agreement, these specialised or sectoral instruments (e.g., dealing with trade or conservation) are linked together, sometimes reinforcing, but at times conflicting or even overlapping, with no or limited coordination (McDermott et al., 2007, Wolfslehner et al., 2020). While the current complexity of forest problems may rule out simple governance solutions, all play a role in forest-relevant decision-making. More importantly, this arrangement makes up the international forest regime, allowing for policy dialogue and cooperation on forests on the global stage (Arts et al., 2010, Rayner et al., 2010).

3.1.5. Pan-European forest governance

Even though no legally binding instruments focus exclusively on European forests, forest-relevant initiatives have also developed on the pan-European level. More specifically, the Ministerial Conference on the Protection of Forests in Europe (MCPFE), nowadays called Forest Europe,[39] has emerged as the leading pan-European political process for forest coordination and participatory decision-making. Forest Europe is a voluntary dialogue process where 46 European countries, the European Commission, several forest sector stakeholders, and international organisations now participate. The Forest Europe process and its periodic Ministerial Conferences have generated several voluntary political commitments,[40] including a definition of what SFM means in the pan-European context (Pülzl and Mayer, 2015, Linser et al., 2018, Lier et al., 2021) and the Pan-European Criteria and Indicators for Sustainable Forest Management, which form the basis for the State of Europe's Forests Report (Forest Europe, 2015, 2020). Moreover, like the Earth Summit, the Forest Europe process has launched attempts to establish a legally binding agreement (LBA) on forests. However, no agreement has been reached as there was a breakdown in negotiations in 2007 and, more recently, during the eighth Ministerial Conference on the Protection of Forests in Europe in 2021.

Another relevant actor on the pan-European stage is the joint United Nations Economic Commission for Europe (ECE) Committee on Forests and the Forest Industry (COFFI) and the FAO European Forestry Commission (EFC).[41] These institutions work together to improve the forest-based sector and contribute to sustainable development in Europe, Central Asia, the

Caucasus, and North America.[42] COFFI and EFC cover many traditional topics related to forest products and services, such as forest resources and forest products markets. They also address issues related to climate change, SFM and the linkages between SFM and the green economy, such as innovation in the forest sector and ecosystem services. For example, they have launched the voluntary Rovaniemi Action Plan for the Forest Sector in a Green Economy (UNECE/FAO, 2013), which has set out to encourage the forest-based sector to become greener.

It can further be noted that the pan-European response to the implementation of the CBD has been the Pan-European Biological and Landscape Diversity Strategy (PEBLDS).[43] Other forest-relevant agreements include the Bern Convention on the Conservation of European Wildlife and Natural Habitats, which includes a list of forest species to be protected, and the European Landscape Convention, which relates in parts to forests. Moreover, while only some regions have developed legally binding agreements that directly affect forests, two sub-regional agreements, the Carpathian Convention and the Alpine Convention, have protocols on forests that are binding to the countries in the respective mountain regions.

The pan-European instruments, both legally and non-legally binding, ministerial conferences, and criteria and indicators (C&I) for SFM demonstrate that the regional arrangements addressing forests vary considerably in the level and type of collaboration and mechanisms employed. They also differ from global forest-relevant governance and make up a distinct part of a multi-level policy system on forests.

3.2. EU forest policy

From a global perspective, while forests cover more than 40 per cent of the total EU land area, the EU accounts for only 5 per cent of the global forest cover. The EU had approximately 180 million hectares (ha) of forest land[44] and other wooded land[45] in 2020; however, in contrast to the global trend, the total forest area has increased by around 10 per cent through natural growth and afforestation since 1990 (Eurostat, 2020). European forests are essential for employment and economic development, providing many wood-based products. For example, forest-based industries in the EU generated a gross value added of 138.6 billion EUR in 2018 (Eurostat, 2020). Furthermore, besides providing a wide range of wood-based products, European forests represent many ecosystems (in terms of species, habitats and ecological processes) that constitute a large part of Europe's biodiversity.

To understand EU forest policy, it is important to realise that no one agency is responsible for forests at the EU level; even on the national level, forest competencies vary significantly between different ministries and agencies responsible for forests. It has been argued that one significant challenge for EU forest governance is rooted in the absence of an instrument or institution that can guide forest-relevant policymaking or, at least, provide a shared voice for the forest-based sector (Wydra, 2013, Kleinschmit et al., 2018, Aggestam and Pülzl, 2020, Aggestam and Giurca, 2021). This argument comes partly from the absence of a common EU forest policy, as forest products (except for cork and some forest-related fruits) have been excluded from existing EU treaties.[46] The formulation and implementation of policies affecting the forest value chain are subject to the principle of subsidiarity and proportionality.[47] and is thus under the competence of EU Member States. However, there is still a long history of EU action on forest-relevant measures (Winkel et al., 2013, Winkel and Sotirov, 2016, Pülzl et al., 2018, Sotirov et al., 2021).

Policy instruments that affect forest-based industries include several EU policy domains, such as climate, agricultural and environmental policies, representing areas where the EU does have explicit competencies. These policy instruments affect things like financing forestry measures under national Rural Development Programmes (RDPs), greenhouse gas (GHG) emissions and carbon accounting, energy efficiency of buildings and international timber trade. Forest governance in the EU consequently takes place across multiple policy domains, institutions, and levels of policymaking that affect the conservation and sustainable use of European forest resources. More importantly, perhaps, many scholars have noted that the mix of EU policy instruments is problematic for forests (Lazdinis et al., 2009, Winkel and Sotirov, 2016, Wolfslehner et al., 2020, Elomina and Pülzl, 2021), mainly as different policy domains frequently have contradictory goals and objectives that remain uncoordinated. One illustration of this problem is conflicting EU targets for renewable energy (e.g., climate change mitigation actions) and biodiversity conservation (e.g., the EU's Natura 2000 network), emphasising important trade-offs between clean energy and conservation policy pathways (Winter et al., 2014, Sotirov et al., 2017). Inconsistent and incoherent EU policy goals can be unproductive and sometimes work against each other (Sotirov and Storch, 2018).

While the treaty on the functioning of the EU (TFEU) does not make any provisions for forests or forestry,[48] it is commonly argued that the EU does have an explicit forest policy. This argument is based mainly on the impact of EU policy instruments on the forest value chain (Aggestam and Pülzl, 2018, 2020, Aggestam and Giurca, 2021), whether direct or indirect. It is also contended in this that if a policy instrument impacts forests, it is de facto a forest policy. The EU has, for example, adopted forest-focused legislation, such as a policy on the protection of forests against atmospheric pollution and the EU Timber Regulation, as well as soft instruments, such as three rounds of the EU Forest Strategy. There are also many forest-relevant instruments, such as the Common Agricultural Policy (CAP) and the European Climate Law, that affect the forest-based sector (Winkel et al., 2013, Aggestam et al., 2017, Pülzl et al., 2018). The absence of a clear definition, or delimiter, for EU forest policy, does, however, make it challenging to provide a concise overview of what it precisely consists of, as this can be dependent on the sub-sector you are interested in or your view on what should be part of the forest value chain (Winkel et al., 2013, Aggestam et al., 2017, Aggestam and Giurca, 2022). This may, in part, be because of the complex police arrangements underlying forest governance but also because of how diverse the forest-based sector is, covering everything from timber to paper and pulp to furniture production as well as renewable energy, construction, public health, and emerging sectors, such as bioplastics and textiles. Perhaps, more importantly, the diverging perspectives on whether there is an EU forest policy allude to the lack of a common framework and language on forests. The absence of a shared understanding has far-reaching implications when characterising EU forest policy and policy research.

Chapter 2 will go more into depth regarding EU policy instruments affecting the forest-based sector. It will also outline the institutional setup related to EU forest policy.

4. What will be covered in this book?

Chapter 2: Forest governance: the EU perspective

Chapter 2 delves into forest policy within the European Union, covering historical perspectives and recent developments in EU forest governance. Although providing a comprehensive overview of every forest-relevant EU policy domain falls outside the scope of this chapter, the primary focus will be to demonstrate the complex interactions between different sectors that influence the use of forest resources. This includes exploring key synergies and trade-offs that continue to present challenges in this area. Additionally, the chapter will introduce the institutional framework that forms the foundation for policymaking on forests in the EU.

Chapter 3: Overview of theories used in forest policy analysis

Chapter 3 ventures into forest policy analysis, offering an overview of prevalent theories in the field. It introduces some of the main theories currently utilised in forest policy research. The chapter initially focuses on classical and mainstream frameworks, such as rational policy analysis and institutional policy analysis. It then explores more contemporary and critical theories, including policy network analysis, the advocacy coalition framework (ACF), and critical policy analysis. This progression is meant to provide an overarching understanding of the development and diversity associated with theoretical approaches in forest policy studies.

Chapter 4: Policy pathways for the European Green Deal and the EU Forest Strategy

Chapter 4 expands on the overview of EU forest governance presented in Chapter 2. This chapter will specifically focus on the European Green Deal, which is being promoted as the cornerstone of European climate policy. The aim is to assess whether and how the European Green Deal has influenced the formulation and execution of the new EU Forest Strategy. In particular, the chapter will examine various policy directions and approaches outlined in the new EU Forest Strategy, exploring its potential trajectories and implications.

Chapter 5: Systemic change for the forest-based sector: implementing the circular bioeconomy

Chapter 5 concentrates on the circular economy concept and its implications for the forest-based sector. It particularly emphasises the crucial role that forest-based industries play in implementing a circular economy. This chapter will use a value chain analysis to evaluate the impact of policies like the circular economy across different sectors within the forest-based industry. Applying a value chain analysis can be especially beneficial for policy analysts and policymakers, as it can help explain the challenges and opportunities associated with introducing new policy instruments.

Chapter 6: Forest ownership and the provision of forest ecosystem services

Chapter 6 addresses the challenges forest management faces due to the increasing demand for multiple forest ecosystem services. A critical aspect of this demand equation relates to how forest owners and stakeholders respond to these changing demands. Consequently, research into forest ownership has gained increased prominence, particularly regarding the emergence of new types of forest owners, the trend toward fragmented ownership structures, and concerns about maintaining a consistent supply of raw materials for forest-based industries. This chapter presents a comprehensive overview of these issues in forest ownership and introduces research on forest actors. It will analyse and discuss how these forest actors' varied interests and objectives intersect with and influence policymaking.

Chapter 7: Forest fires, resilience, and climate change

Recent media coverage of forest fires in countries like Australia and Brazil emphasises the significance of 'fires' as an important topic for this book. Forest fires are crucial for their direct impact on forests, their effects on communities, and their connection to our changing climate. Chapter 7 will approach this socially significant subject from the perspective of framing. It will explore how the EU policy system, tasked with responding to hazards like forest fires, is constructed and implemented. Specifically, the chapter will examine the framing of forest fires within EU policy documents, providing insight into the policy approaches and strategies employed to address fires.

Chapter 8: Nature protection, productivity, and multifunctional forest use

Chapter 8 centres around the complex trade-offs surrounding forest use and the provision of forest ecosystem services (FES). The emphasis is on new forest management paradigms that address the growing need for multiple FES in Europe and globally. The chapter uses the concept of integrated forest management as a lens to demonstrate and discuss the competing demands placed on managed forests, such as the need for timber versus other goods and services, like recreation. Rather than focusing strictly on forest policy analysis, this chapter focuses on managing trade-offs between biodiversity conservation and biomass production in European forests, offering insights into how these often conflicting objectives can be balanced.

Chapter 9: A framework for integrative forest policy analysis

Traditional forest policy analysis approaches often assume that policy systems are well-defined and capable of manipulation and some degree of control. However, the reality of real-life policy problems is markedly different, marked by uncertainties about the future, governed by policymakers with varied or limited capabilities (both legal and technical), and dominated by complexity. These systems are inherently unpredictable. Chapter 9 proposes a dynamic and adaptive approach, equipping prospective forest policy analysts with the means to identify the most suitable analytical tools.

Chapter 10: Conclusions

The concluding chapter of this book synthesises the content and research presented throughout. It will aim to distil and highlight key messages and insights from the preceding chapters. Additionally, the chapter will critically examine the importance of forest policy research within the framework of contemporary EU forest policy. Concluding with recommendations for future research, this chapter provides recommendations for future research in this field.

Online Support Material: Practical toolkit and step-by-step guides for forest policy analysis

The online Support Material of this book is organised into six annexes, interlined with different chapters within the book. These annexes provide a practical, hands-on approach to various analytical methods in forest policy analysis, adopting a step-by-step format. The annexes include:

1. Basic methods of forest policy analysis
2. The pathways to sustainability approach
3. Value chain analysis
4. Actor and network analysis
5. The SWOT and STEEP approach
6. Content, discourse, and frame analysis

The Support Material is designed to be practical and to offer guidelines on how these methods can be applied to analyse different aspects of forest policy. It can be accessed at www.routledge.com/9781032057460

5. Notes

1 This industry covers several sectors that rely on forests for raw materials, such as pulp and paper manufacturing, biomass production and bioenergy, textile production, and furniture manufacturing.
2 COP26 was the 2021 United Nations Climate Change Conference. The 26th United Nations Climate Change Conference was held in Glasgow, Scotland, United Kingdom, between 31 October and 12 November 2021.
3 The CAP is the EU policy providing financial support to Member States farmers. Article 39 of the European Union Treaty sets out the specific objectives of the CAP to (1) increase agricultural productivity by promoting technical progress and ensuring the optimum use of the factors of production, in particular, labour; (2) ensure a fair standard of living for farmers; (3) stabilise markets; (4) ensure the availability of supplies; and (5) ensure reasonable prices for consumers.
4 See https://eustafor.eu/members/romsilva/.
5 Legal or regulatory frameworks refer to the regulations, decisions, directives, regulatory policies, guidelines, recommendations, and procedures made by relevant authorities.
6 Economic tools and incentives designed by the EU to, for example, encourage sustainable forestry practices, reforestation, and forest conservation (direct subsidies), measures for forest management under the CAP (Rural Development Programs), and financial incentives for reducing carbon emissions (Carbon Trading and Emissions Schemes).
7 Regulation 995/2010.
8 EC 2021.
9 See https://ukcop26.org/glasgow-leaders-declaration-on-forests-and-land-use.
10 Regulation 1907/2006.
11 von Carlowitz argued that forestry practices should ensure that forests are used so that the amount of timber harvested does not exceed the forest's capacity to regenerate. He introduced practices like selective cutting (harvesting only mature trees) and reforestation (planting new trees to replace those cut down).
12 Estimates suggest that beef production is responsible for 41 per cent of deforestation, palm oil and soybeans account for 18 per cent while logging for paper and wood in the tropics accounts for 13 per cent (Ritchie, 2021).
13 International refers to forest-related governance or policy initiative that goes beyond national boundaries.
14 See www.cbd.int/.
15 The Aichi Biodiversity Targets underpinned the EU's first Strategy on Biological Diversity.
16 See www.cbd.int/gbf/.
17 See https://unfccc.int/.
18 See https://unfccc.int/land-use-land-use-change-and-forestry-lulucf-under-the-kyoto-protocol.
19 See www.unccd.int/.
20 Agenda 21 is a non-binding action plan of the UN concerning sustainable development.
21 See www.un.org/esa/forests/forum/unffs/index.html.
22 UNFF is a subsidiary body of the United Nations Economic and Social Council (ECOSOC), established by ECOSOC Res. 2000/35, with the primary objective of promoting the management, conservation, and sustainable development of all types of forests.
23 The IAF has five main components: the UN Forum on Forests (UNFF) and its Member States, the UNFF Secretariat, the Collaborative Partnership on Forests (CPF), the UNFF Global Forest Financing Facilitation Network (GFFFN), and the UNFF Trust Fund.
24 The UNs General Assembly adopted this in 2007.
25 See https://forestdeclaration.org.
26 See https://ukcop26.org/glasgow-leaders-declaration-on-forests-and-land-use/.
27 See https://ukcop26.org/the-global-forest-finance-pledge/.
28 See www.un.org/sustainabledevelopment/development-agenda/
29 See www.euflegt.efi.int/home.
30 See https://redd.unfccc.int/.
31 See www.un-redd.org/.

32 See www.euredd.efi.int/.
33 See www.bonnchallenge.org/.
34 Over 70 pledgers from more than 60 countries are restoring 210 million ha of degraded and deforested lands.
35 See https://fsc.org.
36 See www.pefc.org.
37 Cites is implemented through the EU Wildlife Trade Regulations (Regulation, 338/97). The regulation lays down the provisions for import, export, and internal EU trade in specimens of species listed in its four annexes.
38 See www.itto.int.
39 See https://foresteurope.org.
40 See https://foresteurope.org/ministerial-commitments/.
41 See https://unece.org/forests.
42 It can also be noted that the Joint UNECE/FAO Working Party on Forest Statistics, Economics and Management (JWPFSEM) is a subsidiary body of COFFI and EFC.
43 See www.cbd.int/doc/nbsap/rbsap/peblds-rbsap.pdf.
44 FAO characterise a forest as 'Land spanning more than 0.5 hectares with trees higher than 5 meters and a canopy cover of more than 10 per cent, or trees able to reach these thresholds in situ. It does not include land that is predominantly under agricultural or urban land use' (FAO, 2020, p. 4).
45 FAO characterise other wooded land as a 'Land not classified as "Forest", spanning more than 0.5 hectares; with trees higher than 5 meters and a canopy cover of 5–10 per cent, or trees able to reach these thresholds in situ; or with a combined cover of shrubs, bushes and trees above 10 per cent. It does not include land that is predominantly under agricultural or urban land use' (FAO, 2020, p. 4).
46 In the Treaty of Rome (1957), forest products were excluded from the European Community's primary laws on common policies.
47 Article 5(3) of the Treaty on European Union (TEU) and Protocol (No 2) on the application of the principles of subsidiarity and proportionality. The principle of subsidiarity safeguards the ability of EU Member States to make decisions and act on the national level. It authorises EU intervention when the Member States need help to sufficiently achieve an action's objectives.
48 The EU can only propose legislation linked to its shared or exclusive competencies outlined in TFEU13.

6. References

AGGESTAM, F. & GIURCA, A. 2021. The art of the 'green' deal: Policy pathways for the EU Forest Strategy. *Forest Policy and Economics*, 128, 102456.

AGGESTAM, F. & GIURCA, A. 2022. Implementing circular-bioeconomy principles across two value chains of the wood-based sector: A conceptual approach. *Land* [Online], 11.

AGGESTAM, F. & PÜLZL, H. 2018. Coordinating the uncoordinated: the EU Forest Strategy. *Forests*, 9, 125, DOI: https://doi.org/10.3390/f9030125.

AGGESTAM, F. & PÜLZL, H. 2020. Downloading Europe: A regional comparison in the uptake of the EU Forest Action Plan. *Sustainability*, 12.

AGGESTAM, F., WINKEL, G., PÜLZL, H. & SOTIROV, M. 2017. The EU policy framework. *In:* WINKEL, G. (ed.) *Towards a Sustainable European Forest Based Bioeconomy – Assessment and the Way Forward. What Science Can Tell Us 8*. European Forest Institute.

ARTS, B., APPELSTRAND, M., KLEINSCHMIT, D., PÜLZL, H. & VISSEREN-HAMAKERS, I. 2010. Discourses, actors and instruments in international forest governance. *In:* RAYNER, J., BUCK, A. & KATILA, P. (eds.) *Embracing Complexity: Meeting the Challeneges of International Forest Governance*. Vienna: IUFRO World Series Volume 28.

ARTS, B. & BUIZER, M. 2009. Forests, discourses, institutions: A discursive-institutional analysis of global forest governance. *Forest Policy and Economics*, 11, 340–347.

ARTS, B., GIESSEN, L. & VISSEREN-HAMAKERS, I. 2013. International forest policy and Europe: Four pathways of mutual influence. *In:* PÜLZL, H., HOGL, K., KLEINSCHMIT, D., WYDRA, D., ARTS, B., MAYER, P., PALAHI, M., WINKEL, G. & WOLFSLEHNER, B. (eds.) *European Forest Governance: Issues at Stake and the Way Forward.* EFI Series: What Science can tell us; 37–43.

CARLOWITZ, H. C. 2013. Sylvicultura oeconomica. *In:* ROBIN, L., SÖRLIN, S. & WARDE, P. (eds.). *The Future of Nature: Documents of Global Change*. Yale University Press.

CASHORE, B., LEIPOLD, S., CERUTTI, P., BUENO, G., CARODENUTO, S., CHEN, X., DE JONG, W., DENVIR, A., HANSEN, C. P., HUMPHREYS, D., MCGINLEY, K., NATHAN, I., OVERDEVEST, C.,

RODRIGUES, R., SOTIROV, M., STONE, M., TEGEGNE, Y., VISSEREN-HAMAKERS, I., WINKEL, G. & ZEITLIN, J. 2016. Global governance approaches to addressing illegal logging: Uptake and lessons learnt. *In:* KLEINSCHMIT, D., MANSOURIAN, S., WILDBURGER, C. & PURRET, A. (eds.) *Illegal Logging and Related Timber Trade – Dimensions, Drivers, Impacts and Responses: A Global Scientific Rapid Response Assessment Report.* Vienna, Austria: International Union of Forest Research Organizations.

COUNCIL REGULATION 2173/2005. Establishment of a FLEGT licensing scheme for imports of timber into the European Community. *L 347/1.* Brussels: Official Journal of the European Union.

EC 2003. Forest Law Enforcement, Governance and Trade (FLEGT) – Proposal for an EU Action Plan. *COM (2003) 251 final.* Brussels: European Commission.

EC 2021. New EU Forest Strategy for 2030. COM(2021) 572 final. Brussels: European Commission.

EDWARDS, P. & KLEINSCHMIT, D. 2013. Towards a European forest policy: conflicting goals. *Forest Policy and Economics*, 33, 87–91.

EKELUND, H. & HAMILTON, G. 2001. Skogspolitisk historia. Skogstyrelsen: http://shop.skogsstyrelsen.se/shop/9098/art45/4646045-67b381-1695.pdf.

ELOMINA, J. & PÜLZL, H. 2021. How are forests framed? An analysis of EU forest policy. *Forest Policy and Economics*, 127, 102448.

EUROSTAT 2020. Agriculture, forestry and fishery statistics: 2020 edition. *In:* COOK, E. (ed.). Luxembourg: Publications Office of the European Union, 2020.

FAO 2020. Global Forest Resources Assessment 2020: Main report. Rome, Italy: Food and Agriculture Organization of the United Nations.

FOREST EUROPE 2015. State of Europe's Forests 2015. Madrid, Spain: Ministerial Conference on the Protection of Forests in Europe.

FOREST EUROPE 2020. State of Europe's Forests 2020. Bratislava: Ministerial Conference on the Protection of Forests in Europe.

FORESTRY COMMISSION 2017. *The UK Forestry Standard.* Edinburgh: Forestry Commission.

GIURCĂ, A. & DIMA, D. P. 2022. *The Plan B for Romania's Forests and Society.* Transylvania University Press.

GLÜCK, P., ANGELSEN, A., APPELSTRAND, M., ASSEMBE-MVONDO, S., G., A., HOGL, K., HUMPHREYS, D. & WILDBURGER, C. 2010. Core components of the international forest regime. *In:* RAYNER, J., BUCK, A. & KATILA, P. (eds.) *Embracing Complexity. Meeting the Challenges of International Forest Governance.* Vienna: IUFRO World Series 28.

HOSONUMA, N., HEROLD, M., DE SY, V., DE FRIES, R. S., BROCKHAUS, M., VERCHOT, L., ANGELSEN, A. & ROMIJN, E. 2012. An assessment of deforestation and forest degradation drivers in developing countries. *Environmental Research Letters*, 7, 044009.

HUMPHREYS, D. 2004. *Forests for the Future: National Forest Programmes in Europe: Country and Regional Reports from COST Action E19.* Brussels: Office for Official Publications of the European Communities.

KLEINSCHMIT, D., PÜLZL, H., SECCO, L., SERGENT, A. & WALLIN, I. 2018. Orchestration in political processes: Involvement of experts, citizens, and participatory professionals in forest policy making. *Forest Policy Econ.*, 89, 4–15. https://doi.org/10.1016/j.forpol.2017.12.011.

LAZDINIS, M., ANGELSTAM, P. & LAZDINIS, I. 2009. Governing forests of the European Union: Institutional framework for interest representation at the European Community level. *Environmental Policy and Governance*, 19, 44–56.

LIER, M., KÖHL, M., KORHONEN, K. T., LINSER, S. & PRINS, K. 2021. Forest relevant targets in EU policy instruments – can progress be measured by the pan-European criteria and indicators for sustainable forest management? *Forest Policy and Economics*, 128, 102481.

LINSER, S., WOLFSLEHNER, B., BRIDGE, R. J. S., GRITTEN, D., JOHNSON, S., PAYN, T., PRINS, K., RAŠI, R. & ROBERTSON, G. 2018. 25 Years of criteria and indicators for sustainable forest management: How intergovernmental C&I processes have made a difference. *Forests*, 9.

MCDERMOTT, C., O'CARROLL, A. & WOOD, P. 2007. International forest policy – the instruments, agreements and processes that shape it United Nations Forum on Forests Secretariat.

PAUL DRAGOS, A. & ADINA, D. 2003. Land reform and agricultural reform policies in Romania's transition to the market economy: Overview and assessment. *Eastern European Economics*, 41, 49–69.

PÜLZL, H. & MAYER, P. 2015. Assessment of the achievements and added value of the Forest Europe Process. *Final Report.* Vienna: BFW.

PÜLZL, H., WYDRA, D. & HOGL, K. 2018. Piecemeal integration: Explaining and understanding 60 years of European Union forest policymaking. *Forests*, 9, 719.

RAYNER, J., BUCK, A. & KATILA, P. 2010. Embracing complexity: Meeting the challenges of international forest governance. A Global Assessment Report Prepared by the Global Forest Expert Panel on the International Forest Regime. *IUFRO World Series.* Vienna: International Union of Forest Research Organizations (IUFRO).

REGULATION 995/2010. Laying Down the Obligations of Operators Who Place Timber and Timber Products on the Market. *OJ L 295.* Brussels: Official Journal of the European Union.

REGULATION 1907/2006. Registration, Evaluation, Authorisation and Restriction of Chemicals (REACH). OJ L 396. Brussels: Official Journal of the European Communities.

RITCHIE, H. 2021. Cutting down forests: what are the drivers of deforestation? https://ourworldindata.org/what-are-drivers-deforestation.

SLU 2016. *Forest Statistics 2016.* Uppsala: Swedish National Forest Inventory.

SOTIROV, M. & STORCH, S. 2018. Resilience through policy integration in Europe? Domestic forest policy changes as response to absorb pressure to integrate biodiversity conservation, bioenergy use and climate protection in France, Germany, the Netherlands and Sweden. *Land Use Policy*, 79, 977–989. https://doi.org/10.1016/j.landusepol.2017.04.034.

SOTIROV, M., WEISS, G., MARCHETTI, M., GARCIA, S., JUMP, A., GREENWOOD, S., ECOCHARD, L., HILY, E., BASTRUP-BIRK, A., PARVIAINEN, J., SARVASOVA, Z., SANTOPUOLI, G., VIZZARRI, M., BROTONS, L., HERMOSO, V., MIKUSINSKI, G., GROSSMANN, C., SCHABEL, A., SCHMACK, A. & BLUM, M. 2017. Natura 2000 and forests – assessing the state of implementation and effectiveness. EFI procurement reference number: 4042–01–2016. University of Freiburg, Freiburg, 1 March. Draft Final Report. University of Freiburg, Freiburg, 1 March.

SOTIROV, M., WINKEL, G. & ECKERBERG, K. 2021. The coalitional politics of the European Union's environmental forest policy: Biodiversity conservation, timber legality, and climate protection. *Ambio*, 50, 2153–2167.

SPANOS, K., GAITANIS, D., SKOUTERI, A., PETRAKIS, P. V. & MELIADIS, I. 2018. Implementation of forest policy in Greece in relation to biodiversity and climate change. *Open Journal of Ecology*, 8, 174–191.

UNDP 2019. New York Declaration on Forests Endorser Perspectives Report: Progress, Challenges and Enabling Conditions for Action. *In:* DESANTIS, N., ERVIN, J., CRAIG, M., MARX, M., PAZ, C. & KANTCHEVA, N. (eds.). New York: United Nations Development Programme.

UNECE/FAO 2013. Rovaniemi Action Plan for the Forest Sector in a Green Economy. *ECE/TIM/SP/35.* Geneva: United Nations Economic Commission for Europe.

WINKEL, G., AGGESTAM, F., SOTIROV, M. & WEISS, G. 2013. Forest Policy in the European Union. *In:* PÜLZL, H., HOGL, K., KLEINSCHMIT, D., WYDRA, D., ARTS, B., MAYER, P., PALAHÍ, M., WINKEL, G. & WOLFSLEHNER, B. (eds.) *European Forest Governance: Issues at Stake and the Way Forward.*

WINKEL, G. & SOTIROV, M. 2016. Whose integration is this? European forest policy between the gospel of coordination, institutional competition, and a new spirit of integration. *Environ. Plann. C.*, 34, 496–514, https://doi.org/10.1068/c1356j.

WINTER, S., BORRASS, L., GEITZENAUER, M., BLONDET, M., BREIBECK, R., WEISS, G. & WINKEL, G. 2014. The impact of Natura 2000 on forest management: a socio-ecological analysis in the continental region of the European Union. *Biodiversity and Conservation*, 23, 3451–3482.

WOLFSLEHNER, B., PÜLZL, H., KLEINSCHMIT, D., AGGESTAM, F., WINKEL, G., CANDEL, J., ECKERBERG, K., FEINDT, P., MCDERMOTT, C., SECCO, L., SOTIROV, M., LACKNER, M. & LAZYA ROUX, J. 2020. European forest governance post-2020. *From Science to Policy 10.* Joensuu, Finland: European Forest Institute. https://doi.org/10.36333/fs10

WYDRA, D. 2013. The Legal Context of European Forest Policymaking. *In:* PÜLZL, H., HOGL, K., KLEINSCHMIT, D., WYDRA, D., ARTS, B., MAYER, P., PALAHI, M., WINKEL, G. & WOLFSLEHNER, B. (eds.) *European Forest Governance: Issues at Stake and the Way Forward.* EFI Series: What Science can tell us, 29–36.

2 Forest governance

The EU perspective

1. The state of forests in Europe

The most recent iteration of the State of Europe's Forests report argues that the total forest land in Europe is increasing. Forests were estimated to cover approximately 227 million hectares (ha), covering 34.8% of the total land area, while 27 million ha (or around 4% of the total land area) was reported as being other wooded land in 2020 (Forest Europe, 2020). Most countries have between 30 and 45% of their land area covered by forests. This is for the pan-European region but without the Russian Federation.[1] The EU reported having approximately 180 million ha of forest and other wooded land in the same year (Eurostat, 2020). There is, however, considerable variation across the pan-European region (Figure 2.1), reflecting a diversity in climatic conditions, soil types, topography, and altitude. Overall, 46% of European forests are considered coniferous trees, while 37% is broadleaf forest, with the remaining 17% being mixed stands (Forest Europe, 2020). The larger share of coniferous trees is mainly due to Northern Europe's predominance of boreal forests. Forest coverage also varies considerably among the EU Member States, where six countries – Sweden, Spain, Finland, France, Italy, and Germany.[2] – account for over two-thirds of the EU's total forest area, including other wooded land.

What is a forest?

This is a straightforward question; however, no single definition or classification is valid across all countries. National ministries or agencies dealing with forests apply varied definitions depending on their respective legislative frameworks, national priorities, and forest characteristics. For example, in Finland, the classification of forestry land is based on the annual forest increment, which is a definition that focuses on the productivity and growth potential of the forest land. In contrast, forests in Austria are defined based on crown cover, minimum size, and their fulfilment of certain functions, such as for forestry purposes. This definition emphasises the physical characteristics and functional aspects of forests.

These variations in definitions have implications for international reporting and policymaking on forests. For example, Turkey has recently reclassified large areas previously considered 'sparse woody vegetation' to forests. This resulted in a reduction of 10 million ha of other wooded land (OWL) being reported for the State of European Forests report in 2020 (Forest Europe, 2020). These variations can make it difficult to compare countries,

DOI: 10.4324/9781003198987-2

not only regarding the use of statistics, but it may create barriers in international policy-making as there is no common language.

Even though forest definitions and classifications vary across countries, efforts to develop international forestry statistics have resulted in a shared classification scheme set up by the Food and Agriculture Organization of the United Nations (FAO). The FAO is the main centre for collecting and analysing data and disseminating international forestry statistics. The FAO classification has been adopted by Eurostat, the main provider of statistical information in the EU, and Forest Europe and its reporting on the state of forests (Forest Europe, 2015, 2020). However, one problem with the FAO classification is that it does not distinguish between forest types, which implies that you cannot monitor changes in the forest, such as differentiating between a planted or natural forest (Chazdon et al., 2016).

Based on statistical information from the FAO, one of the main publications is the Global Forest Resource Assessment (FRA), an international forest resources assessment published every five to ten years since 1946 (FAO, 2010, 2015, 2020). This is the main source of globally synthesised information on forests. In its terms and definitions, forests are characterised as 'Land spanning more than 0.5 hectares with trees higher than 5 meters and a canopy cover of more than 10 per cent, or trees able to reach these thresholds in situ. It does not include land that is predominantly under agricultural or urban land use' in the FRA (FAO, 2018, p. 4).

2. Forest governance in the European Union

Forests play an important role in European culture and history, both from an environmental and socio-economic perspective, as a source of jobs, food, biodiversity, and recreation. Even more, for the later point, the relevance of forests in relation to human well-being and health has increased significantly in the wake of the COVID-19 pandemic (Samuelsson et al., 2021, Ciesielski et al., 2022). Forest-based products and services, including non-wood forest products (NWFP), are thus not only an inherent and integrated element of many economic sectors but also an integral part of social and cultural activities. The wide spectrum of sectors and private and public interests emphasises the relevance of forests. However, from the perspective of governing forests in Europe, the EU does not have a common forest policy. Aside from cork, forest products are excluded from existing EU treaties, which means that the formulation and implementation of EU forest policy is subject to the principle of subsidiarity. As such, the EU cannot draft legally binding (hard) forest policies on forests. This has effectively resulted in EU forest governance using non-binding (soft) policy instruments. There are, however, exceptions. In areas where the EU has competencies, such as agriculture, trade, and the environment, it has been able to address forests more explicitly. The downside of this approach is a significant number of conflicting policy objectives (e.g., climate change mitigation versus biodiversity conservation) and policy incoherence as a defining feature of EU forest governance. The following chapter will demonstrate some of the complex interactions between different sectors affecting forest use, covering unresolved synergies and trade-offs. It will also introduce forest-focused and forest-relevant EU policy domains.

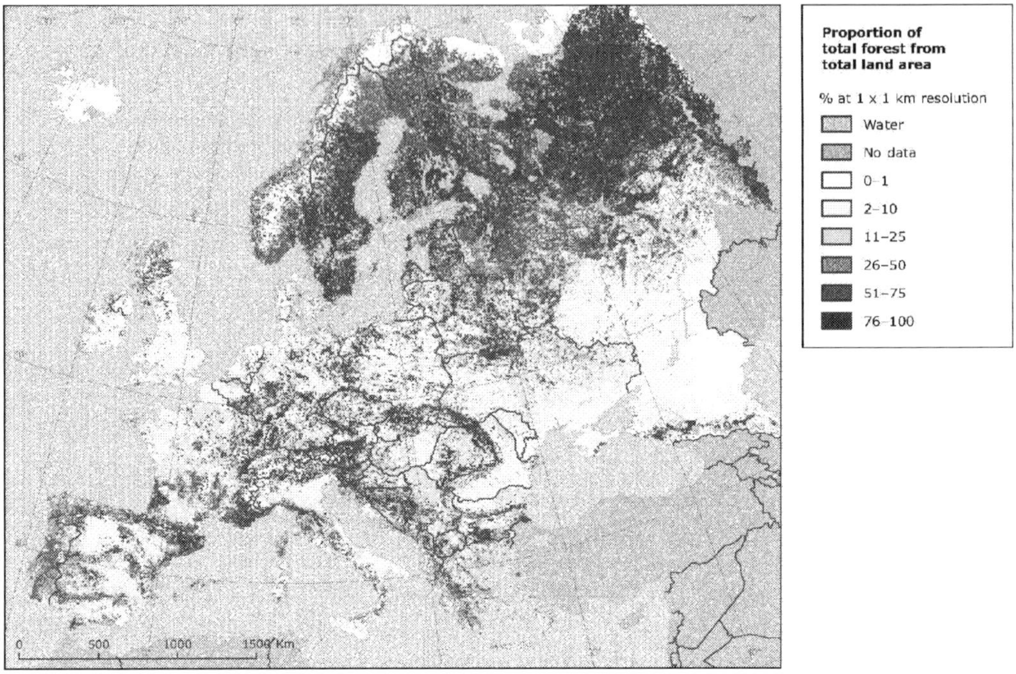

Figure 2.1 Forest map of Europe.

Source: European Forest Institute (Päivinen et al., 2001, Schuck et al., 2002, Kempeneers et al., 2012)

> The EU does not have a common forest policy. This means forest products, such as timber and other wood-based products) are excluded from existing EU treaties, and the formulation and implementation of policies affecting forests are subject to the principle of subsidiarity (Article 5(2) of the EEC treaty).[3] Forests are, therefore, explicitly managed by EU Member States. The only exception is for cork and some forest-related fruits explicitly included in the Treaties of the European Union.[4] For example, cork oak forests are included in Annex I of the EU habitats directive (Directive, 92/43/EEC) and supportive measures have been included in the afforestation programmes of the Common Agriculture Policy (CAP).

2.1. *Past and ongoing EU policy developments on forests*

Early EU policies affecting forests can be found from the 1960s and onwards. These initial EU initiatives have mainly been anchored in agricultural and rural development policy and financial support, covering measures that directly and indirectly subsidised investments in forestry (Arts et al., 2013, Lazdinis et al., 2019), such as afforestation, fire prevention, and forest roads. The first explicit call for an EU forest policy came in 1978 when the European Commission proposed a Council Resolution on a set of forest policy principles. The 1978 resolution suggested, among other things, the need for better coordination of national forest policies to achieve increased timber production, environmental conservation, and public access to forests for recreation (EC, 1978). While the resolution was never adopted, it is interesting to note that several of the policy problems emphasised in 1978 remain the same today, such as the need for the different sectors of forestry to identify 'common interests' and to 'speak with one voice' at the national and EU level (p. 31), improving public awareness on forests, and having harmonised forest statistics. These topics are still high on the EU policy agenda.

Since there has never been a consensus around an EU forest policy among EU Member States, it was not until 1998 that we find what is commonly referred to as the EU Forest Strategy. This voluntary policy instrument was adopted in 1998 and provided general guidelines for an EU forest policy designed to coordinate other EU forest-relevant policies (EC, 1998). Since being adopted, the Strategy has undergone two iterations, with a second Strategy adopted in 2013 (EC, 2013) and a third Strategy adopted in 2021 (EC, 2021a). The second Strategy served as an updated and integrative framework that addressed the increasing demands on forests while tackling changes in societal and policy priorities relevant to the forest sector. For the third Strategy leading up to 2030, the European Commission has outlined two key objectives, namely, the (1) protection, restoration, and sustainable management of forests, and (2) ensuring the multifunctionality of EU forests. Furthermore, the new Strategy sets out a pledge to plant at least 3 billion additional trees by 2030, fully respecting ecological principles with long-term planning and monitoring (EC, 2021a).

EU policy frameworks can consist of **soft policy documents** (non-binding documents) or **hard policy documents** (legally binding documents). Formulating an agreement or policy in legally binding terms requires national acceptance and legislative approval. In contrast, soft policy documents are voluntary and do not require legislative implementation at the Member States' level. Soft policy documents also imply the absence of enforcement mechanisms.

With the adoption of the European Green Deal in 2019 and a new EU Biodiversity Strategy in 2020, the relevance of EU forest policy has arguably increased (EC, 2019b, 2020a). For instance, as a flagship initiative of the Green Deal, the third EU Forest Strategy is meant to support EU targets to reduce GHG emissions by at least 55% by 2030 (EC, 2021a, 2021g). Moreover, building on SFM, the commitments and actions proposed in the third EU Forest Strategy have been designed to contribute to the measures laid out in the European Commission's 'Fit for 55' package (EC, 2021e) as part of the EU's climate law. It is further relevant to stress that while the EU can only address forests through soft policy instruments, other policy domains have adopted legally binding instruments that directly concern forests, such as nature restoration and climate action.

For the forest-based sector, there are also many relevant EU directives, regulations, and policy tools affecting forest use outside the realm of forests and forestry. This includes the updated bioeconomy strategy and action plan to develop a sustainable and circular bioeconomy across the EU (EC, 2018a, 2018d). The bioeconomy strategy aims to improve and scale up the use of renewable resources. The new circular economy action also sets out to minimise waste generation and increase the life cycle of products, materials, and resources (EC, 2020c). While many more EU policy instruments are directly relevant to forest-based industries (e.g., REACH and the EU Timber Regulation), these examples demonstrate that many instruments affect how forests and wood can be utilised. This includes objectives and targets concerning the decarbonisation of the building sector to maximising the potential of carbon storage in forests to compensating for land use and forestry sector emissions by CO_2 removals to financing measures that support SFM to halting the loss of biodiversity and ecosystem services.

The discourse on forest use occurs within a complex institutional environment, set against a backdrop of different sectoral perspectives, policy goals, and implementation tools (Aggestam and Pülzl, 2018, Lazdinis et al., 2019). This effectively means that forest use and EU forest governance have to be considered through the lens of several EU policy domains, including competing interests across multiple stakeholder groups, countries, and sectors with varied effects and relevance for the forest-based sector (Korhonen et al., 2018, Purkus et al., 2018, Watanabe et al., 2018).

2.2. EU policymaking and legislative procedures

When countries join the EU, they give up some of their power to make decisions. In other words, the decisions made at the EU level have legal effects in every Member State. Before considering EU forest policy in more detail, it is relevant to briefly consider how the EU works. Learning about the EU will help in understanding EU forest governance.

2.2.1. EU institutions

The **Council of the European Union** is the main decision-making body of the EU. Its members are government ministers from each of the EU Member States, and it is where the national interests of the EU Member States are represented. The Council is divided into ten configurations, each dealing with a particular subject matter. The Presidency of the Council of the EU is held by each Member State for six months, following a pre-established rota.

The **European Commission** is the executive wing of the EU. The Commission has the sole right to propose legislation, although the European Parliament can request that the Commission bring forward a proposal (as set out under Article 225 of the TFEU). The Commission sets objectives and annual priorities for action, drafts proposals for new laws, and manages the implementation of EU policies and the spending of EU funds. The Commission refers to the College of Commissioners as a political body and the institution. The **College of Commissioners** consists of 27 Commissioners, one from each Member State. Each Commissioner has a Cabinet that supports the Commissioner in delivering their programmes. The Commission is further organised into **Directorates-Generals** (DGs) responsible for different policy domains (see 5.1.2).[5] In addition, a number of executive agencies manage programmes set up by the Commission, while service departments are set up to deal with specific administrative issues. Even though the EU does not have any competence on forests, DG Environment, DG Agriculture and Rural Development, DG Regional Policy, DG Research, etc., are all involved in developing and monitoring measures that affect the forest-based sector.[6]

The **European Parliament** is the only elected institution of the EU, and its members are elected every five years. The Parliament is led by its President, elected by the members for 2.5 years, who represents the Parliament and oversees its procedures. Following the implementation of the Lisbon Treaty, nearly all EU legislation is co-decided between the European Parliament and the Council under the ordinary legislative procedure. In principle, this puts the European Parliament on an equal footing with the Council. Moreover, Parliament adopts the EU budget jointly with the Council. It can also be noted that the Parliament is divided into several specialised standing committees that address different topics (e.g., foreign affairs and development). There are 20 parliamentary committees, each consisting of between 25 and 88 Members of the European Parliament. Committees that address forests include, for example, the Committee on Agriculture and Rural Development, the Committee on the Environment, Public Health and Food Safety, and the Committee on Regional Development.

The **Committee of the Regions** is an assembly of regional and local authorities and representatives from all EU Member States. It addresses opinions on their behalf to the Council and the Commission. In other words, the Committee gives regions and cities a formal say in EU policymaking. This helps to ensure that the position and needs of regional and local authorities are considered (e.g., the Commission, the Council of the European Union and the European Parliament must consult the Committee when developing legislation that may affect local and regional governments. For example, the Committee issued an opinion on the EU Forest Strategy for 2030 (EC, 2022c). It also has a say in the allocation of structural funds affecting forests.

2.2.2. Who is responsible for what?

The Lisbon Treaty[7] sets out the divisions of competencies between the EU and its Member States. The treaty outlines the areas of exclusive competency, shared competency, and supporting competence (see Table 2.1):

- **Exclusive competencies**: Policy domains where only the EU can legislate and adopt legally binding decisions. EU Member States can adopt legislation in these domains if allowed by the EU.
- **Shared competencies**: Policy domains where the EU and Member States can formulate legislation; however, Member States can only exercise their competence if the EU has not done so.
- **Supporting competencies**: Policy domains where the EU may support, coordinate, or supplement the actions taken by Member States; however, the EU cannot adopt legislation in these domains.

Table 2.1 Competences in the EU.

Exclusive competence	*Shared competence*	*Supporting competence*
Customs Union	Internal Market	Human health
Conservation of marine biological resources (a part of fisheries policy)	Social Policy	Culture
Establishing rules for the effective functioning of the Internal Market	Economic, social, and territorial cohesion	Tourism
European Monetary Union	Public health	Industry
Common Commercial Policy	Agriculture and fisheries	Civil protection
	Environment	Administrative cooperation
	Transport	Education, vocational training, youth and sport
	Consumer protection	
	Area of Freedom, Security and Justice	
	Energy	
	Trans-European networks	

2.2.3. Policymaking procedures

The legislative procedure in the EU contains several steps, and pushing for specific priorities is possible before, during or after each stage. To push for specific priorities or issues within the legislative process, it is important to understand the procedures and know who plays what role and when. For example, the Commission usually makes proposals for a legislative act, while the EP may also adopt legislative initiatives.

The **ordinary legislative procedure** is the most common form of policymaking in the EU. It applies to most of the policy areas that are contained within the TFEU. Under the ordinary legislative procedure, the Parliament has the same rights and policymaking authority as the Council. For example, Parliament has veto power as compared to other procedures. This legislative procedure has up to three stages once the Commission has issued a proposal: (1) first reading by the European Parliament and then the Council, (2) second reading by the European Parliament and then the Council, and (3) a conciliation committee and third reading. The process can, however, be concluded or stopped after the first or second reading (see Figure 2.2).

There is also the **consent procedure**, where the European Commission makes a proposal that the EP must reject or accept. Acceptance requires an absolute majority of votes. It can also be noted that the EP cannot make any amendments to the proposal, and a rejection of a proposed act constitutes a veto by the EP (e.g., the Council cannot take any decisions without consent from Parliament). Some policy areas covered by the consent procedure include EU membership or withdrawal, association agreements, agreements establishing specific institutional frameworks, and agreements with significant budgetary implications.

Finally, there is the **consultation procedure**, when the European Commission and the Council are only required to consult and ask the EP for an opinion on legislative proposals. This implies that the Council can enact legislation Parliament does not approve (e.g., the EP can only delay the process). This procedure is mainly used in a few areas, such as internal market exemptions and competition law.

> If you would like to learn more about how the EU works, see *How the EU Really Works* (Costa and Brack, 2019).

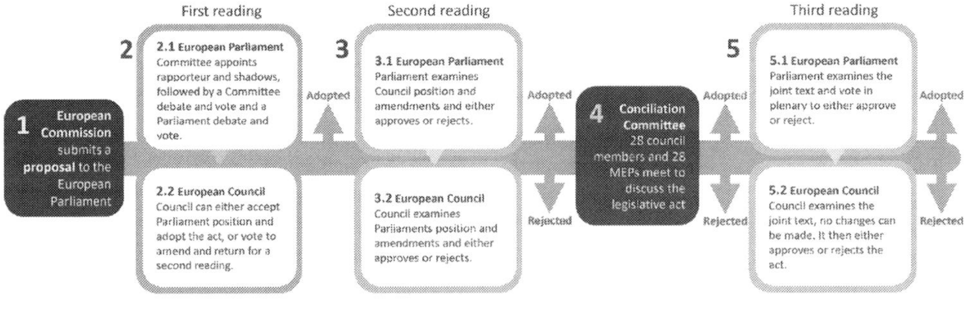

Figure 2.2 EU policymaking procedure.

Source: own figure

2.3. *How does EU forest governance work?*

2.3.1. Forests and forestry

EU forest-relevant issues go through the **Working Party on Forestry** (WPF),[8] which is part of the preparatory bodies under the Council of Ministers[9] and the **Standing Forestry Committee** (SFC)[10]:

- **Working Party on Forestry** (WPF) handles international forest processes and internal dossiers within the EU and comprises relevant delegates and experts. The WPF is mainly involved in discussions on international forest fora and processes such as the UNFF, Forest Europe, the UN Food and Agriculture Organization (FAO) Committee on Forestry (COFO), and the UNECE Committee on Forests and the Forest Industry (COFFI). For example, it is involved in negotiating a legally binding agreement (LBA) on forests under the Forest Europe process and the Forest Law Enforcement, Governance and Trade (FLEGT) Regulation. This includes, among other things, the EU Member States and the Commission coordinating positions prior to major international meetings through the WPF.
- **Standing Forestry Committee** (SFC) assist the Commission in preparing legislative proposals and policy initiatives, coordinating with the EU Member States (e.g., exchange of views), and providing expertise to the Commission when preparing to implement measures. For instance, the SFC brings together EU Member States representatives chaired by the Commission. It mainly acts as an ad hoc consultation forum that provides expertise in developing forest-relevant measures, such as rural development and the environment. It also provides a venue for exchanging information and technical expertise among and between the EU Member States and the Commission. While the Commission chairs the Committee, the SFC has members representing all the EU countries.[11] Following the adoption of the new Forest Strategy, the governance structure of the SFC is set to be reviewed, such as potentially merging the SFC and the Working Group on Forest and Nature.

> Following the adoption of the new (or third) EU Forest Strategy, the **governance structure** of the SFC was set to be reviewed. The new Strategy put forward that the Commission would 'propose an updated governance' (p. 23) to merge the SFC and the Working Group on Forest and Nature (EC, 2021a, 2021g). While the proposed merger between the SFC and the Working Group has not taken place, the Commission has set up a new stakeholder platform – the **Forest and Forestry Stakeholder Platform** – to assist the Commission services (e.g., DG Agri, DG Environment and DG Clima) in implementing the new Strategy. The new platform builds on the Civil Dialogue Group on Forestry and Cork, which has now ceased, and the Working Group on Forest and Nature.

Coordination also takes place through the informal **EU Forestry Directors-General meetings**. For instance, Presidencies of the EU periodically organise informal meetings of the Director-Generals responsible for forestry. These meetings are mainly organised to, for example, discuss the role of forests in current policies and strategies, most recently on the new EU Forest Strategy. The **Inter-Service Group on Forestry** is another intra-institutional Commission mechanism established in 2001 which sets out to improve coordination between Commission services responsible for forest-relevant policies. It is a mechanism that allows the EC to obtain the opinion of other Directorates-Generals (DGs) and is meant to improve the coherence between policy areas affecting forests.

Overall, given the legal status of forests and the EU Forest Strategy, most activities carried out by the EC concern coordination activities, such as formal and informal information sharing through the SFC and the Inter-Service Group on Forestry. The EU Forest Strategy outlines further allocation of responsibility for implementing measures (e.g., accelerating the uptake of forest-related innovations) but are yet to be designed (and implemented) in close cooperation with the EU Member States.

2.3.2. Forest-based industries

Forest-based industries can engage the Commission through formal and ad hoc cooperation and communication mechanisms, primarily in the context of existing committees. For some examples, the following groups provide the Commission with relevant technical expertise:

- **Expert Group on Forest and Forestry Stakeholder Platform**[12] is a new forum for regular information exchange between forest and forestry-related stakeholders that operate on a European level. The platform held its first meeting in 2023 and it involves EU-based stakeholder organisations operating in the field of forests and forestry aiming to assist the Commission services in preparing relevant policy initiatives. It also sets out to deliver opinions on specific matters and to monitor policy developments in the fields of forests and forestry, with a particular focus on the implementation of the new EU Forest Strategy.
- **Expert Group on Forest-based Industries and Sectorally Related Issues**,[13] which was set up in 2014, involves representatives from the forest-based industries, relevant Member State authorities and groupings holding relevant information. This is the main body for cooperation between forest-based industries and the Commission. It provides legal, economic and technical advice on issues, in particular, concerning the preparation and implementation of policies and legislation that may affect the forest-based industries.
- **Advisory Group on Construction Products**[14] focus on the woodworking and construction sectors (e.g., the construction industry is a big driver of sawnwood consumption) and mainly assist the Commission in issues related to the Construction Products Regulation (Regulation, 305/2011).

Several committees and groups can be considered relevant to forest-based industries. Table 2.2 provides some additional examples.[15]

Table 2.2 Expert Groups and other entities.

Group	*Lead*	*Role*
Commission Expert Group/Multi-Stakeholder Platform on Protecting and Restoring the World's Forests, including the EU Timber Regulation and the FLEGT Regulation[16]	DG Environment	Advice and assistance in the preparation of legislative proposals and policy initiatives, the preparation of delegated acts and the implementation of Union legislation.
EU Biodiversity Platform (including several sub-groups, such as the Working Group on Green Infrastructure and Restoration)	DG Environment	Assist in the implementation of the EU Biodiversity Strategy for 2030.
Expert Group on Forest Fires	DG Environment	Reviewing forest fire trends in Europe and determining adequate and effective responses.

(Continued)

Table 2.2 (Continued)

Group	*Lead*	*Role*
Informal Green Public Procurement Advisory Group	DG Environment	Promote green public procurement in the EU and assist in the preparation of legislation.
Expert group on Forestry statistics	Eurostat	Facilitate the exchange of information, experiences and good practices between statistical institutions.
Commission expert group 'Industrial Forum'	DG Grow	Exchange best practices aiming at the transformation of industrial ecosystems.
Working Group on Technical Harmonization (including on motor vehicles)[17]	Council of the EU	Examines draft legislation that aims to remove obstacles to intra-EU trade in goods.
Working Party on Civil Protection (PROCIV)[18]	Council of the EU	Prevention of, preparedness for, and response to natural and anthropogenic disasters.

2.4. *International forest governance and policy*

Several international mechanisms, including legally and non-legally binding policy instruments, have been developed to facilitate SFM and foster the protection of forests. International refers to forest-relevant governance or policy initiatives beyond national and EU boundaries. The sections below briefly introduce some relevant processes and instruments on the international level.

2.4.1. Pan-European forest governance

The **Ministerial Conference on the Protection of Forests in Europe** (MCPFE), nowadays known as **Forest Europe**,[19] is a voluntary process that involves 46 countries, including the EU, the Russian Federation, Switzerland and Norway. The Forest Europe process has generated several political commitments[20] and a definition of what SFM means in the pan-European context (Lier et al., 2021, Linser et al., 2018). This includes a set of criteria and indicators that form the basis for the State of Europe's Forests Report. It can also be noted that the Forest Europe process has attempted to establish a legally binding agreement (LBA) on forests for the pan-European region. However, the discussions for this LBA broke down during negotiations in 2007 and, more recently, for the eighth Ministerial Conference on the Protection of Forests in Europe in 2021. Currently, Germany holds the chairmanship of the Forest Europe process and the Liaison Unit is located in Bonn (LUBO) and acts as a secretariat.

The joint **United Nations Economic Commission for Europe (ECE) Committee on Forests and the Forest Industry** (COFFI) and the **FAO European Forestry Commission** (EFC)[21] work together to strengthen the forest sector and contribute to sustainable development in the pan-European context. The current programme of work pursued by the joint UNECE/FAO Forestry and Timber Section covers many traditional topics related to forest products and services, such as forest resources and forest products markets. It also addresses issues related to climate change, SFM in general and linkages between SFM and the green economy, such as innovation in the forest sector and ecosystem services.

Other forest-relevant agreements include the Pan-European Biological and Landscape Diversity Strategy (PEBLDS),[22] the Bern Convention on the Conservation of European Wildlife and Natural Habitats includes a list of forest species to be protected, and the European Landscape Convention relates in parts to forests. In addition, two sub-regional agreements, the Carpathian Convention and the Alpine Convention, have protocols on forests that are binding to several countries in the respective mountain regions.

2.4.2. Global forest policy

The Rio Summit, held in 1992, was a significant turning point in international environmental policy as it represented a moment when three global conventions (the Rio Conventions) at the United Nations Conference on Environment and Development (UNCED) were adopted. However, at this time, no agreement could be reached on forests nor in the years following the summit. This means no global convention on forests (Pülzl, 2010). There are nevertheless legally binding conventions that do relate directly to the forest value chain, such as the International Tropical Timber Agreement (ITTA), the United Nations Convention on Biological Diversity (CBD), and the United Nations Framework Convention on Climate Change (UNFCCC). In addition, there are other legally binding agreements for the international trade of forest products, such as the World Trade Agreement (WTA) and the Convention on International Trade in Endangered Species of Wild Fauna and Flora (CITES).[23]

International Tropical Timber Agreement

The International Tropical Timber Agreement (ITTA)[24] is an intergovernmental, legally binding agreement to provide a framework for cooperation between tropical timber producers and consumers. The first ITTA was signed in 1983 and entered into force in 1985. We are presently in the third iteration of the ITTA, negotiated under the UN Conference on Trade and Development in 2006 and entered into force in 2011. It is valid until 2021. The ITTA is administrated by the International Tropical Timber Organization (ITTO), and its decision-making body is the International Tropical Timber Council (ITTC).

Convention on Biological Diversity

The Convention on Biological Diversity (CBD)[25] opened for signatures at the Rio Summit and entered into force in 1993. It is a legally binding treaty under the UN, and it is the only global framework addressing forest biological diversity and genetic resources, their conservation and sustainable use. Notably, the Kunming-Montreal Global Biodiversity Framework (KM-GBF) recently replaced the 2020 Aichi Biodiversity Strategic Goals and Targets included in the Strategic Plan for Biodiversity 2011–2020. Forest-related targets outlined within the GBF agreement are expected to become increasingly important. National biodiversity strategies, plans, or programs and the expanded work program on forest biological diversity are the main policy tools for global forest biodiversity conservation. The expanded program of work on forest biological diversity on forest biodiversity contains 12 goals, 27 objectives, and 130 actions arranged around three priorities: (1) Conservation, sustainable use, and access and benefit-sharing; (2) institutional and socio-economic enabling environment (governance); and (3) knowledge, assessment, and monitoring.

United Nations Framework Convention on Climate Change

The United Nations Framework Convention on Climate Change (UNFCCC),[26] which also opened for signature at the Rio Summit and entered into force in 1994. It is the first global environmental treaty addressing climate change and setting targets for reducing greenhouse gas (GHG) emissions. Relating to forests, the Kyoto Protocol and Paris Agreement set out a legally binding obligation for accounting, monitoring, and reporting of GHG emissions/sinks from the Land Use, Land Use Change, and Forestry (LULUCF) sector.[27]

After the Rio Summit and the breakdown in negotiations on a global forest convention, the Intergovernmental Panel on Forests (IPF) was established in 1995 under the United Nations Commission on Sustainable Development (CSD). The IPF was later replaced by the Intergovernmental Forum on Forests (IFF) in 1997 and, most recently, by the United Nations Forum on Forests (UNFF)[28] in 2000. This is presently the main intergovernmental fora[29] for international forest policy, however, governments negotiating under the UNFF have not found a consensus for a legally binding convention on forests.

Despite the absence of a binding agreement, and as part of the International Arrangement on Forests (IAF) that came out of the Rio Summit,[30] states have adopted two non-legally binding policy documents related directly to forests. These are the 'Non-Legally Binding Authoritative Statement of Principles for a Global Consensus on the Management, Conservation and Sustainable Development of All Types of Forests' (or the Forest Principles) and Chapter 11 'Combating Deforestation' in Agenda 21, a non-binding action plan of the UN regarding sustainable development. In addition, under the UNFF, states have adopted the 'UN Non-Legally Binding Instrument on All Types of Forests' (commonly known as the Forest Instrument), the first time countries have agreed on a global common definition of SFM.[31] Moreover, in 2014, following the Climate Summit held in New York, the voluntary and non-legally binding New York Declaration on Forests (or the New York Declaration) was released as a pledge to halve the rate of deforestation by 2020 and to end it by 2030.[32] It can also be noted that states have adopted the UN Strategic Plan for Forests (UNSPF) for the 2020–2030 period in 2017. The Strategic Plan features six Global Forest Goals and 26 associated targets to be reached by 2030, which are voluntary.

The Declaration on Forests and Land Use and the Global Forest Finance Pledge

Under the umbrella of the UNFCCC and COP26 in Glasgow, 124 countries have signed a Declaration on Forests and Land Use,[33] (or the Glasgow Declaration) on 2 November 2021. The Declaration aims to end and reverse deforestation by 2030 and achieve a sustainable land-use transition. In contrast to the New York Declaration, substantial progress has also been made regarding mobilising forest finance. As part of issuing a Global Forest Finance Pledge,[34] the EU, Canada, United Kingdom, Norway, South Korea, and the United States of America have announced that they will provide 12 billion USD of public climate finance between 2021 and 2025. This financial pledge aims to support action on restoring degraded land, tackling wildfires, and advancing the rights of indigenous peoples and local communities. It can also be noted that the national governments of 28 countries have committed to stopping deforestation associated with the global trade of food and other agricultural products, such as palm oil, soya and cocoa. These are areas that, in part, drive global deforestation, such as forest loss generated by the demand for agricultural land. Major financial companies have likewise committed to ending investments in deforestation-related activities.

While recent political and private sector commitments are encouraging, earlier efforts have not been successful. For example, a recent assessment of the New York Declaration, which set out similarly ambitious targets to end deforestation by 2030, argues that the Declaration failed to deliver on crucial pledges (UNDP, 2019). In fact, deforestation has increased, and restoration has been progressing more slowly than expected since the New York Declaration was signed. The assessment also argues that forest finance is only a small part of what is needed to restore and save our forests. However, what may become a deciding factor for the Glasgow Declaration is the scale at which national governments have taken up the Declaration. For instance, the New York Declaration was only backed up by 40 countries; in contrast, 143 countries signed the Glasgow Declaration, accounting for over 90% of the world's forests (Uhlich and Burck, 2022). In contrast to the New York Declaration, this is coupled with significant financial pledges. While it is too early to tell, these developments may strengthen the existing commitment to stop deforestation by 2030. However, more generally, if the mitigation benefits of forests are to be realised, then supportive policies, such as the Glasgow Declaration, are needed, including in the rules and guidelines for implementing the Paris Agreement.

3. What is EU forest-relevant policy?

It is challenging to provide a precise overview of EU forest-relevant policy for two reasons. First, as noted earlier, the underlying legislative complexity of EU forest policy means the EU can only propose legislation linked to its shared or exclusive competencies, as outlined in the TFEU.[35] This is due to the need for provisions for forests or forestry, aside from cork, in the TFEU. It is further complicated because the EU has adopted forest-focused and forest-related legislation, such as the EU Timber Regulation, the Common Agricultural Policy (CAP) and the European Green Deal, directly affecting the forest-based sector. Second, the industries that use forests and wood represent a wide range of products, sectors and enterprises (Aggestam and Pülzl, 2018, Rivera León et al., 2016), ranging from sawnwood, construction, recreation, furniture, energy, paper and pulp, plastics and textiles (Figure 2.3). This implies that the soft and hard policy instruments that are relevant depend on the given sector (Aggestam and Pülzl, 2018, 2020, Aggestam and Giurca, 2021). So, without providing a clear sectoral delimitation, this means a wide range of instruments can be considered as relevant (see Figure 2.3).

The present section provides a brief overview of different policy domains and instruments that affect the forest-based sector, covering forest-focused and forest-related policy. It will be a partial overview, as the main purpose is to demonstrate the various policy instruments affecting the forest-based sector. It is also a time-fixed overview as the EU policy landscape continuously changes.

3.1. Forest-focused policy

Even though the EU lacks explicit competence in forests, it has been argued that the EU has an implicit forest policy, mainly because it has adopted policies that affect forests and the forest-based sector directly (Aggestam and Pülzl, 2018).

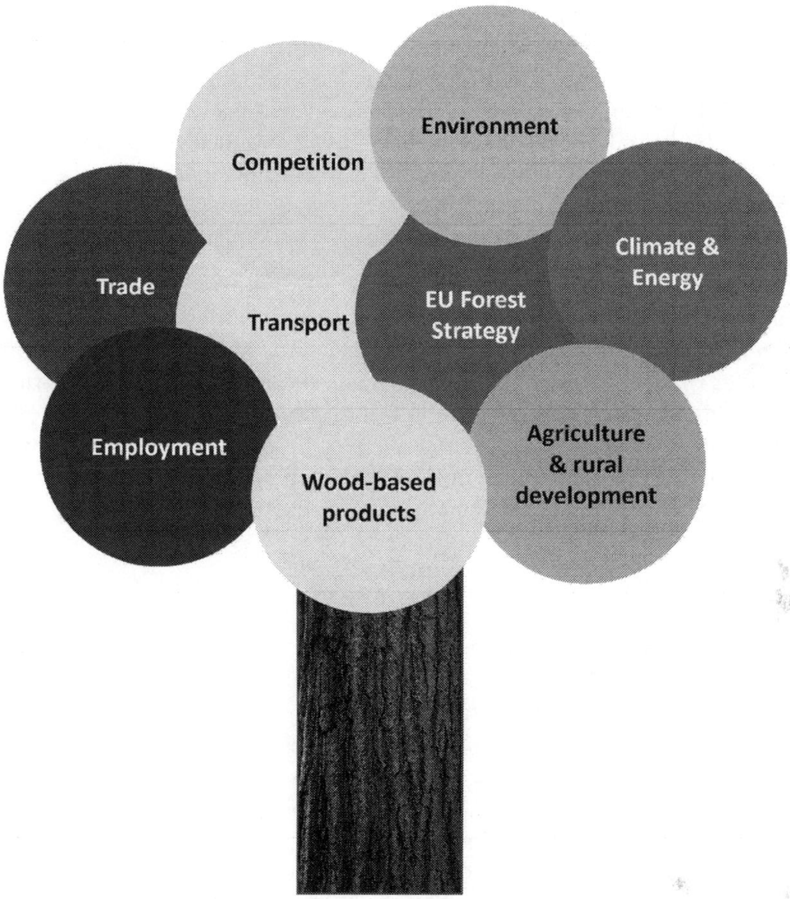

Figure 2.3 Forest-relevant policy domains.

Source: own figure

3.1.1. The EU Forest Strategy

The three forest strategies the European Commission has developed are at the heart of EU policymaking on forests. These strategies have been reinforced by two action plans supporting overall implementation until 2020. Even though they are voluntary instruments, not prioritised by industry (Aggestam and Pülzl, 2018), they can be seen as strategically important documents for the forest-based sector.

3.1.1.1. A FORESTRY STRATEGY FOR THE EUROPEAN UNION: 1998–2011

The first EU Forest Strategy, adopted in 1998, provided general guidelines for an EU forest policy designed to coordinate other forest-relevant policies (EC, 1998). The Strategy employed key principles related to SFM and addressed, among other things, competitiveness, job creation, forest protection, and the delivery of forest ecosystem services through a multi-functional approach. In addition, the first Strategy explicitly noted the domains of EU competence and relevant processes and platforms through which coordination should take place, examples being the SFC and consultative committees on forests, which later became the Civil Dialogue Group on Forestry and Cork and the Expert Group Forest-based industries and Sectorally Related Issues. The policy tool for implementing the first Strategy was the EU Forest Action Plan, covering the 2007 to 2011 period (EC, 2006, Pelli et al., 2012). This plan was operationalised through voluntary cooperation between EU Member States (no enforcement capabilities), with some coordinating actions implemented by the EC.

> The evaluation of the Forest Action Plan demonstrated that if an EU Member State did not assign any direct relevance to the non-binding and voluntary EU Forest Strategy, it did not matter whether the Strategy already aligned with relevant policy priorities at the EU level. In other words, no EU Member States chose to implement activities contrary to their national interests or agendas. There was a general lack of activities attributable to the Strategy, mainly due to the absence of binding targets, varying commitments to action, and the lack of direct funding (Aggestam and Pülzl, 2020, Pelli et al., 2012).

3.1.1.2. A NEW EU FOREST STRATEGY FOR FORESTS AND THE FOREST-BASED SECTOR: 2013–2020

The second EU Forest Strategy was adopted in 2013 and followed the same path as the first Strategy (EC, 2013). It responded to emerging challenges for forests and the forest-based sector and served as an updated and integrative framework that addressed the increasing demands on forests while also accounting for changes in societal and policy priorities. Like the Forest Action Plan, the primary tool for implementation was the voluntary Forest Multi-Annual Implementation Plan (Forest MAP), which set out several relevant activities (EC, 2015, Wolfslehner et al., 2019). In contrast to the first iteration, the second Strategy recognised the complex and fragmented forest policy environment; however, its priority areas still tried to cover various conflicting policy activities and goals. These ranged from applying the 'cascading use of wood' principle to using forest biomass for energy and carbon sequestration and halting global deforestation (Wolfslehner et al., 2019, EC, 2015, 2018c).

Regarding forest-based industries, the second Strategy only make general references to the industry (e.g., referring to general strategies). Interestingly, it does not refer to climate- and energy-related legislation, only the Kyoto Protocol and related strategies. It singled out the Natura 2000 legislation, the overarching EU Environmental Action Programme, the Biodiversity Strategy and international agreements. However, for instance, environmental legislation concerning waste management and wood-based products was not addressed.

> While the second Strategy and the Forest MAP recognised the complex and fragmented forest policy environment, it was significantly hampered by its lack of legal liability and direct connection to EU financial policy instruments (Aggestam and Pülzl, 2020, Pelli et al., 2012). Research has also shown that the instruments that the Strategy covers are not considered relevant to forest-based industries. For example, from an industries perspective, it was mainly the EU Timber Regulation (Regulation, 995/2010, 607/2012, 363/2012) and

the Forest Law Enforcement, Governance and Trade (FLEGT) Regulations (Regulation, 1024/2008, Regulation, 2173/2005) that were considered to be relevant (Aggestam and Pülzl, 2018). This would suggest that the EC and the forest-based industries have very different perspectives on what would constitute relevant forest and wood policy.

3.1.1.3. THE NEW EU FOREST STRATEGY FOR 2030: 2021–2030

It is important to note that the third EU Forest Strategy came out of a policy discussion that included the new EU Biodiversity Strategy (EC, 2020a), an EU taxonomy for sustainable activities[36] (Regulation, 2020/852), the 2050 Climate Change Mitigation Strategy (EC, 2018b), as well as the Adaptation Strategy (EC, 2021f) and Circular Economy Action Plan (EC, 2020b). These initiatives have been pushed as important components of the European Green Deal, providing a background for the most recent iteration of the EU Forest Strategy (EC, 2021a, 2021g).

The third Strategy has been designed to protect and restore EU forests. The EC has characterised the objective of the third Strategy as 'a vision and concrete actions to improve the quantity and quality of EU forests and strengthen their protection, restoration and resilience. Furthermore, it aims to adapt Europe's forests to the new conditions, weather extremes and high uncertainty brought about by climate change' (Forest Strategy website).[37] It is further argued that the third Strategy will support the socio-economic functions of forests for rural areas and boost the forest-based bioeconomy (EC, 2021a, 2021c, 2021d, 2021g). This will be achieved through:

- forest restoration and reinforced SFM for climate adaptation and forest resilience;
- sustainable use of wood-based resources for bioenergy;
- re- and afforestation of biodiverse forests;
- protecting primary and old-growth forests;
- skill development and empowering people for a sustainable forest bioeconomy;
- financial incentives for forest owners and managers;
- sustainable forest bioeconomy for long-lived wood products;
- supporting a non-wood bioeconomy.

The third Strategy also focus on (1) developing an EU-integrated forest monitoring framework, (2) implementing and enforcing the Community acquis, (3) developing a more coherent EU forest governance framework, and (4) improving the research and innovation agenda on forests. It puts forward a range of upcoming actions, such as a future proposal on new EU SFM indicators, thresholds, and ranges; an EU voluntary closer-to-nature forest management certification scheme; the further development of payment for ecosystem services; and a proposal on EU Forest Observation, reporting and data collection. It also clearly recognises the importance of forest owners in implementing all the outlined actions, such as through educational programmes. In parallel, it can be noted that a roadmap for planting three billion trees across Europe by 2030 accompanies the third Strategy (EC, 2021a, 2021d). This measure, which is driven mainly by the new Biodiversity Strategy (EC, 2020a), will have to follow clearly defined ecological standards to meet the guiding principle of 'plant and grow the right tree in the right place, for the right purpose' (EC, 2021d, p. 3).

The third Strategy is more expansive than its predecessors; however, the lack of other policy domains is evident, focusing mainly on the environment and climate while neglecting other domains highlighted by the European Green Deal, such as the EU's social and growth objectives. Aside from emphasising the role of wood in the construction and building sector (p. 5), the third Strategy does not address the importance of the economy of wood. For instance, the Strategy primarily considers the substitution benefits of using wood products, not the other wide range of wood-producing functions of forests.

The Strategy acknowledges previous challenges in the forest-based sector, such as limited funding allocated to forests in earlier rural development funds (see p. 17). This recognition addresses criticisms of past strategies (Wolfslehner et al., 2019, Aggestam and Giurca, 2021, Gordeeva et al., 2022). Nevertheless, the Strategy's main objectives remain focused on reforestation and afforestation (see p. 15), supported by a roadmap (EC, 2021a, 2021d). This emphasis suggests that afforestation will remain a key approach for EU Member States, consistent with trends over the last two programming periods (Pelli et al., 2012, Alliance Environnement et al., 2018, Wolfslehner et al., 2019). As part of the third Strategy, the EC has also proposed a legislative framework for enhanced forest monitoring and strategic plans (EC, 2021a, 2021g, Aggestam and Giurca, 2021). This upcoming legislation, pending adoption, aims to improve forest monitoring capabilities.

Interestingly, regarding EU forest governance, the Strategy proposed merging the SFC with the Working Group on Forest and Nature. While this merger was not finalised, concerns were raised about creating a single expert group with a broader scope (EUSTAFOR, 2021, Gordeeva et al., 2022). A key challenge was seen as managing the increasing range of issues within one group, which sparked fears that forest-specific issues might not receive adequate priority.

Furthermore, despite criticism of previous Strategies for failing to address the incoherence in EU forest policy (Aggestam and Pülzl, 2018, 2020, Gordeeva et al., 2022), the new Strategy does not propose any significant changes to improve governance and inter-service communication within the European Commission. Instead, the Strategy focuses on enhancing communication between relevant ministries in EU Member States, particularly those responsible for the environment and forests. This approach implies that the third Strategy may not substantially increase coherence between EU forest-relevant policies (Aggestam and Giurca, 2021). Introducing mechanisms for effective policy coordination at the EU level could have been a strategic move to reconcile conflicting policy objectives affecting forests. However, the Strategy appears to maintain the status quo of its predecessors, potentially overlooking opportunities for more integrated and effective forest policy coordination at the EU level.

3.1.2. Timber production and illegal logging

In addition to the three forest strategies, two forest-focused legislations have been developed by the EC to promote legally produced wood and reduce illegal logging outside the Union:

The **EU Timber Regulation** (EUTR) was enacted in 2013 as part of the FLEGT Action Plan. The EUTR specify the obligations of operators and traders[38] who place wood and wood products on the EU market for the first time. The EUTR applies to imports from countries outside the EU, excluding Norway, Iceland and Liechtenstein. For example, it prohibits selling illegally harvested wood or wood products derived from such wood and requires operators and traders to exercise 'due diligence'. The due diligence obligations[39] imply that operators and traders must be able to specify from where they have bought their wood-based products and to whom they have supplied their products. Products covered by the EUTR are listed in the Annex of the Regulation.

More recently, the **regulation on deforestation-free products** (Regulation, 2023c), repealing the EUTR, aims to limit EU-driven deforestation and forest degradation. The new regulation obliges enterprises to demonstrate that their goods are legally produced and deforestation-free, following the due diligence rules. It implies obligations for enterprises trading in, for example, palm oil, soy, cocoa, coffee, rubber, charcoal, and wood. This may have significant global implications as palm oil and soya account for more than two-thirds of global deforestation (FAO, 2020).

The **Forest Law Enforcement, Governance and Trade (FLEGT) Regulation** aims to reduce illegal logging by strengthening sustainable and legal forest management, improving governance and promoting trade in legally produced wood outside the EU. This includes the EU FLEGT Action Plan, which sets out seven measures to support timber-producing countries, promote trade in legal timber, promote environmentally and socially beneficial public procurement policies, support private sector initiatives (e.g., FLEGT licensing) and address the problem of conflict timber (e.g., shaping EU development aid). FLEGT essentially help improve forest governance in third countries and reduce levels of illegal wood harvesting and related trade, mainly through the prevention of the import of illegal wood into the EU from FLEGT partner countries. FLEGT uses trade deals in the form of Voluntary Partnership Agreements (VPAs) between wood-producing countries and the EU to ensure that wood from those countries is verifiably legally logged and traded. In other words, the VPAs are meant to ensure that wood can be traced back to its origin.

The first review of the EUTR found that implementation of the regulation has been slow and insufficient. For example, for national enforcement bodies, it has been noted that more resources are needed to effectively enforce the EUTR (Regulation, 995/2010). The evaluation of the FLEGT Action Plan highlights that it has only been partially effective in raising awareness of the problem of illegal logging, improving global forest governance, and reducing the demand for illegal timber in the EU. It also raises concerns as 'deficient planning and monitoring systems, inadequate coordination and insufficient human and financial resources have hampered effectiveness and efficiency' in implementing the Action Plan (p. 3). There is furthermore little evidence that VPAs help implement the EUTR (EC, 2016). It can also be noted that the EC recently adopted a proposal for a regulation on deforestation-free products (EC, 2021h). The proposal is part of a broader plan of action to tackle deforestation and forest degradation and, if accepted, may have implications for the EUTR and FLEGT.

3.2. Forest-related policy

3.2.1. Agriculture and rural development

It is important to consider the Common Agricultural Policy (CAP) as around 90% of EU funding for forests comes from the European Agricultural Fund for Rural Development (EAFRD) (Regulation, 1305/2013). The CAP fundamentally shapes agricultural practices, rural development, and land use within the EU. This makes it an important instrument from a forestry perspective. Briefly, the EU reached an agreement on the reform of the new CAP, which entered into force in 2023 (Regulation, 2021/2117, Regulation, 2021/2115, Regulation, 2021/2116). For

2021 to 2022, a transitional regulation was in force, which meant that RDPs were conditionally extended until the end of 2022 (Regulation, 2020/2220). The CAP is financed through two funds as part of the EU budget: the European Agricultural Guarantee Fund (EAGF), which provides direct support and funds market measures, and the European Agricultural Fund for Rural Development (EAFRD), which finances rural development. Financing for forests in the EU comes mainly from the rural development pillar, providing the financial incentives for SFM, with a broad set of possible measures that EU Member States can prioritise and select.

The new programming period started in 2023, covering ten key objectives that are addressed through the EU Member States strategic plans[40] for the 2023–2027 period (EU, 2023).[41] It can be noted that the two pillars of the CAP have been combined under a single CAP strategic plan,[42] and the reformed CAP aims to provide more targeted support to smaller farms (the redistribution income support mechanism), a new 'green' architecture, a performance-based approach whereby EU Member States will have to report their achievements each year (Regulation, 2021/2115, Regulation, 2021/2116, Regulation, 2021/2117). Most EU Member States have scheduled support for forestry under the CAP, including an increase in forest areas, agroforestry, restoring forests after natural disasters and adverse weather, and improving forest infrastructure, as well as support for SFM (EU, 2023).

> Forestry has an important role in supporting SFM, the EU Forest Strategy, and other EU policy objectives and priorities, e.g., environment and climate-related objectives. A recent evaluation demonstrated that the RDPs have been quintessential in the uptake of forestry measures by EU Member States (Alliance Environnement et al., 2018). However, the design and implementation of the forestry measures under RDPs are also entirely in the hands of the EU Member States and relevant regions (e.g., views diverge significantly regarding the priorities for rural development). This means that programming approaches across the RDPs vary significantly. The result is a mixed picture of EU-added value and impacts associated with forestry measures across the EU (Alliance Environnement et al., 2018, Wolfslehner et al., 2019).

3.2.2. Wood-based products

There is no policy domain for 'products' nor specifically for wood-based products. This section refers to the many EU rules and regulations determining how EU forest-based industries can produce and label products. Relevant examples include regulations related to consumer product safety, such as the EU's **General Products and Safety Directive** (2001/95/EC). Standards have been developed under this directive to ensure consumers' safety. EU regulations regarding food packaging and food safety have likewise been adopted. For instance, regulations on **Food Contact Materials** provide a harmonised EU legal framework by laying down common rules for packaging materials and articles intended to come into contact with food (Regulation, 1935/2004). This regulation was amended in 2019 (Regulation, 2019/1381). Other standard-setting processes include instruments, such as the **EU Ecolabels** for tissue paper, converted paper products, newsprint paper, printed paper, copying and graphic paper, wooden furniture, and wooden floor coverings (Council Regulation, 880/92). The EU Ecolabel is a voluntary scheme established by the Commission in 1992 for products and services that meet specific environmental standards.

Another regulation relevant to the forest-based industries is the **Registration, Evaluation, Authorisation and Restriction of Chemicals** (REACH Regulation, 1907/2006), which provides a comprehensive framework for chemical manufacture and use in Europe. REACH is important for most manufacturers as it introduced an integrated registration and authorisation system for all chemical substances and products containing chemical substances produced or supplied in the EU. Similarly, the **Biocidal Product Regulation** (ECHA Regulation, 528/2012) concerns the placing on the market and use of biocidal products. This relates to substances used to treat wood-based materials against harmful organisms like pests or bacteria. Products that have been treated require authorisation before they can be placed on the EU market. Interlinked with the ECHA regulation, the **Classification, Labelling and Packaging Regulation** (CLP Regulation, 1272/2008), amended in 2010 and 2015, ensures that chemical hazards are communicated to EU workers and consumers. The CLP classification and labelling method are based on the United Nations' Globally Harmonised System. According to the CLP Regulation, suppliers of chemicals must classify, label, and package substances and mixtures.

The listed products-related regulations are non-exhaustive and primarily serve to illustrate that different regulations will affect forest-based industries in different ways, whereby, for example, paper and paperboard manufacturers may be more affected by REACH and regulations on Food Contact Materials whereas woodworking (e.g., sawnwood) may be more affected by the ECHA regulation. In other words, the degree of wood processing (primary to tertiary) and the product type (e.g., packaging materials) determines the applicability and effect of the respective legislative frameworks on different enterprises' value and supply chains.

3.2.3. Construction

The legal framework guiding construction products is the **Construction Products Regulation** (CPR Regulation, 305/2011). This regulation sets out the conditions for marketing construction products and the related use of CE marking,[43] as well as criteria for assessing construction products' performance. Certain obligations are also put on operators. For example, manufacturers must provide a performance declaration and technical documentation and affix the CE marking. Importers and distributors must check that the product is accompanied by technical documentation and bears the CE marking. The CPR does not set product requirements for construction products, and the EU Member States are responsible for safety, environmental and energy requirements applicable to construction works.

It can be noted that the **CPR regulation** has undergone a review[44] and a proposal for a revised CPR regulation was adopted in 2022 (EC, 2022d), which is set to simplify rules for construction products on the EU market. The proposal was released as part of the Circular Economy Package, including a proposal for a regulation on Eco-design for Sustainable Products (EC, 2022b) and an Eco-design and Energy Labelling Working Plan 2022–2024 (EC, 2022a) that aims to make products in the EU market more sustainable. The new CPR regulation would introduce sustainability elements as an integral part of product requirements for construction products. If the new harmonised standards for marketing construction products are adopted, it would imply that the EC would broaden its role regarding the standardisation of construction products.

The revision of the CPR is also supposed to play a role in the foreseen **Strategy for a Sustainable Built Environment**[45] that will address the sustainability performance of construction products. However, progress has yet to be made on the Strategy to date. Moreover, as part of the European Green Deal, the Commission has published a **renovation wave strategy** (EC, 2020d). This Strategy contains an action plan with concrete regulatory, financing and enabling measures to boost building renovation. It is foreseen that a revision of the **Energy Performance of Buildings Directive** will follow.

> The construction sector is significantly interlinked with energy policy. For example, to boost the energy performance of buildings, the EU has established a legislative framework that includes the **Energy Performance of Buildings Directive** and the **Energy Efficiency Directive** (Directive, 2010/31/EU, and 2012/27/EU). Both directives have been amended as part of the Clean Energy for all Europeans package in 2018 and 2019. The revision of these directives resulted in, among other things, the establishment of a set of standards and accompanying technical reports to support the Energy Performance of Buildings Directive called the energy performance of buildings standards (EPB standards). These are managed by the European Committee for Standardisation (CEN).[46]

3.2.4. Environment

The following sub-sections on the environment will cover a broad range of policy areas such as biodiversity and conservation, highlighting important EU directives like the Birds Directive and the Habitats Directive and strategies, including the Biodiversity Strategy for 2030. It will also introduce legislative aspects from air quality to waste management. Furthermore, it explores the emerging bioplastics sector, its environmental implications, and the critical area of sustainable finance, particularly the EU taxonomy for sustainable economic activities.

3.2.4.1. BIODIVERSITY AND CONSERVATION

Several EU policy instruments are crucial in protecting and enhancing biodiversity across various ecosystems, including forest ecosystems. The **Birds Directive** (Directive, 2009/147/EC) focuses on conserving wild bird species and their habitats. It establishes a Special Protection Areas (SPAs) network to safeguard important bird habitats, including bird species that depend on forest ecosystems. The **Habitats Directive** (Directive, 92/43/EEC) focuses on conserving natural habitats and species of community interest. It establishes the Natura 2000 network, which includes Special Areas of Conservation (SACs). Forest habitats, such as old-growth forests, are included in the Natura 2000 network. In addition, the Biodiversity Strategy for 2030 (EC, 2020a) is a comprehensive framework aimed at halting biodiversity loss and restoring ecosystems. It sets ambitious targets, including protecting 30% of land and sea areas, restoring degraded ecosystems, and mainstreaming biodiversity considerations into various policies, including forestry. The **LIFE Programme** (Regulation, 2021/783) is the main funding instrument for nature conservation and environmental projects. It supports initiatives related to preserving and restoring biodiversity, including forest conservation projects, promoting SFM practices, and protecting forest-dependent species.

The new **Nature Restoration Law** (EC, 2022e), following the goals set under the European Green Deal (EC, 2019b), is considered a recent landmark instrument and key element of the EU's Biodiversity Strategy for 2030 (EC, 2020a). It contains targets to restore at least 20% of degraded habitats by 2030 and all ecosystems in need of restoration by 2050. It can be noted that the text of the provisional agreement on the nature restoration regulation, which was revised by the European Parliament and by the Council of Europe in 2023, has been formally adopted by Council and will enter into force in 2024 (Council of the EU, 2023); however, the adopted text has been significantly revised. For example, measures to improve the sustainability of farming were voted down and certain amendments have weakened the prospective enforcement mechanisms of the regulation, undermining the EU Member States accountability in fulfilling biodiversity commitments. Nevertheless, the regulation sets out measures to enhance forest biodiversity, including indicators related to deadwood, forest structure, connectivity, carbon stock, and species diversity. Several articles in the regulation can be considered directly relevant for forests and the forest-based sector, such as on energy from renewable resources, on the restoration of forest ecosystems, and on the planting of three billion additional trees.

EU biodiversity and conservation policies are intrinsically linked to forests. These policies are crucial in influencing forest growth, the forest carbon sink, the trade-offs between wood harvest and carbon sequestration, and forest biodiversity conservation. Other key aspects include guiding responsible timber harvesting practices and enhancing forests' resilience to climate change's impacts, including addressing the increased risks of pests, diseases, and extreme weather events.

3.2.4.2. ENVIRONMENTAL PROTECTION

For forest-based industries, environmental policy is not about biodiversity or conservation but legislative acts concerned with environmental protection, such as air quality, acidifying substances, wastewater treatment and waste incineration. For example, the **Industrial Emission Directive** (IED Directive, 2010/75/EU) represents a key EU instrument regulating industrial installation pollutant emissions. The IED sets EU-wide emission limit values for selected pollutants and promotes using techniques that reduce emissions, such as applying Best Available Techniques (BATs) reference documents (BREFs).[47] The BAT documents are the reference for setting the permit conditions for installations covered by the IED. BREFs have been adopted for the production of pulp, paper and board, large combustion plants, waste treatment, surface treatment using organic solvents, including wood and wood products preservation with chemicals and wood-based panels, all of which can be considered as being relevant for the forest-based sector. In other words, the IED sets the conditions that industries need to fulfil to receive permits, and the EU Member States take the necessary measures to ensure that no installation is operated without a permit. For example, an important element of the permits is the requirement to adopt the BATs to achieve a high level of environmental protection.

There is similarly a body of legislation concerned with ambient air quality and emissions to air, such as the **Air Quality Framework Directive** (96/62/EC), the **Ambient Air Quality and Cleaner Air Directive** (2008/50/EC) and the **National Emission Ceilings Directive** (Directive, 2016/2284). The Air Quality Framework Directives specify all pollutants for which the EU

Member States must introduce air quality standards and objectives. While the Ambient Air and Cleaner Air Directive set standards for 12 air pollutants, such as particulate matter (PM10 & PM2.5), ozone, benzene and lead, the National Emission Ceilings Directive set the total emission limit for five pollutants responsible for acidification, eutrophication and ground-level ozone pollution (sulphur dioxide, nitrogen oxides, VOCs, PM2.5, and ammonia). In setting legally binding standards on air pollution emissions and controlling emissions of harmful substances, these EU regulations imply compliance costs for forest-based industries.

Waste is another fundamental part of the environmental policy domain that affects industrial operations. The **Waste Framework Directive** (2008/98/EC) establishes basic concepts and definitions for waste management. For example, it introduces the extended producer responsibility, the polluter pays principle, and recycling and recovery targets. These principles directly affect forest-based industries as the extended producer responsibility includes accepting returned products and the waste that remains after those products have been used, including the subsequent waste management and financial responsibility for such activities. The polluter pays principle, which states that waste management costs shall be borne by the original waste producer or by the current or previous waste holders, also leads to substantive obligations for the forest-based industry. Other waste-relevant legislation covers landfills (Directive, 1999/31/EC), wastewater treatment (Directive, 91/271/EEC) and packaging waste (Directive, 94/62/EC).

The **Environmental Liability Directive** (2004/35/EC) establishes a framework, based on the polluter pays principle, to prevent or remedy environmental damage to protected species and natural habitats, water and soil. As listed in the directive, companies carrying out dangerous activities fall under strict liability. Companies pursuing other activities not directly covered by the directive are liable for fault-based damage to protected species or natural habitats. This can create enforcement costs for enterprises if a causal link between activities and damages has been established.

> EU measures to protect the environment and human health through controlling pollutants and waste affect industrial operations that use wood, and generate significant compliance costs for the forest-based sector (Rivera León et al., 2016). This is arguably offset by the environmental benefits these instruments generate. However, this is also a continuously changing policy landscape. For example, the EU is presently updating most of the above-noted instruments, including the IED[48] and the Ambient Air Quality Directives,[49] as part of the measures set out in the European Green Deal.

3.2.4.3. BIOPLASTICS

Forest-based industries are generally seen as covering four sectors, namely, woodworking, furniture, pulp and paper manufacturing and printing. However, sectoral boundaries are increasingly being blurred with the development of new wood-based products and services. Several emerging sectors present significant opportunities for forest-based industries. For example, wood industry by-products (e.g., wood polymers and extractives) and bio-based products, in general, can be utilised to produce many new types of materials, products and services, such as the use of lignins and tannins in composites manufacturing, adhesives and foams. Cellulose-based plastics are bioplastics manufactured using cellulose or derivatives of cellulose. They are manufactured using softwood as the dominant raw material. However, while the global demand for bio-based plastics has increased, it still only represents a small market. For instance, in Europe, where bioplastics are being promoted, only 1% of the plastic produced annually is bioplastics (Bajpai, 2019).

In terms of relevant environmental EU laws affecting the production of cellulose-based plastics, the recent directive on the reduction of the impact of certain plastic products on the environment (the **Single Used Plastics Directive**) is the first attempt at the European level to tackle the problem of plastic marine litter in a structured manner (Directive, 2019/904). The Single Use Plastics Directive prohibits, among other things, single-use/disposable products, such as cotton buds, cutlery, plates and straws.

> While the Single Used Plastics Directive is a laudable and timely initiative, it has severe implications for forest-based industries involved in the production of bioplastics. More specifically, the guidelines on single-use plastic products (EC, 2021b) changed the definition of 'plastic' to include cellulose-based fibre as plastic. This development may hamper current and future investments into sustainable bio-based materials. Furthermore, it may not align with other EU policy initiatives. For example, the EU Bioeconomy Strategy has set out to promote the development and use of bio-based plastics (EC, 2012, EC, 2018d). This latter point highlights the incoherence found between EU policy instruments.

3.2.4.4. SUSTAINABLE FINANCE

The EU Taxonomy Regulation (Regulation, 2020/852) is a framework designed to assess the environmental sustainability of economic activities. It serves as a guide to identify which enterprises and investments can be considered climate-friendly, providing a pathway for organisations to gain recognition for their green investments. To be classified as environmentally sustainable under this regulation, economic activities must meet four key criteria: (1) make a substantial contribution to at least one of six defined environmental objectives, (2) not cause significant harm to any of these six objectives, (3) adhere to minimum social safeguards as outlined in Article 18 of the regulation, and (4) satisfy specific technical screening criteria relevant to each environmental objective. The six environmental objectives covered are climate change mitigation, climate change adaptation, the sustainable use and protection of water and marine resources, the transition to a circular economy, pollution prevention and control, and the protection and restoration of biodiversity and ecosystems. Thus, this taxonomy is a strategic tool in guiding investments towards more sustainable and ecologically responsible practices.

The taxonomy regulation acknowledges the role of forest management in climate change mitigation, particularly highlighting the role of forests as carbon sinks. However, this focus on forests as carbon sinks means that the regulation must fully account for other wood-based products' carbon sequestration potential. An exception is made for forest bioenergy, which is considered sustainable under the Renewable Energy Directive. Additionally, the regulation's criteria for defining SFM are not uniform but instead defer to national forestry laws, the guidelines established by the Forest Europe process, and recognised certification programmes, such as the Programme for the Endorsement of Forest Certification (PEFC). This approach allows for flexibility in applying SFM standards based on regional and national contexts.

Representatives from the forest-based industry have raised concerns that the criteria for SFM under the taxonomy regulation could impose additional administrative burdens on forest owners seeking to certify their management practices as sustainable.[50] Additionally, this regulation significantly impacts sustainable financial products, requiring financial market participants and advisors to disclose the environmental objectives of their investments, which could influence the sustainable management of forest resources. Criticism has also been directed at the taxonomy regulation for categorising nuclear and gas activities as transitional and thus sustainable. Even more, environmental NGOs have actively challenged the inclusion of bioenergy and forestry projects in the regulation, arguing that incentives for using woody biomass for energy production could adversely affect forest ecosystems.[51] Furthermore, there are claims that the current sustainability principles for SFM, as defined by Forest Europe, are insufficient and perpetuate the status quo.[52] This ongoing and active debate suggests that future amendments to the taxonomy regulation may be on the horizon (Begemann et al., 2023).

> The EU Taxonomy Regulation is a framework for assessing the environmental sustainability of economic activities, including forest management, by setting criteria for climate-friendly enterprises and investments while facing significant debate and potential amendments regarding its treatment of forest bioenergy and sustainable forest management standards.

3.2.5. Employment

Several EU policies and measures influence employment. Among these is the **Working Time Directive**, which aims to protect workers' health and safety by establishing minimum standards applicable throughout the EU concerning working hours, including limits to weekly working hours and daily rest periods (Directive, 2003/88/EC). There is also a legal framework on health and safety at work (the European Framework Directive), including a directive on the protection of workers from the risks related to exposure to carcinogens or mutagens at work (Directive, 89/391/EEC, Directive, 2004/37/EC, Directive, 98/24/EC). The basis for these directives and associated acts is to prevent risks in the workplace and promote safer and healthier working conditions (e.g., setting noise exposure levels), emphasising the well-being and health of workers. They also aim to improve productivity and competitiveness.

These directives are relevant for certain sectors of forest-based industries that are more labour-intensive and that have riskier working environments. Examples concern carrying and handling heavy loads, working with chemicals that can damage our health, and risks from loud noise and vibrations. For example, three lists of indicative occupational exposure limits have been published since 2000 concerning risks related to chemical agents at work. For forest-based industries, this implies having to take necessary preventive measures to reduce or eliminate risks, such as risks associated with having chemical agents in a workplace.

> National governments oversee setting their employment standards; however, the EU complements policy initiatives taken by the EU Member States by setting minimum standards in employment law. For forest-based industries, while these types of directives push up

production costs (Rivera León et al., 2016), they also guarantee better safety and health protection for employees. This is particularly important as, for example, the forestry industry has some of the highest incidences of fatal or major injuries as compared to other sectors (EC, 2008).

3.2.6. Transport

Transport policies are relevant for the entire forest-based value chain. This includes legislation regulating marine fuel sulphur content (Directive, 2012/33/EU, Directive, 2016/802) and carbon dioxide emissions from maritime transport (Regulation, 2015/757). The **Sulphur Directive** limits the maximum sulphur content of gas oils, heavy fuel oil in land-based applications, and marine fuels. The directive also contains additional fuel-specific requirements for ships calling at EU ports. This regulation may, as such, affect maritime transport costs. There is also a regulation on waste shipment (Regulation, 1013/2006) covering procedures for transboundary shipments. The **Waste Shipment Regulation** implements the provisions of the Basel Convention. It includes a ban on the export of hazardous wastes to non-OECD countries and the export of waste for disposal. The regulation was amended in 2014 (Regulation, 660/2014) and set out a control system for the movement of waste covering almost all types of waste shipped, including national and transit transports, except radioactive waste and a few other types of waste (Directive, 2015/719, Directive, 2002/7/EC, Regulation, 661/2009, Commission Regulation, 1230/2012).

Road freight transport is the backbone of logistics for the forest-based sector, including transport operations relating to forestry. The EU road transport policy, covering emission performance standards, maximum authorised dimensions and weights and legal obligations for drivers, consequently sets out some important operational conditions for the sector as a whole. For example, **road-haulage specifications** affect the dimensions and maximum weights authorised for transport, common rules for certain types of carriage of goods, speed limitations and testing of exhaust emissions (e.g., overloaded vehicles can impact road safety and cause road damage). The EU has also adopted a **Fuel Quality Directive** (Directive, 2009/30/EC) and a **Renewable Energy Directive** (RED II, Directive (2018/2001)), which provides technical and environmental specifications for fuels used in transport. For example, the fuel and energy directives regarding indirect land use change (ILUC) emissions requirements have been harmonised, which may affect forest land use patterns (e.g., increased bioenergy demand) and wood production. Also, as the use of wood for energy is intensifying (in part as RED II incentivises the burning of wood), it may push up the price of wood (as a raw material) due to increased competition over biomass (Rivera León et al., 2016). For example, a recent study that analysed the trade-offs between renewable energy targets and forest conservation in Europe, demonstrates that subsidised energetical use of biomass by RED III endangers forest carbon sinks and biodiversity. The study also argues that European forests are insufficiently protected by the Natura 2000 network to counteract increased use of forest biomass for bioenergy production (Stubenrauch et al., 2022).

The **Fit for 55 package** has generated numerous legislative proposals, such as an amending directive for RED II and calls for a new Fuel Quality Directive (e.g., the proposed Euro 7/VII emission standards (EC, 2022f) to set stricter standards for petrol and diesel cars, vans, lorries and buses). These changes are set to achieve climate neutrality in the EU by 2050 as part of the European Green Deal. It is, for example, expected that the use of renewable energy in the transport sector will be amended and that there will be a new 13% GHG intensity reduction target. This implies that the transport sector will see legislative changes soon and, in extension, the forest-based sector. It is, however, too early to speculate on what the impact will be.

3.2.7. Trade

The EU applies trade defence instruments in accordance with the World Trade Organization. This affects the operations of any enterprise, including those in the forest-based sector. The Commission monitors the application of trade defence instruments and follows up on the enforcement of measures. This includes **anti-dumping** and **anti-subsidy measures** (Regulation, 1225/2009, Regulation, 597/2009) to re-establish a competitive environment for EU industries when harmed by dumped or subsidised imports. The EU furthermore applies **common tariffs for all forest products** (Regulation 2658/87, 2015/1754). The tariff level is zero for most wood products, except for some manufactured wooden and furniture goods. However, this also relates to tariff barriers and duties that apply to exports and protectionist subsidies for rival goods from forest-based industries outside the EU.

Moreover, if you want to import wood or wood products into the EU, you must observe specific phytosanitary requirements. The **phytosanitary regulations**, covering a range of food safety and animal and plant health-related regulations, set criteria for goods entering the EU to prevent the entry and spread of diseases and pests into new areas (Regulation, 2019/2072). In this case, forest-based industries may be affected when there are requirements for phytosanitary certificates for regulated articles. In particular, manufacturers of wood packaging are affected by specific rules targeting their products (e.g., wooden pallets and other wooden packaging) that incorporate the standard for phytosanitary treatment and related marking (e.g., ISPM 15).[53] For instance, any wood packaging for export must meet the requirements of the country where it was manufactured and the destination country's requirements.

A general concern for the forest-based industries in the EU is that competition with third countries is negatively affected by unfair trading practices that make other countries more competitive. EU trade-related regulations may create an uneven playing field that restricts wood products trade and decreases the competitiveness of the forest-based sectors. For example, some sectors face both duties on imported raw materials and semi-finished products as fixed by the EU and tariffs on exports of finished products as fixed by foreign countries.

3.2.8. Climate and energy

Several policy instruments on climate change and energy affect the forest-based value chain. At the EU level, climate and energy are commonly bundled together, such as in the **2030 climate and energy framework** (EC, 2014) or, more recently, the **European Green Deal** (EC, 2019b) and the **Fit for 55 package**.[54] The Fit for 55 package is a set of proposals for new legislation that refers to the EU's target of reducing net GHG emissions by at least 55% by 2030 (EC, 2020e).

The **2030 climate and energy policy framework** builds on the 2020 climate and energy package (EC, 2014). The framework puts forward a governance framework based on national plans to assess progress over time and binding targets to increase the share of renewable energy and energy efficiency following the review of the Energy Efficiency Directive by 2030. In addition, the framework highlights the significant interlinkages between climate change and energy, including other policy areas such as environment and transport. More recently, the EU has set out to be climate-neutral by 2050. This objective is at the heart of the **European Green Deal**, covering nearly all EU policies, and aligns with commitments to global climate action under the Paris Agreement (EC, 2018b, EC, 2019b). As part of the European Green Deal, the Commission has proposed the first **European Climate Law** to enshrine the 2050 climate-neutrality target into law (EC., 2020).

Land Use, Land Use Change and Forestry (LULUCF) is another important component of ongoing emission-reduction efforts that relate directly to the forest-based sector. LULUCF has been fully included in the EU's climate targets for 2030 (Regulation, 2018/841), and it sets a binding commitment for the EU Member States to ensure that accounted emissions from land use are compensated by an equivalent accounted removal of CO2 from the atmosphere. It can be noted that the Commission has amended the existing LULUCF legislation with a delegated act setting forest reference levels that each country must apply between 2021 and 2025 (Regulation, 2021/268). Forest reference levels are benchmarks for accounting net emissions from existing EU forests. The revised LULUCF regulation furthermore accounts for emissions from biomass in energy production, which was not accounted for under previous EU law. The new provisions imply that more robust accounting rules and governance for forest management will be used to record and account for EU Member States' 2030 climate commitments. As it presently stands, the LULUCF regulation will allow EU Member States to reduce the amount of carbon dioxide they remove from the atmosphere over the next five years.

The **Renewable Energy Directive** (Directive, 2009/28/EC) demonstrates the significant inter-linkages between climate and energy legislation. It also highlights the connection between biomass production, wood mobilisation, and forest-based industries. For instance, the non-binding **Biomass Action Plan** was developed to promote the production and use of biomass as an alternative source of energy (EC, 2005, 2010). The recast directive requires the EU to fulfil at least 32% of its total energy needs with renewable energy by 2030 (Directive, 2018/2001). This includes new provisions for enabling self-consumption of renewable energy, an increased 14% target for the share of renewable fuels in transport by 2030, and strengthened criteria for ensuring bioenergy sustainability. There may also be further revisions to the Renewable Energy Directive (RED III) to align it with the **energy system integration and hydrogen strategies** adopted in 2020. The legislative proposal for RED III suggests phasing out electricity production from primary forest biomass by 2026. It also aims to avoid using forest biomass from highly biodiverse forests for electricity production (EC, 2021i).

While most EU energy legislation does not target specific sectors, it does affect the availability and the price of energy, which is as important for the forest-based sector as any other industry. For example, energy legislation affects direct energy costs when forest-based industries produce their electricity and indirectly when they buy and sell it on the grid. Relevant policy instruments include:

- The **Energy Efficiency Directive** (2012/27/EU) established a common framework of binding measures to promote energy efficiency as part of the 20–20–20 targets set by the 2020 climate and energy package. In 2018, as part of the **Clean Energy for all Europeans package** (EC, 2019a), the Directive on Energy Efficiency was amended to update the policy framework to 2030 and beyond (Directive, 2018/2002, Directive, 2018/844). The key element of the amended directive is an energy efficiency target of at least 32.5% for 2030.

- The **Energy Performance of Buildings Directive** (Directive, 2010/31/EU) is the main instrument that aims to improve the energy performance of buildings in the EU. New elements were also introduced into the buildings directive as part of the **Clean Energy for all Europeans package** (EC, 2019a) in 2018 (Directive, 2018/844), and further revisions are expected with the introduction of the **renovation wave strategy** (EC, 2020d) and objectives to decarbonise buildings.
- The **Fourth Energy Package** consists of the Electricity Directive (2019/944/EU) and the Electricity Regulation (2019/943/EU), the Risk-Preparedness Regulation (2019/941/EU) and the Agency for the Cooperation of Energy Regulators (ACER) Regulation (2019/942/EU). This is part of the Clean Energy for All Europeans package to redesign the EU electricity market.
- The **Energy Taxation Directive** (Council Directive, 2003/96/EC) sets minimum rates for energy taxation related to energy-bearing products (mineral oils, coal, natural gas and electricity) and the CO_2 emissions of specific energy sources. It also helps reduce market distortions resulting from national energy legislation and the measures adopted under different national energy legislation and those adopted under the ETS Directive (REF). It can be noted that the ETS Directive is expected to be revised as a key part of the Fit for 55 package and European Green Deal. The current proposal represents a major update, outlining a new ETS target of a 61% reduction by 2030 compared to 2005.[55] Given the prominent role that carbon removals will play in the future, it is expected that this will impact the forest-based sector.

EU legislation on climate and energy has significant implications for the forest-based sector, principally as the EU has pledged to become carbon neutral by 2050. This would suggest that there will be increasing pressure on forest management (e.g., ensuring that forests act as a carbon sink) and industry (e.g., accounting for emissions). Current policy frameworks are furthermore not without criticism. For example, RED II has been criticised for not providing full transparency on how EU Member States should set forest management reference levels. Also, incentives for burning wood, both through RED III and LULUCF regulations, increase demand for wood. However, it does not ensure that wood used for energy is climate-friendly (e.g., bioenergy production can be increased without accounting for carbon dioxide emission increases).

3.3. What is relevant EU forest-based policy?

The hard and soft EU policy instruments introduced in this chapter indicates that a broad range of EU policies affects the forest-based sector. Revisiting the initial discussion, it is further evident that these policy instruments have different effects across different segments of the forest-based sector. This diversity in impact largely depends on the specific position of an enterprise within the value and supply chain, which is inherently linked to the type of wood product being manufactured. For instance, a furniture manufacturing company is governed by different regulations and directives than other wood processing industries like sawmills. One illustration of this is the influence exerted by the Single Use Plastics Directive on the production of bioplastics. For instance, the inclusion of biodegradable and compostable plastics under its scope means that products made from forest-based materials must follow to the same restrictions and standards as conventional plastics. These regulatory variations across forest value chains highlights a critical point: determining what constitutes relevant forest-based policies is highly dependent on the stakeholders' perspective and the specific sector under consideration.

This critical point can be exemplified by the third EU Forest Strategy, which has faced criticism for failing to address a range of EU policy instruments that are important to the forest-based sector. While the Strategy focuses on several forestry-focused policy areas, it overlooks crucial policy instruments that, for example, hold significance for forest-based industries (Aggestam and Pülzl, 2018, Lier et al., 2021, Gordeeva et al., 2022). This oversight encompasses legislation related to energy production and use, industrial emissions, air quality, waste management, environmental liability, phytosanitary requirements, road safety, waste shipment, and marine fuels. Additionally, it neglects areas crucial for maintaining operational standards, such as workplace health and safety, product safety, human health protection, packaging, and trade-related legislation, including trade defence and tariffs (Söderberg and Eckerberg, 2013, Gordeeva et al., 2022, Stubenrauch et al., 2022).

Expanding further on the complexities of EU forest-based policy:

- **Climate change.** Climate-oriented policy instruments, such as the European Green Deal, focus on reducing carbon emissions and promoting renewable resources, which is expected to increase demand for forest products. For instance, legislation on climate change, like the EU's commitments under the Paris Agreement, emphasises the role of forests in carbon sequestration and biomass production.
- **Bioeconomy and circular economy.** The EU's push towards a bioeconomy, where renewable biological resources replace fossil-based materials, positions forests as critical natural resources, such as using forest resources to produce bio-based products. The circular economy concept, which focuses on reducing, reusing, and recycling materials, encourages the development of innovative wood products and efficient recycling processes, which is expected to enhance the sector's sustainability.
- **Trade and market regulations.** EU trade policies and agreements can affect the forest-based sector in terms of market access, tariffs, and phytosanitary standards. These factors play a role in shaping the international competitiveness of EU forest products. Market regulations and standards, including those for product safety and consumer protection, influence how forest products are produced, processed, and marketed within and outside the EU.
- **Socio-economic considerations.** EU labour laws and health and safety regulations impact employment practices within the forest-based sector. These regulations ensure worker safety, particularly in high-risk activities such as logging and sawmilling. It can also be noted that the social dimension of forest policies, including the rights of indigenous and local communities and the involvement of various stakeholders in forest management, is also increasingly recognised in EU policymaking.
- **Innovation and technological progress.** The EU's emphasis on research and innovation, as seen in programs like Horizon 2020, influences research and technological innovations in the forest-based sector. This includes developing new wood-based materials, sustainable forestry practices, and more efficient processing technologies. In addition, advances in digital technologies, such as earth observations and remote sensing for forest monitoring, as well as blockchain for traceability in the timber supply chain, are also influenced by EU policies promoting digitalisation and innovation.

These examples highlight that to understand the network of policy instruments at play, it is imperative to consider all policy domains that influence forest-based value chains, especially those dependent on wood as a raw material. This includes sectors like construction, textiles, bioplastics, and chemical production. Furthermore, evolving policy developments in areas like nature restoration, the nature restoration, bioeconomy and the circular economy necessitate a more integrative view of the forest-based value chain and context-specific interconnections. For instance, it is not only about producing carbon-neutral biomass, timber, and non-wood products but also about the approaches employed in their production and processing. Moreover, the transition towards a greener, circular or bioeconomy encompasses SFM, nature conservation, air quality, energy efficiency, sustainable trade practices, and public health. All these aspects are integral to existing EU policy objectives, highlighting the need for a shared understanding of what constitutes relevant EU forest-based policy.

Want to find out more?

For those seeking more knowledge on EU forest policy, please see the references provided below or some of the many references throughout the text.

- European forest governance post-2020 (Wolfslehner et al., 2020).[56]
- *European Forest Governance: Status Quo and Optimising Options with Regard to the Paris Climate Target* (Stubenrauch et al., 2022).
- Towards a sustainable European forest-based bioeconomy – assessment and the way forward (Winkel, 2017).
- *European Forest Governance: Issues at Stake and the Way Forward* (Pülzl et al., 2013).

The set of questions listed below may also help guide further inquiry into specific areas of interest:

1. How can varying national definitions of a forest impact international forest policy and management?
2. Why does the EU not have a common forest policy, and what are the implications of this for forest management across EU Member States?
3. What role does the EU play in international forest governance initiatives, and how does this impact global forest management practices?
4. How do EU forest-relevant policies impact forest-based industries, such as timber, paper, and biomass production?
5. What economic and environmental trade-offs are involved in managing forests for conservation and industrial use in the EU?
6. What emerging trends or challenges might influence the future direction of EU forest policy?

4. Assignment I: Understanding EU forest policy

Description

This assignment requires students to critically analyse the current state of forest policy within the European Union (EU), as outlined in Chapter 2. Students will explore the complexities of

EU forest governance, including the absence of a common EU forest policy, the role of the EU in international forest governance, and the interplay between different policy domains affecting forests.

Purpose

The aim is to improve the student's understanding of the multi-sectoral EU forest governance system, develop analytical skills in policy analysis, and encourage critical thinking about potential alternatives for EU forest policy.

Outline of task

- Read Chapter 2 to understand the key aspects of EU forest governance.
- *Identify and analyse key themes*: Focus on the EU forest strategy, forest ownership dynamics, and multifunctional forest use. Pay attention to the cross-sectoral policies influencing European forests.
- *Additional materials*: Read other relevant publications on EU forest policy and governance.
- *Draft a policy analysis*: Write a detailed analysis covering the strengths and weaknesses of a selected number of EU forest-focused and forest-related policies. Discuss how these policies impact forest management and conservation.
- *Propose alternatives*: Based on your analysis, suggest potential improvements or alternative approaches to EU forest policy. Consider aspects like sustainability, socio-economic impacts, and biodiversity conservation.

Suggested additional reading

- EU Forest Strategy documents and related policy papers (available on EUR-Lex[57] and other EU websites).
- Pülzl, H., Wydra, D. & Hogl, K. 2018. Piecemeal integration: Explaining and understanding 60 years of European Union forest policymaking. *Forests*, 9, 719.
- Aggestam, F. & Pülzl, H. 2020. Downloading Europe: A regional comparison in the uptake of the EU Forest Action Plan. *Sustainability*, 12.

Deliverable

- Prepare a structured essay with a clear introduction, analysis, and conclusion (length: 2000–2500 words).

5. Notes

1 See www.eea.europa.eu/data-and-maps/figures/country-groupings-in-europe-2.
2 Sweden had the largest area of forests and other wooded land, an estimated 30.3 million ha. The second largest area was recorded in Spain (28.0 million ha), followed by Finland (23.2 million ha), France (18.1 million ha), Italy and Germany (both at 11.4 million ha).
3 The principle of subsidiarity rules out Union intervention when the EU Member States themselves can effectively deal with an issue at the central, regional or local levels. See www.europarl.europa.eu/factsheets/en/sheet/7/the-principle-of-subsidiarity.
4 See https://en.wikipedia.org/wiki/Treaties_of_the_European_Union.
5 All EU departments and agencies are at https://ec.europa.eu/info/departments (accessed 29 November 2022).
6 All Committees can be found at www.europarl.europa.eu/committees/en/envi/home/highlights (accessed 29 November 2022).

7 The Lisbon Treaty amended the Maastricht Treaty, Treaties of Rome, and other documents to simplify and streamline the institutions that govern the EU.
8 See www.consilium.europa.eu/en/council-eu/preparatory-bodies/working-party-forestry/.
9 Committees or Working Parties may only be set up by the Council or the Committee of Permanent Representatives (Coreper).
10 See https://ec.europa.eu/transparency/regexpert/index.cfm?do=groupDetail.groupDetail&groupID=3550.
11 The governments of EU countries nominate members of the Committee.
12 See https://ec.europa.eu/transparency/expert-groups-register/screen/expert-groups/consult?lang=en&groupId=3888&fromMeetings=true&meetingId=49468.
13 See https://ec.europa.eu/transparency/regexpert/index.cfm?do=groupDetail.groupDetail&groupID=1356.
14 See https://ec.europa.eu/transparency/expert-groups-register/screen/expert-groups/consult?do=groupDetail.groupDetail&groupID=1329.
15 The Commission has set up a register listing Commission expert groups and similar entities (see https://ec.europa.eu/transparency/regexpert/).
16 See https://ec.europa.eu/transparency/expert-groups-register/screen/expert-groups/consult?do=groupDetail.groupDetail&groupID=3282.
17 The working group deals with drafting proposals for legislation on type-approval of agricultural and forestry vehicles (see www.consilium.europa.eu/en/council-eu/preparatory-bodies/working-party-technical-harmonisation/#).
18 See www.consilium.europa.eu/en/council-eu/preparatory-bodies/working-party-civil-protection/).
19 See https://foresteurope.org.
20 See https://foresteurope.org/ministerial-commitments/.
21 See https://unece.org/forests.
22 See www.cbd.int/doc/nbsap/rbsap/peblds-rbsap.pdf.
23 Cites are implemented through the Wildlife Trade Regulations (Regulation, 338/97).
24 See www.itto.int/.
25 See www.cbd.int/.
26 See https://unfccc.int/.
27 See https://unfccc.int/land-use-land-use-change-and-forestry-lulucf-under-the-kyoto-protocol.
28 See www.un.org/esa/forests/forum/unffs/index.html.
29 UNFF is a subsidiary body of the United Nations Economic and Social Council (ECOSOC), established by ECOSOC Res. 2000/35 with the main objective of promoting the management, conservation and sustainable development of all types of forests.
30 The IAF has five main components: the UN Forum on Forests (UNFF) and its Member States, the UNFF Secretariat, the Collaborative Partnership on Forests (CPF), the UNFF Global Forest Financing Facilitation Network (GFFFN), and the UNFF Trust Fund.
31 The UN General Assembly adopted this in 2007.
32 See https://forestdeclaration.org.
33 See https://ukcop26.org/glasgow-leaders-declaration-on-forests-and-land-use.
34 See https://ukcop26.org/the-global-forest-finance-pledge.
35 See https://eur-lex.europa.eu/legal-content/EN/TXT/PDF/?uri=CELEX:12012E/TXT.
36 The EU taxonomy is a classification system that has established a list of environmentally sustainable economic activities. It is part of the EU's plans to achieve net zero emissions. For example, it includes projects aimed at accelerating logging.
37 See https://ec.europa.eu/environment/strategy/forest-strategy_en.
38 '*Operators*' are businesses that initially place wood or wood products on the EU market, and '*traders*' are businesses trading in wood or wood products already placed on the EU market.
39 Due diligence obligations depend on an EC benchmarking system that assesses countries and the level of risk of deforestation and forest degradation driven by the respective commodities. Minimum inspection standards are set for competent authorities in line with the benchmarking system.
40 Country factsheets provide an overview of the agricultural sector and rural development at EU and MS levels (EU, 2023a).
41 EU Member States are free to choose the interventions it considers most effective to meet its specific objectives based on assessing its own needs.
42 25% of the direct payments will be allocated to eco-schemes, including providing agroforestry and carbon farming incentives.
43 See https://europa.eu/youreurope/business/product-requirements/labels-markings/ce-marking/index_en.htm.

44 The Commission presented an implementation report on the CPR in 2016 and engaged stakeholders in consultation through relevant technical platforms. The evaluation of the CPR was published in 2019, with a report on the relevance of the tasks of the European Organisation for Technical Assessment (EOTA).
45 See www.europarl.europa.eu/legislative-train/theme-a-european-green-deal/file-strategy-for-a-sustainable-built-environment.
46 See www.cen.eu.
47 See https://eippcb.jrc.ec.europa.eu/reference/.
48 See https://environment.ec.europa.eu/publications/proposal-revision-industrial-emissions-directive_en (accessed 22 November 2022).
49 See https://environment.ec.europa.eu/topics/air/air-quality/revision-ambient-air-quality-directives_en (accessed 22 November 2022).
50 See www.forestindustries.se/news/news/2021/04/eu-taxonomy/.
51 See https://forestdefenders.eu/environmental-coalition-sues-the-european-commission-over-inclusion-of-biomass-and-forestry-activities-in-eu-sustainable-finance-taxonomy/.
52 See www.birdlife.org/news/2022/09/06/eu-green-taxonomy-burning-biodiversity/.
53 See www.fao.org/3/mb160e/mb160e.pdf.
54 See www.consilium.europa.eu/en/policies/green-deal/fit-for-55-the-eu-plan-for-a-green-transition (accessed 25 November 2022).
55 See https://climate.ec.europa.eu/eu-action/eu-emissions-trading-system-eu-ets/revision-phase-4–2021 2030_en.
56 See https://efi.int/sites/default/files/files/publication-bank/2020/efi_fstp_10_2020.pdf.
57 See https://eur-lex.europa.eu/.

6. References

AGGESTAM, F. & GIURCA, A. 2021. The art of the 'green' deal: Policy pathways for the EU Forest Strategy. *Forest Policy and Economics*, 128, 102456.

AGGESTAM, F. & PÜLZL, H. 2018. Coordinating the uncoordinated: the EU Forest Strategy. *Forests*, 9, 125, DOI: https://doi.org/10.3390/f9030125.

AGGESTAM, F. & PÜLZL, H. 2020. Downloading Europe: A regional comparison in the uptake of the EU Forest Action Plan. *Sustainability*, 12.

ALLIANCE ENVIRONNEMENT, ORÉADE-BRÈCHE SARL, IEEP & EFI. 2018. Evaluation study of the forestry measures under rural development. Luxembourg: Publications Office of the European Union.

ARTS, B., GIESSEN, L. & VISSEREN-HAMAKERS, I. 2013. International forest policy and Europe: Four pathways of mutual influence. *In:* PÜLZL, H., HOGL, K., KLEINSCHMIT, D., WYDRA, D., ARTS, B., MAYER, P., PALAHI, M., WINKEL, G. & WOLFSLEHNER, B. (eds.) *European Forest Governance: Issues at Stake and the Way Forward.* EFI Series: What Science can tell us; 37–43.

BAJPAI, P. 2019. *Biobased Polymers: Properties and Applications in Packaging*. Amsterdam, The Netherlands, Elsevier.

BEGEMANN, A., DOLRIIS, C. & WINKEL, G. 2023. Rich forests, rich people? Sustainable finance and its links to forests. *Journal of Environmental Management*, 326, 116808.

CHAZDON, R. L., BRANCALION, P. H. S., LAESTADIUS, L., BENNETT-CURRY, A., BUCKINGHAM, K., KUMAR, C., MOLL-ROCEK, J., VIEIRA, I. C. G. & WILSON, S. J. 2016. When is a forest a forest? Forest concepts and definitions in the era of forest and landscape restoration. *Ambio*, 45, 538–550.

CIESIELSKI, M., TKACZYK, M., HYCZA, T. & TACZANOWSKA, K. 2022. Was it really different? COVID-19-pandemic period in long-term recreation monitoring – A case study from polish forests. *Journal of Outdoor Recreation and Tourism*, 100495.

COMMISSION REGULATION 1230/2012. Type-approval requirements for masses and dimensions of motor vehicles and their trailers and amending Directive 2007/46/EC. *L 353/31*. Brussels: Official Journal of the European Union.

COSTA, O. & BRACK, N. 2019. *How the EU Really Works*. Routledge.

COUNCIL DIRECTIVE 2003/96/EC. Restructuring the Community framework for the taxation of energy products and electricity. *L 283/51*. Brussels: Official Journal of the European Union.

COUNCIL OF THE EU 2023. Nature restoration: Council and Parliament reach agreement on new rules to restore and preserve degraded habitats in the EU. www.consilium.europa.eu/en/press/press-releases/2023/11/09/nature-restoration-council-and-parliament-reach-agreement-on-new-rules-to-restore-and-preserve-degraded-habitats-in-the-eu/: Council of the EU.

COUNCIL REGULATION 880/92. Community eco-label award scheme. *No L 99 / 1.* Brussels: Official Journal of the European Communities.

DIRECTIVE 89/391/EEC. The introduction of measures to encourage improvements in the safety and health of workers at work. *L183/1.* Brussels: Official Journal of the European Union.

DIRECTIVE 91/271/EEC. Urban waste water treatment. *L 135/40.* Brussels: Official Journal of the European Communities.

DIRECTIVE 92/43/EEC. The conservation of natural habitats and wild fauna and flora. *L 206/7.* Official Journal of the European Union: European Council.

DIRECTIVE 94/62/EC. Packaging and packaging waste. *L 365/10* Brussels: Official Journal of the European Communities.

DIRECTIVE 96/62/EC. Ambient air quality assessment and management. Brussels: Official Journal of the European Communities.

DIRECTIVE 98/24/EC. Protection of the health and safety of workers from the risks related to chemical agents at work. *L 131/11.* Brussels: Official Journal of the European Communities.

DIRECTIVE 1999/31/EC. Landfill of waste. *L 182/1.* Brussels: Official Journal of the European Communities.

DIRECTIVE 2002/7/EC. Laying down for certain road vehicles circulating within the Community the maximum authorised dimensions in national and international traffic and the maximum authorised weights in international traffic. *L 67/47.* Brussels: Official Journal of the European Communities.

DIRECTIVE 2003/88/EC. Organisation of working time. *L 299/9.* Brussels: Official Journal of the European Union.

DIRECTIVE 2004/35/EC. Environmental liability with regard to the prevention and remedying of environmental damage. *L 143/56.* Brussels: Official Journal of the European Union

DIRECTIVE 2004/37/EC. Protection of workers from the risks related to exposure to carcinogens or mutagens at work *L 158/50.* Brussels: Official Journal of the European Union.

DIRECTIVE 2008/50/EC. Ambient air quality and cleaner air for Europe Official Journal of the European Communities: European Council.

DIRECTIVE 2008/98/EC. On waste and repealing certain directives. Official Journal of the European Communities: European Council.

DIRECTIVE 2009/28/EC. Promotion of the use of energy from renewable sources and amending and subsequently repealing Directives 2001/77/EC and 2003/30/EC. *L 140/17.* Official Journal of the European Union: European Parliament and Council.

DIRECTIVE 2009/30/EC. Specification of petrol, diesel and gas-oil and introducing a mechanism to monitor and reduce greenhouse gas emissions. Brussels: Official Journal of the European Union.

DIRECTIVE 2009/147/EC. Conservation of wild birds. Brussels: Official Journal of the European Union.

DIRECTIVE 2010/31/EU. Energy performance of buildings. *L 153/13.* Brussels: Official Journal of the European Union.

DIRECTIVE 2010/75/EU. Industrial emissions (integrated pollution prevention and control). Official Journal of the European Communities: European Council.

DIRECTIVE 2012/27/EU. Energy efficiency. Brussels: Official Journal of the European Union.

DIRECTIVE 2012/33/EU. Sulphur content of marine fuels. *L 327/1.* Brussels: Official Journal of the European Union.

DIRECTIVE 2015/719. Laying down for certain road vehicles circulating within the Community the maximum authorised dimensions in national and international traffic and the maximum authorised weights in international traffic. *L 115/1.* Brussels: Official Journal of the European Union.

DIRECTIVE 2016/802. Reduction in the sulphur content of certain liquid fuels. *L 132/58* Brussels: Official Journal of the European Union

DIRECTIVE 2016/2284. Directive on the reduction of national emissions of certain atmospheric pollutants, amending Directive 2003/35/EC and repealing Directive 2001/81/EC. Brussels: Official Journal of the European Union.

DIRECTIVE 2018/844. Amending Directive 2010/31/EU on the energy performance of buildings and Directive 2012/27/EU on energy efficiency. *L 156/75.* Brussels: Official Journal of the European Union.

DIRECTIVE 2018/2001. Promotion of the use of energy from renewable sources. *L 328/82* Brussels: Official Journal of the European Union

DIRECTIVE 2018/2002. Energy efficiency. *L 328/210* Brussels: Official Journal of the European Union.

DIRECTIVE 2019/904. Reduction of the impact of certain plastic products on the environment. *L 155/1.* Brussels: Official Journal of the European Union.

EC 1978. Proposal for a Council resolution concerning the objectives and principles of forestry policy. *COM (1978) 621*. Brussels: European Commission.
EC 1998. A forestry strategy for the European Union. *COM 1998/649 final*. Brussels: European Commission.
EC 2005. Biomass Action Plan. *COM(2005) 628 final*. Brussels: European Commission.
EC 2006. EU Forest Action Plan. *COM(2006) 302 final*. Brussels: European Commission.
EC 2008. Occupational safety and health in Europe's forestry industry. *E-Facts 29*. Brussels: European Commission.
EC 2010. Sustainability requirements for the use of solid and gaseous biomass sources in electricity, heating and cooling. *COM(2010)11 final*. Brussels.
EC 2012. Innovating for Sustainable Growth: A Bioeconomy for Europe. *COM(2012) 60 final* Brussels: European Commission.
EC 2013. A new EU Forest Strategy: for forests and the forest-based sector. *COM(2013) 659 final*. Brussels: European Commission.
EC 2014. A policy framework for climate and energy in the period from 2020 to 2030. *COM(2014) 015 final*. Brussels: European Commission.
EC 2015. Multi-annual Implementation Plan of the new EU Forest Strategy. *SWD(2015) 164*. Brussels: European Commission.
EC 2016. Evaluation of the EU Action Plan for Forest Law Enforcement Governance and Trade (FLEGT). *SWD(2016) 275*.
EC 2018a. Bioeconomy: the European way to use our natural resources Action plan 2018. Brussels: European Commission.
EC 2018b. A Clean Planet for all: A European strategic long-term vision for a prosperous, modern, competitive and climate neutral economy. *COM(2018) 773 final*. Brussels: European Commission.
EC 2018c. Progress in the implementation of the EU Forest Strategy 'A new EU Forest Strategy: for forests and the forest sector'. *COM(2018) 811 final*. Brussels: European Commission.
EC 2018d. A sustainable Bioeconomy for Europe: Strengthening the connection between economy, society and the environment. *COM(2018) 673 final*. Brussels: European Commission.
EC 2019a. *Clean energy for all Europeans*. Luxembourg: Publications Office of the European Union.
EC 2019b. The European Green Deal. *COM(2019) 640 final*. Brussels: European Commission.
EC 2020a. EU Biodiversity Strategy for 2030: Bringing nature back into our lives. *COM(2020) 380 final*. Brussels: European Commission.
EC 2020b. A new Circular Economy Action Plan: For a cleaner and more competitive Europe. *COM(2020) 98 final*.
EC 2020c. A new Circular Economy Action Plan: For a cleaner and more competitive Europe. *COM(2020) 98 final*. Brussels: European Commission.
EC 2020d. A Renovation Wave for Europe – greening our buildings, creating jobs, improving lives. *COM(2020) 662 final*. Brussels: European Commission.
EC 2020e. Stepping up Europe's 2030 climate ambition Investing in a climate-neutral future for the benefit of our people. *COM(2020) 562 final*. Brussels: European Commission.
EC 2021a. Annex to the New EU Forest Strategy for 2030: Roadmap of the Commission's action to implement the pledge to plant 3 billion additional trees by 2030 in the EU. *COM(2021) 572 final*. Brussels: European Commission.
EC 2021b. Commission notice – Commission guidelines on single-use plastic products in accordance with Directive (EU) 2019/904 of the European Parliament and of the Council on the reduction of the impact of certain plastic products on the environment. *C 216/1*. Brussels: Official Journal of the European Union.
EC 2021c. Commission staff working document: Stakeholder Consultation and Evidence Base *SWD(2021) 652 final*. Brussels: European Commission.
EC 2021d. Commission staff working document: The 3 billion tree planting pledge for 2030. *SWD(2021) 651 final*. Brussels: European Commission.
EC 2021e. 'Fit for 55': delivering the EU's 2030 Climate Target on the way to climate neutrality. *COM(2021) 550 final*. Brussels: European Commission.
EC 2021f. Forging a climate-resilient Europe – the new EU Strategy on Adaptation to Climate Change. *COM(2021) 82 final*. Brussels: European Commission.
EC 2021g. New EU Forest Strategy for 2030. *COM(2021) 572 final*. Brussels: European Commission.
EC 2021h. Proposal for a regulation on the making available on the Union market as well as export from the Union of certain commodities and products associated with deforestation and forest degradation and repealing Regulation (EU) No 995/2010. *COM(2021) 706 final*. Brussels: European Commission.

EC 2021i. Proposal for amending Directive (EU) 2018/2001 of the European Parliament and of the Council, Regulation (EU) 2018/1999 of the European Parliament and of the Council and Directive 98/70/EC of the European Parliament and of the Council as regards the promotion of energy from renewable sources, and repealing Council Directive (EU) 2015/652. *COM(2021) 557 final.* Brussels: European Commission.

EC 2022a. Ecodesign and Energy Labelling Working Plan 2022–2024. *C 182/1.* Brussels: Official Journal of the European Union.

EC 2022b. On making sustainable products the norm. *COM(2022) 140 final.* Brussels: European Commission.

EC 2022c. Opinion of the European Committee of the Regions – EU Forest Strategy for 2030. *C 301/61.* Brussels: Official Journal of the European Union.

EC 2022d. Proposal for a Regulation laying down harmonised conditions for the marketing of construction products, amending Regulation (EU) 2019/1020 and repealing Regulation (EU) 305/2011. *COM(2022) 144 final.* Brussels: European Commission.

EC 2022e. Proposal for a regulation on nature restoration. *COM(2022) 304 final.* Brussels: European Commission.

EC 2022f. Proposal for a Regulation on type-approval of motor vehicles and engines and of systems, components and separate technical units intended for such vehicles, with respect to their emissions and battery durability (Euro 7) and repealing Regulations (EC) No 715/2007 and (EC) No 595/2009. *COM(2022) 586 final.* Brussels: European Commission.

EC. 2020. Proposal for a regulation establishing the framework for achieving climate neutrality and amending Regulation (EU) 2018/1999 (European Climate Law). *COM/2020/80 final.* Brussels: European Commission.

EU 2023. Approved 28 CAP Strategic Plans (2023–2027) – Summary overview for 27 Member States: Facts and figures. Brussels: European Commission.

EUROSTAT 2020. Agriculture, forestry and fishery statistics: 2020 edition. *In:* COOK, E. (ed.). Luxembourg: Publications Office of the European Union, 2020.

EUSTAFOR. 2021. *The New EU Forest Strategy for 2030: Position of European Forest Owners and Managers* [Online]. https://eustafor.eu/uploads/20211004_The-New-EU-Forest-Strategy-for-2030-Position-of-European-Forest-Owners-and-Managers.pdf.

FAO 2010. Global Forest Resources Assessment 2010. *FAO Forestry Paper 163.* Rome: Food and Agriculture Organization of the United Nations.

FAO 2015. Global Forest Resources Assessment 2015. Rome: Food and Agriculture Organization of the United Nations.

FAO 2018. FRA 2020 Terms and definitions. *Forest Resources Assessment working paper 188.* Rome: Food and Agriculture Organization of the United Nations.

FAO 2020. Global Forest Resources Assessment 2020: Main report. Rome, Italy: Food and Agriculture Organization of the United Nations.

FOREST EUROPE 2015. State of Europe's Forests 2015. Madrid, Spain: Ministerial Conference on the Protection of Forests in Europe.

FOREST EUROPE 2020. State of Europe's Forests 2020. Bratislava: Ministerial Conference on the Protection of Forests in Europe.

GORDEEVA, E., WEBER, N. & WOLFSLEHNER, B. 2022. The New EU Forest Strategy for 2030: An Analysis of Major Interests. *Forests*, 13, 1503.

KEMPENEERS, P., SEDANO, F., SEEBACH, L., STROBL, P. & SAN-MIGUEL-AYANZ, J. 2012. Data fusion of different spatial resolution remote sensing images applied to forest-type mapping. *IEEE Transactions on Geoscience and Remote Sensing*, 49, 4977–4986.

KORHONEN, J., GIURCA, A., BROCKHAUS, M. & TOPPINEN, A. 2018. Actors and Politics in Finland's Forest-Based Bioeconomy Network. *Sustainability*, 10.

LAZDINIS, M., ANGELSTAM, P. & PÜLZL, H. 2019. Towards sustainable forest management in the European Union through polycentric forest governance and an integrated landscape approach. *Landscape Ecology*, 34, 1737–1749.

LIER, M., KÖHL, M., KORHONEN, K. T., LINSER, S. & PRINS, K. 2021. Forest relevant targets in EU policy instruments – can progress be measured by the pan-European criteria and indicators for sustainable forest management? *Forest Policy and Economics*, 128, 102481.

LINSER, S., WOLFSLEHNER, B., BRIDGE, R. J. S., GRITTEN, D., JOHNSON, S., PAYN, T., PRINS, K., RAŠI, R. & ROBERTSON, G. 2018. 25 Years of criteria and indicators for sustainable forest management: How intergovernmental C&I processes have made a difference. *Forests*, 9.

PÄIVINEN, R., LEHIKOINEN, M., SCHUCK, A., HÄME, T., VÄÄTÄINEN, S., KENNEDY, P. & FOLVING, S. 2001. Combining earth observation data and forest statistics. *EFI Research Report 14.* Joensuu: European Forest Institute.

PELLI, P., AGGESTAM, F., WEISS, G., INHAIZER, H., KEENLEYSIDE, C., GANTIOLER, S., BOGLIO, D. & POLÁKOVÁ, J. 2012. Ex-post evaluation of the EU Forest Action Plan. http://ec.europa.eu/agriculture/evaluation/market-and-income-reports/forest-action-plan-2012_en.htm.

PÜLZL, H. 2010. *Die Politik des Waldes* [*The Politics of Forests*]. Vienna: Böhlau.

PÜLZL, H., HOGL, K., KLEINSCHMIT, D., WYDRA, D., ARTS, B., MAYER, P., PALAHI, M., WINKEL, G. & WOLFSLEHNER, B. 2013. European forest governance: Issues at stake and the way forward. What Science Can Tell Us 2. European Forest Institute. www.efi.int/publications-bank.

PURKUS, A., HAGEMANN, N., BEDTKE, N. & GAWEL, E. 2018. Towards a sustainable innovation system for the German wood-based bioeconomy: Implications for policy design. *Journal of Cleaner Production*, 172, 3955–3968.

REGULATION 305/2011. Laying down harmonised conditions for the marketing of construction products. Brussels: Official Journal of the European Union.

REGULATION 363/2012. Commission Delegated Regulation on the procedural rules for the recognition and withdrawal of recognition of monitoring organisations as provided for in Regulation (EU) No 995/2010 of the European Parliament and of the Council laying down the obligations of operators who place timber and timber products on the market. Brussels: Official Journal of the European Union.

REGULATION 528/2012 Making available on the market and use of biocidal products (BPR). Brussels: Official Journal of the European Communities.

REGULATION 607/2012. Commission Implementing Regulation on the detailed rules concerning the due diligence system and the frequency and nature of the checks on monitoring organisations as provided for in Regulation (EU) No 995/2010 of the European Parliament and of the Council laying down the obligations of operators who place timber and timber products on the market Text with EEA relevance. *L 177/16.* Brussels: Official Journal of the European Union.

REGULATION 660/2014. Amending Regulation (EC) No 1013/2006 on shipments of waste. *L189/135.* Brussels: Official Journal of the European Union.

REGULATION 661/2009. Type-approval requirements for the general safety of motor vehicles, their trailers and systems, components and separate technical units intended therefor. *L 200/1.* Brussels: Official Journal of the European Union.

REGULATION 995/2010. Laying down the obligations of operators who place timber and timber products on the market. *OJ L 295.* Brussels: Official Journal of the European Union.

REGULATION 1013/2006. Shipments of waste. Brussels: Official Journal of the European Union.

REGULATION 1024/2008. Commission Regulation laying down detailed measures for the implementation of Council Regulation (EC) No 2173/2005 on the establishment of a FLEGT licensing scheme for imports of timber into the European Community. *L 277/23.* Brussels: Official Journal of the European Union.

REGULATION 1272/2008. Classification, labelling and packaging of substances and mixtures (CLP Regulation). Brussels: Official Journal of the European Union.

REGULATION 1305/2013. Support for rural development by the European Agricultural Fund for Rural Development (EAFRD) and repealing Council Regulation (EC) No 1698/2005. *L 347/487.* Brussels: Official Journal of the European Union.

REGULATION 1907/2006. Registration, Evaluation, Authorisation and Restriction of Chemicals (REACH). *OJ L 396.* Brussels: Official Journal of the European Communities.

REGULATION 1935/2004. Materials and articles intended to come into contact with food. Brussels: Official Journal of the European Union.

REGULATION 2015/757. Monitoring, reporting and verification of carbon dioxide emissions from maritime transport, and amending Directive 2009/16/EC. *L 123/55* Brussels: Official Journal of the European Union.

REGULATION 2018/841. Inclusion of greenhouse gas emissions and removals from land use, land use change and forestry in the 2030 climate and energy framework, and amending Regulation (EU) No 525/2013 and Decision No 529/2013/EU. *L 156/1.* Brussels: Official Journal of the European Union.

REGULATION 2019/1381. Transparency and sustainability of the EU risk assessment in the food chain and amending Regulations (EC) No 178/2002, (EC) No 1829/2003, (EC) No 1831/2003, (EC) No 2065/2003, (EC) No 1935/2004, (EC) No 1331/2008, (EC) No 1107/2009, (EU) 2015/2283 and Directive 2001/18/EC. *L 231/1.* Brussels: Official Journal of the European Union.

REGULATION 2019/2072. Commission Implementing Regulation establishing uniform conditions for the implementation of Regulation (EU) 2016/2031 of the European Parliament and the Council, as regards protective measures against pests of plants, and repealing Commission Regulation (EC) No

690/2008 and amending Commission Implementing Regulation (EU) 2018/2019. *L 319/1.* Brussels: Official Journal of the European Union.
REGULATION 2020/852. Establishment of a framework to facilitate sustainable investment, and amending Regulation (EU) 2019/208. *L 198/13.* Brussels: Official Journal of the European Union.
REGULATION 2020/2220. Laying down certain transitional provisions for support from the European Agricultural Fund for Rural Development (EAFRD) and from the European Agricultural Guarantee Fund (EAGF) in the years 2021 and 2022. *L 437/1.* Brussels: Official Journal of the European Union.
REGULATION 2021/268. Commission Delegated Regulation amending Annex IV to Regulation (EU) 2018/841 of the European Parliament and of the Council as regards the forest reference levels to be applied by the Member States for the period 2021–2025. *L 60/21.* Brussels: Official Journal of the European Union.
REGULATION 2021/783. Establishing a Programme for the Environment and Climate Action (LIFE), and repealing Regulation (EU) No 1293/2013. *L 172/53.* Brussels: Official Journal of the European Union.
REGULATION 2021/2115. Establishing rules on support for strategic plans to be drawn up by Member States under the common agricultural policy (CAP Strategic Plans) and financed by the European Agricultural Guarantee Fund (EAGF) and by the European Agricultural Fund for Rural Development (EAFRD) and repealing Regulations (EU) No 1305/2013 and (EU) No 1307/2013. *L 435/1.* Brussels: Official Journal of the European Union.
REGULATION 2021/2116. Financing, management and monitoring of the common agricultural policy and repealing Regulation (EU) No 1306/2013. *L 435/187* Brussels: Official Journal of the European Union
REGULATION 2021/2117. Establishing a common organisation of the markets in agricultural products, (EU) No 1151/2012 on quality schemes for agricultural products and foodstuffs, (EU) No 251/2014 on the definition, description, presentation, labelling and the protection of geographical indications of aromatised wine products and (EU) No 228/2013 laying down specific measures for agriculture in the outermost regions of the Union. *L 435/262.* Brussels: Official Journal of the European Union.
REGULATION 2173/2005. Council Regulation on the establishment of a FLEGT licensing scheme for imports of timber into the European Community. *L 347/1.* Brussels: Official Journal of the European Union.
RIVERA LEÓN, L., BOUGAS, K., AGGESTAM, F., PÜLZL, H., ZOBOLI, E., RAVET, J., GRINIECE, E., VERMEER, J., MAROULIS, N., ETTWEIN, F., VAN BRUSSELEN J. & GREEN, T. 2016. An assessment of the cumulative cost impact of specified EU legislation and policies on the EU forest-based industries. Brussels: DG GROW.
SAMUELSSON, K., BARTHEL, S., GIUSTI, M. & HARTIG, T. 2021. Visiting nearby natural settings supported wellbeing during Sweden's 'soft-touch' pandemic restrictions. *Landscape and Urban Planning*, 214, 104176.
SCHUCK, A., VAN BRUSSELEN, J., PÄIVINEN, R., HÄME, T., KENNEDY, P. & FOLVING, S. 2002. Compilation of a calibrated European forest map derived from NOAA-AVHRR data. *EFI Internal Report 13.* Joensuu: European Forest Institute.
SÖDERBERG, C. & ECKERBERG, K. 2013. Rising policy conflicts in Europe over bioenergy and forestry. *Forest Policy and Economics*, 33, 112–119.
STUBENRAUCH, J., GARSKE, B., EKARDT, F. & HAGEMANN, K. 2022. European Forest governance: Status quo and optimising options with regard to the Paris Climate Target. *Sustainability*, 14, 4365.
UHLICH, T. & BURCK, J. 2022. *Fact Sheet: Glasgow Leaders' Declaration on Forests and Land Use.* German Federal Ministry for Economic Cooperation and Development, G7 | G20 Track 2 Dialogue.
UNDP 2019. New York Declaration on Forests Endorser Perspectives Report: Progress, Challenges and Enabling Conditions for Action. *In:* DESANTIS, N., ERVIN, J., CRAIG, M., MARX, M., PAZ, C. & KANTCHEVA, N. (eds.). New York: United Nations Development Programme.
WATANABE, C., NAVEED, N. & NEITTAANMÄKI, P. 2018. Digital solutions transform the forest-based bioeconomy into a digital platform industry – A suggestion for a disruptive business model in the digital economy. *Technology in Society*, 54, 168–188.
WINKEL, G. 2017. Towards a sustainable European forest-based bioeconomy- assessment and the way forward. *What Science Can Tell Us 8.* Joensuu: European Forest Institute.
WOLFSLEHNER, B., AGGESTAM, F., PULZL, H., HENDRIKS, K., LINDNER, M., SOTIROV, M., KULIKOVA, E., SECCO, L., LOVRIC, M., NABUURS, G.-J., HURMEKOSKI, E., PETTENELLA, D., ARETS, E., BAULENAS SERRA, E., BOZZANO, M., DERKS, J., GATTO, P., LINSER, S., MASIERO, M., MAVSAR, R. & GIESSEN, L. 2019. Study on progress in implementing the EU Forest Strategy. Final report. Luxembourg: Publications Office of the European Union. DOI:10.2762/160685.
WOLFSLEHNER, B., PÜLZL, H., KLEINSCHMIT, D., AGGESTAM, F., WINKEL, G., CANDEL, J., ECKERBERG, K., FEINDT, P., MCDERMOTT, C., SECCO, L., SOTIROV, M., LACKNER, M. & LAZYA ROUX, J. 2020. European forest governance post-2020. *From Science to Policy 10.* Joensuu, Finland: European Forest Institute. https://doi.org/10.36333/fs10

3 Overview of theories and approaches used in forest policy research

1. Introduction

Forest policy research has emerged as a distinct field of study in the latter half of the twentieth century, reflecting growing concerns over forest management practices' social, economic, and environmental impacts. Understanding the processes by which forest policies are created (e.g., for what and by whom) and their impacts was recognised as essential, not only for improving policymaking per se but also for the promotion of sustainable forest management practices. Forest policy researchers started to draw on theoretical frameworks and models from social sciences to attain some degree of understanding. In its early stages, the focus was on practical aspects of forest management, such as sustainable timber production (Nocentini et al., 2021), which resulted in limited use of theoretical analysis as most work was primarily normative and oriented towards policy advice (Arts, 2012, Arts and van de Graaf, 2009). Since this initial stage, forest policy research has substantially transformed, becoming increasingly theory-driven.

This chapter will focus on different theoretical frameworks and models from social sciences, all of which have been applied to analyse forest policy processes. It will not be focusing on methodological approaches other than as examples. Instead, it will explore the conceptual tools that forest policy researchers have utilised to understand and make sense of the interactions between actors, institutions, society, and forests. By examining some of the theoretical perspectives used to analyse forest policy, this chapter will provide readers with a short overview of the multidisciplinary field of policy research and the various approaches through which policy processes are and can be examined. For example, Weible and Sabatier (2018) have highlighted the importance of using theories of the policy process, such as the Punctuated Equilibrium Theory, Multiple Streams Framework, Policy Feedback Theory, and the Advocacy Coalition Framework (ACF), to explain how policies are developed and implemented. Another example comes from Zittoun and Peters (2016), who have emphasised the role of social constructionist theories in policy analysis. Constructionist theories focus on discursive practices and power relations and how they shape policy ideas, values, and norms. These examples demonstrate theoretical frameworks' role in identifying actors involved in policymaking, factors that influence their behaviour, and instruments and mechanisms that facilitate or prevent policy change and uptake.

One such framework is institutional theory (Meyer and Rowan, 1977, DiMaggio and Powell, 1991, Scott, 1995), which has been utilised to explain formal and informal rules that govern policymaking, including the roles of various actors and their relationships (Cashore, 2002). ACF is another framework that underscores the role of advocacy coalitions[1] in shaping policy ideas and facilitating policy change (Sabatier and Jenkins-Smith, 1996). Policy network analysis examines actor relationships in the policy process, explaining how various actors and their interactions influence policy outcomes (Rhodes, 1997). These are additional

DOI: 10.4324/9781003198987-3

examples of theoretical frameworks used to improve our understanding of policymaking processes in the context of European and international forest policy. Even more, today, several scientific journals focus, in part or entirely, on forest policy analyses within the broader field of policy sciences, such as Forest Policy and Economics,[2] Forests,[3] and the Journal of Forest Research.[4]

The present chapter will zoom in on theory-informed policy analysis as a research method in forest policy. It offers an overview of mainstream theories in forest policy sciences and introduces the use of theory in forest policy research, highlighting its significance to understanding forest policies and their impact on the environment and socio-economic conditions. The chapter will introduce various policy theories and briefly review their applicability to forest policy research. The chapter concludes by emphasising the importance of theory in forest policy analysis and the potential to generate valuable insights for policy practitioners and decision-makers.

> It is crucial to adopt policy-oriented approaches appropriate to the context to enhance the effectiveness of forest policy research. However, researchers and practitioners in this field often encounter limited resources, as has been pointed out by scholars such as Browne et al. (2019), Clarke et al. (2016) and Coveney (2010).
>
> Moreover, it has been noted that existing resources frequently lack a comprehensive overview of the theoretical orientations and analytical approaches in policy research (Breton and De Leeuw, 2011). This highlights the need for better guidance on relevant approaches in forest policy research.

1.1. The emergence of forest policy analysis

Forest policy analysis is rooted in historical developments. It has evolved gradually, with several significant milestones playing an essential role in its emergence as a distinct field of study. One such milestone was the development of forestry sciences in the eighteenth century, which established a scientific basis for forest management and silviculture (Nocentini et al., 2021, Arts, 2012, Arts and van de Graaf, 2009). This led to the development of forest management practices prioritising timber production; however, it disregarded the ecological and social values of forests (Lowood, 1990, Heilbron, 1990). Another milestone, as argued by Bas Arts (2012), was the introduction of forestry curricula at universities across the globe, including policy courses at both the undergraduate and postgraduate levels. In essence, this has enabled foresters to gain knowledge of policy theories.

Establishing the world's first national reserves and parks in the late nineteenth century has arguably also played a critical role in modern forest conservation efforts and associated research. In contrast to forest management and silviculture research, the conservation movement that followed the first national forest reserves and parks recognised the need for conservation strategies encompassing ecological and social factors. For example, while many early national parks were established as protected areas for scientific study, they later became sites for outdoor recreation, education, and nature conservation (US Forest Service, 1968). These developments have likely affected the application of theoretical frameworks in forest policy research.

During the 1960s and 1970s, the environmental movement gained significant momentum; interlinked with concerns over pollution and deforestation, environmental interests began to shape public policy (Hays, 1981). In other words, policy scientists began to show a greater interest in forest issues due to the growing importance of environmental policy (Arts, 2012). This marked another turning point in forest policy analysis, as policymakers and researchers began to recognise the significance of ecological and social factors in forest management decisions. Furthermore, this aligns with the emergence of multiple-use forestry, which aims to balance timber production with other forest values such as recreation, wildlife habitat, and water quality (Glück, 2000).

Early efforts in forest policy analysis were primarily concerned with understanding the policy processes that shape forest management decisions. Researchers in the field sought to identify actors involved in decision-making, including interest groups, agencies, and political institutions, and to analyse how these key actors interacted and influenced policy outcomes (Sabatier and Jenkins-Smith, 1996, Cashore, 2002, Aggestam and Vogelpohl, 2009, Crewett, 2015, Baycheva-Merger et al., 2018). In more recent years, researchers have increasingly started to emphasise holistic approaches to forest management, highlighting not only the economic and political dimensions of decision-making but also the ecological and social impacts of different management strategies (Leban et al., 2016, Borrass et al., 2017, Nagel et al., 2017, Aggestam et al., 2020). The development of policy analysis tools and frameworks has, as such, enabled policymakers and researchers to evaluate the impacts of different forest policies and management strategies, providing evidence-based approaches to forest management decision-making (Krott, 2005, Krott, 2012, Arts, 2012, Arts and van de Graaf, 2009). Moreover, the evolution of forest policy research demonstrates that it is complex and multifaceted.

2. Theoretical frameworks and approaches in forest policy research

As emphasised above, European and international forest policy studies draw on several theoretical frameworks. The frameworks that will be covered in this chapter include:

- Institutional policy analysis: Exploring the role of institutions, such as legal frameworks, regulatory bodies, and stakeholder groups, in shaping forest policy outcomes. This framework has been used to explain variations in forest policy outcomes across European countries.
- Policy network analysis: Examining the relationships between actors involved in forest policymaking, such as government agencies, interest groups, and civil society organisations. This framework can aid in identifying key actors and their interests, as well as the power dynamics within a policy network.
- Advocacy coalition framework (ACF): Analysing the formation and evolution of policy coalitions, including their beliefs, resources, and interactions with other coalitions.
- Critical policy analysis: Reviewing the role of power, ideology, and discourse in shaping policy outcomes. It can be particularly valuable in uncovering hidden interests and values that underlie policy decisions.
- Rational policy analysis: Identifying the most efficient means of achieving specific policy goals, considering costs, benefits, and trade-offs. This approach has been commonly employed in European forest policy research.

It is important to acknowledge that these frameworks do not exist in isolation. Researchers often employ a mix of analytical frameworks to achieve a more nuanced understanding of forest policy development and its outcomes. The underlying concept here is that employing multiple (or integrative) theoretical perspectives makes identifying various factors driving policy changes feasible, thus facilitating a more comprehensive depiction of policymaking and relevant implementation procedures. Furthermore, while this chapter confines its focus to the frameworks mentioned above, the policy research community employs a wide range of additional theoretical and methodological approaches, such as content, discourse, and frame analysis (Goffman, 1974, Gilliam and Bales, 2001, Dewulf et al., 2004, Elomina and Pülzl, 2021), the pathways to sustainability approach (Leach et al., 2007, Beland Lindahl et al., 2016, Aggestam and Giurca, 2021), and policy modelling (Solberg, 1986, Toppinen and Kuuluvainen, 2010, Banerjee and Alavalapati, 2014), to mention a few. This overview has limitations and serves primarily as a brief introduction to applying forest policy research tools and methods.

It is equally important to recognise that distinct theoretical and methodological assumptions regarding the nature of the world underpin most analytical approaches. These underlying assumptions, embedded within various theoretical frameworks, fundamentally shape researchers' perceptions and interpretations of social, political, economic, and ecological factors that impact the policymaking process. Consequently, the choice of theory and methodology not only shapes the researchers' perspectives but also influences the nature of data collection and analysis, thereby impacting the conclusions derived from the scientific work.

This diversity can challenge researchers as it can affect the degree of comparability between studies, emphasising the subjective nature of relevant research findings. For example, if there is not a shared consensus on what 'forest policy' means (as explored in Chapter 2), how can findings from forest policy research be trusted? Moreover, it underscores that integrating approaches is not always feasible due to inherent theoretical incompatibilities. For example, the disparity between theories arguing for rational actors and those acknowledging decision-making influenced by bounded rationality highlights this challenge (more on this below). These considerations and variations highlight why it is important to consider underlying assumptions and perspectives when deciding upon the appropriate analytical approach. It also stresses the need for a shared understanding within the respective fields of policy work to ensure that results are understood and accepted.

2.1. Institutional policy analysis

As a theoretical and empirical approach, institutional policy analysis explains how institutions shape policy outcomes. Institutions are commonly defined as the formal and informal rules, norms, and decision-making processes that govern the behaviour of organisations and individuals (Gormley, 1987). Formal institutions include laws, regulations, and administrative procedures, while informal institutions include social norms and cultural practices. Institutional policy analysis draws on insights from political science, sociology, law, and other disciplines, and it has its roots in new institutional economics (DiMaggio and Powell, 1991, Brousseau and Glachant, 2008). One of the key insights from institutional policy analysis is that institutions matter for policy implementation and outcomes. Institutions and institutional arrangements can facilitate or create barriers to implementing policies by shaping the distribution of costs and benefits between different groups. For example, a well-designed regulatory framework can ensure that a policy intervention is implemented effectively, while a poorly designed framework is a common reason for poor implementation.

During the early stages of forest policy research, the focus rested mainly on technical and economic aspects of forest management, neglecting the institutional dimensions. However, recognising the limitations of purely technical approaches, researchers began incorporating institutional analysis into forest policy research to explore the influence of institutions, including property rights, governance structures, and decision-making processes, on policy outcomes (Poteete and Ostrom, 2004, Williamson, 1991). One of the key applications of institutional policy analysis in forest policy research has been to examine the institutional arrangements for forest governance, including the roles and responsibilities of different actors, such as governments and private sector actors. Researchers have used institutional policy analysis to understand how forest-based actors interact, make decisions and identify the factors that shape their behaviours and decisions (Gormley, 1987, Polski and Ostrom, 1999). For example, Ostrom (1990) has argued that the success of common-pool resource management, such as forests, depends on the design of institutions that incentivise cooperation among users.

Another area where institutional policy analysis has been applied in forest policy research is in the examination of the effectiveness of different types of policy interventions, such as certification schemes, community-based forest management, and public-private partnerships (Cashore, 2002) For example, Cashore et al. (2004) used institutional policy analysis to understand the institutional context of forest certification schemes and to identify opportunities for improvement. New institutional economics and new institutionalism have also influenced forest policy research. These theoretical perspectives have highlighted the importance of understanding formal and informal institutions as integral to policymaking (North, 1990, Hood, 1990).

In conclusion, by examining institutions and institutional arrangements, it is possible to understand better the interplay between institutions, actors, and policy outcomes. This knowledge enables researchers to identify opportunities for policy reform, improved governance structures, and enhanced collaboration among stakeholders, which can, for example, help pave the way for SFM. In other words, institutional policy analysis is a helpful tool for researchers who want to understand how institutions influence policy outcomes and how they can be improved.

Case example: Forests, discourses, and institutions

The case study Bas Arts and Mareleen Buizer presented in their 2009 paper, '*Forests, discourses, institutions: A discursive-institutional analysis of global forest governance*', offers a valuable contribution to forest policy research. The authors apply the policy arrangement approach to analyse global forest politics from the 1980s, providing new insights into the complex interplay between discursive and institutional dimensions of policymaking. By examining the relationship between discourses, coalitions, rules of the game, power, and social practices over time, the authors demonstrate how discursive changes have led to shifts in the global forest policy arrangement. The paper also highlights the emergence of new partnerships, certification programs, voluntary instruments, and the empowerment of non-state actors because of sustainable development, biodiversity, and governance discourses. The paper illustrates forest policy's dynamic and multidimensional nature and the importance of examining discursive and institutional factors to improve our understanding of policy processes.

2.2. *Rational policy analysis*

Rational policy analysis represents a systematic and theoretical framework that primarily aims to guide the decision-making process in public policy formulation. Its objective is to provide a structured and evidence-based approach to selecting optimal policy options, specifically those that maximise societal benefits. This approach, sometimes termed the 'economic model', fundamentally rests on the assumption that both individuals and organisations function as 'rational actors' that strive to maximise their utility and that policymakers possess access to complete and accurate information (Montgomery, 1987). It presupposes a sense of rationality in individual behaviour, where decisions are logically derived. However, it is crucial to acknowledge that these foundational assumptions have been criticised, given that individuals do not consistently exhibit logical behaviour, and varied factors influence their decision-making processes. Furthermore, reality often deviates from the assumption of perfect information accessibility, potentially impairing the capacity to make objective decisions (Hoefer, 2021).

Even with these critiques, the rational approach centres on establishing clearly defined policy objectives and identifying the most efficient strategies for their attainment. The rational model is grounded in several assumptions. First, it posits that human behaviour is rooted in stable and consistent values that guide their decision-making processes. Second, it presupposes that individuals possess an adequate information base to make informed choices, thus enabling them to optimise gains and rationally minimise losses efficiently. Rational policy analysis finds substantial application within the public policy domain, where policy alternatives concerning environmental regulations, infrastructure investments, and health and safety measures, among others, are evaluated through rational policy analysis. By integrating elements like cost–benefit analysis and risk assessment (Shreve and Kelman, 2014, Christine and Luuk, 2021), rational policy analysis provides the necessary tools for policy evaluation that are anchored in understanding the associated costs and benefits.

One cornerstone of rational policy analysis is the practice of cost–benefit analysis, which constitutes a quantitative technique employed to estimate the economic ramifications of various policy options (Hanley, 2001, Arrow et al., 2014, Boardman et al., 2018). The cost–benefit analysis encompasses both monetised and non-monetised aspects of benefits and costs, intending to facilitate a comprehensive assessment of the policy's economic footprint. By quantifying and comparing these dimensions, policymakers are empowered to analyse the inherent trade-offs, thereby enabling the formulation of choices aligned with overarching societal objectives. Risk assessment is another aspect of rational policy analysis, a methodology designed to evaluate the uncertainties intertwined with policy outcomes while aiding decision-makers in managing associated risks in the policymaking process (Kunreuther, 2002). By reviewing diverse policy alternatives and their potential impacts, policymakers can assess associated risks and make decisions to optimise policy outcomes. For instance, risk analysis may provide policymakers with the insights necessary to pinpoint the drivers of policy-related risks and to make informed decisions by weighing potential benefits against established uncertainties (Ninan and Inoue, 2013, Pascual et al., 2017, Rakatama et al., 2017, Lazaridou et al., 2021).

Rational policy analysis has been effectively applied within forest policy research to assess several issues, including the economic impacts of forest management practices and policies. In forest conservation, for instance, rational policy analysis has been employed to systematically evaluate the costs and benefits of establishing protected areas (Kniivilä et al., 2002, Naidoo and Ricketts, 2006, Brander et al., 2020). SFM practices, such as timber production and the utilisation of non-timber forest products, have also undergone cost–benefit analyses to assess their economic viability relative to implementation costs (Croitoru, 2007). Furthermore, forest certification programs, such as the Forest Stewardship Council (FSC), have been subjected to rational policy analysis to ascertain their economic and environmental contributions towards the promotion of SFM (Rametsteiner and Simula, 2003, Wolff and Schweinle, 2022). These applications of rational policy analysis have collectively contributed to advancing our understanding of the economic and environmental implications of forest management practices and policies.

An additional application of rational policy analysis is examining the impact of forest policy instruments. Taxation measures imposed on forest-related activities, such as logging or carbon emissions, have undergone extensive cost–benefit analysis to assess their economic repercussions and as a way to promote sustainable practices (Palmer and Bulkan, 2010, Kniivilä et al.,

2002). Similarly, subsidies aimed at incentivising reforestation or adopting SFM practices have been analysed using rational policy analysis to determine their effectiveness and the economic considerations underpinning them (Rämö et al., 2023, Kotecký, 2015). Furthermore, regulations governing forestry practices, encompassing aspects like biodiversity conservation, water resource management, and sustainable logging, have all been evaluated to clarify their economic costs and benefits (Rakatama et al., 2017, Kniivilä et al., 2002, Brander et al., 2020).

Integrating risk analysis into forest policy evaluation has enhanced decision-making processes. Risk analysis techniques have been employed to evaluate the uncertainty associated with different forest policy outcomes, considering factors such as market fluctuations, climate change impacts, and natural disasters and their potential effects on policy effectiveness (Bonazountas et al., 2005, Hanewinkel et al., 2011, Yousefpour et al., 2012). Identifying key drivers of policy risks in forest management has been achieved through risk analysis, facilitating a better understanding of potential consequences and enabling the development of strategies to mitigate risks.

The application of rational policy analysis within the context of forest policy research has considerably advanced our understanding of the economic and environmental impacts associated with a diverse array of forest management practices and policies. By integrating quantitative techniques such as cost–benefit analysis, risk analysis, and others, both researchers and policymakers have been better equipped to navigate the forest policy landscape. This approach has enabled them to weigh trade-offs and make well-informed choices conducive to promoting SFM.

Case example: Forest management and species richness

Forestry practices inevitably alter forest characteristics, affecting biodiversity. An investigation by Chaudhary et al. (2016) is an illustrative example, wherein an economic trade-off analysis was conducted to demonstrate the effects associated with different forest management strategies and their ensuing impacts on species richness. The study reveals that the consequences of forest management are markedly divergent among various approaches; it is observed that the pursuit of equivalent economic wood value through disparate means results in divergent levels of biodiversity loss. A comprehensive understanding of the impacts of distinct forest management practices necessitates an integrative assessment that can effectively encompass environmental and economic dimensions. Though the study may not conform precisely to the conventional structure of a cost–benefit analysis, it demonstrates the capacity to quantitatively assess impacts in terms of economic and environmental factors, thereby facilitating an enhanced understanding of relevant trade-offs. This contextual framework underscores the inherent value of undertaking a cost–benefit analysis in forest management decisions. Such analytical paradigms can help policymakers make more informed decisions that factor in economic objectives and biodiversity conservation.

2.3. Network analysis

Policy network analysis represents a flexible framework for analysing factors affecting the interactions and relationships between actors involved in policymaking and policy implementation. This methodological approach systematically analyses social networks and affiliations connecting different actors, covering governmental entities, interest groups, non-governmental organisations (NGOs), and individual actors (Marin and Mayntz, 1991, Bressers and O'Toole, 1998, Besussi,

2006). Its fundamental purpose is to explain how these relationships influence the development and implementation of public policies, including forest policy.

A defining feature of a policy network analysis lies in mapping the relationships between actors and uncovering how these interactions affect the policymaking process. This analytical approach commonly requires data collection regarding the connections among actors, including the frequency and intensity of their interactions. Subsequently, this data is used to construct visual representations of policy networks. These visual aids (Figure 3.1) demonstrate the network's key figures and power brokers to show how various actors influence and shape the policymaking landscape (Knoke and Yang, 2008, Compston, 2009).

Another important dimension of policy network analysis concerns the analysis of power dynamics and vested interests within the policymaking process (e.g., logging companies can hold significant economic power that they use to influence policymakers) (Nepstad et al., 2014). The approach tries to disentangle how distinct actors wield their influence within the policymaking process and sets out to identify the main drivers driving policy outcomes. In other words, through the review of actor-to-actor relationships, policy network analysis tries to explain how different actors promote their interests and perspectives in the policymaking process, ultimately striving to outline the distribution of power and influence affecting the network as well as policy outcomes (Borgatti and Foster, 2003).

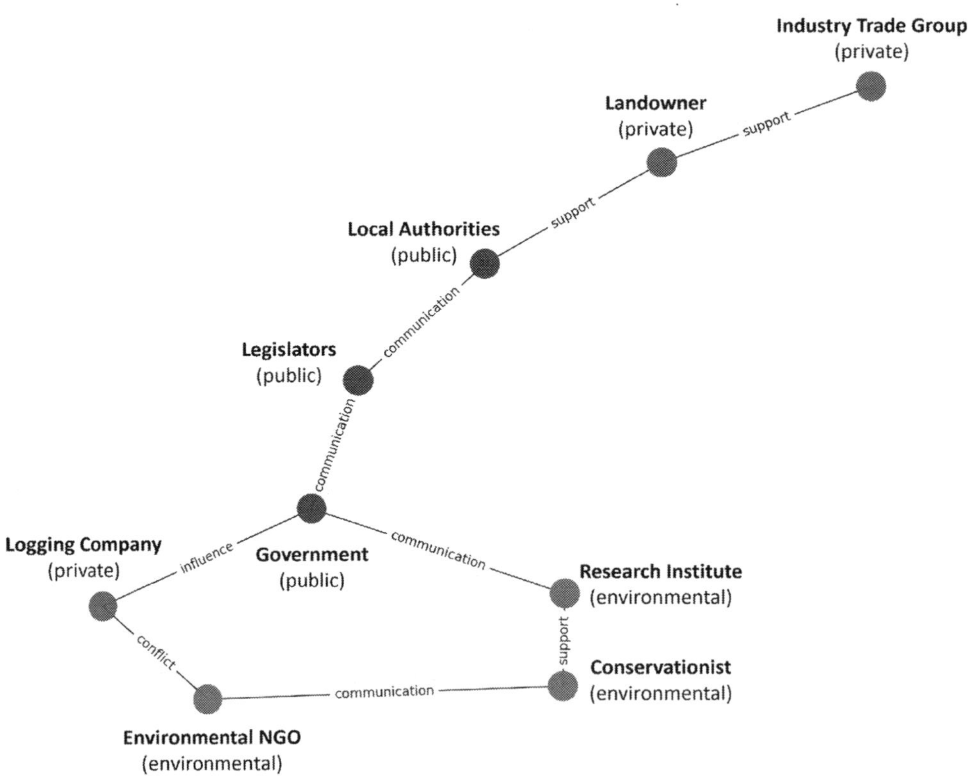

Figure 3.1 Simplified example of a social network analysis.

Source: own figure

As a research tool, policy network analysis can be a versatile and adaptable methodology applicable to various policy domains and sectors, covering environmental policy, healthcare policy, educational policy, and more (Wasserman and Faust, 1994, Scott, 2017). It can provide researchers with a comprehensive and nuanced overview and understanding of actor relationships and interactions, if applied correctly. This can provide valuable insights into the policy formulation and implementation process, such as by which means actors influence or affect policymaking. On a more practical level, accounting for power dynamics and interconnections among actors, a policy network analysis can identify areas for potential collaboration and cooperation between actors.

In the context of forest policy research, policy network analysis has found meaningful applications in the study of relationships between actors involved in forest certification schemes, such as the Forest Stewardship Council (FSC) and the Programme for the Endorsement of Forest Certification (PEFC) (Gale and Cadman, 2014), or to assess the level of stakeholder engagement in the development of forest policies (Baulenas et al., 2021), or to study forest biodiversity conservation (Borg et al., 2015) and processes related to the international trade of wood and non-wood forest products (Lovrić et al., 2018). These studies have helped clarify the relationships and interactions between actors intricately involved in forest policy formulation and implementation. Governmental bodies, NGOs, industry stakeholders, and individuals all occupy roles within this complex web of policymaking, and policy network analysis assists in breaking down and explaining their interplay. Another notable application relates to identifying important players (or power brokers) and influential figures within the forest policy network. By outlining the connections between actors, a policy network analysis can reveal those who exert the most significant impact on the policymaking process and show the strategies through which they wield their influence. This knowledge can offer valuable insights concerning policy outcomes and identify opportunities for constructive collaboration among actors, such as in developing more efficient policy tools and incentives.

Case example: Forest and water policy integration

Using qualitative interviews and social network analysis, Baulenas et al. (2021) explored two case studies in Germany and Spain. The research aimed to shed light on the dynamics of policy integration, using social network analysis and emphasising the role and impact of policy networks. The paper reveals interesting regional variations in policy integration. For example, in Catalonia, efforts to integrate forest and water policies were driven by shared perceptions and collaborative fact-finding activities, resulting in the inclusion of forest-water considerations in climate change legislation. On the other hand, in Baden-Württemberg, the lack of a policy broker and contextual factors led to more fragmented interactions and technical policy outputs. This study demonstrates the application of social network analysis to understand the complexities of policy integration. It demonstrates how policy networks and their dynamics can influence the outcomes of forest-water policy integration efforts in distinct regional contexts.

2.4. Advocacy coalitions

The Advocacy Coalition Framework (ACF) is an actor-based analytical framework that analyses policymaking and policy change. Developed in the 1980s, by Paul Sabatier and Hank

Jenkins-Smith, ACF was conceived to examine complex policy change processes, with a specific focus on the role that belief systems, advocacy coalitions (or networks)[5] and policy learning have on policy development and outcomes (Sabatier, 1987, Sabatier, 1988, Sabatier and Jenkins-Smith, 1996). ACF offers an alternative perspective to the conventional understanding of the policy process as a linear cycle. It acknowledges the complexity of policymaking environments, which often feature diverse actors and multiple levels of government. The framework accommodates decisions made under considerable uncertainty, acknowledges the long time it can take to see policy outcomes, and includes a range of policy processes, including political and technical aspects (Jenkins-Smith and Sabatier, 1994, Weible and Sabatier, 2018).

At the heart of the ACF lies the premise that actors engaged in policymaking do not operate in isolation; instead, they form coalitions characterised by shared beliefs and preferences. Furthermore, these coalitions are not static but exhibit patterns of dynamic evolution in response to shifting policy developments and external pressures over time (Figure 3.2).

Key components of the ACF are: (1) **Policy belief systems**, often referred to as policy cores. These are core beliefs (e.g., religious principles and norms), values, and causal explanations related to a particular policy issue or problem. Actors, such as government officials, interest groups, and experts, are, in turn, affiliated with specific belief systems, which arguably guide their preferences and actions. (2) **Advocacy coalitions** that represent actors with similar policy belief systems that collaborate and coordinate to influence policy decisions. These networks are commonly

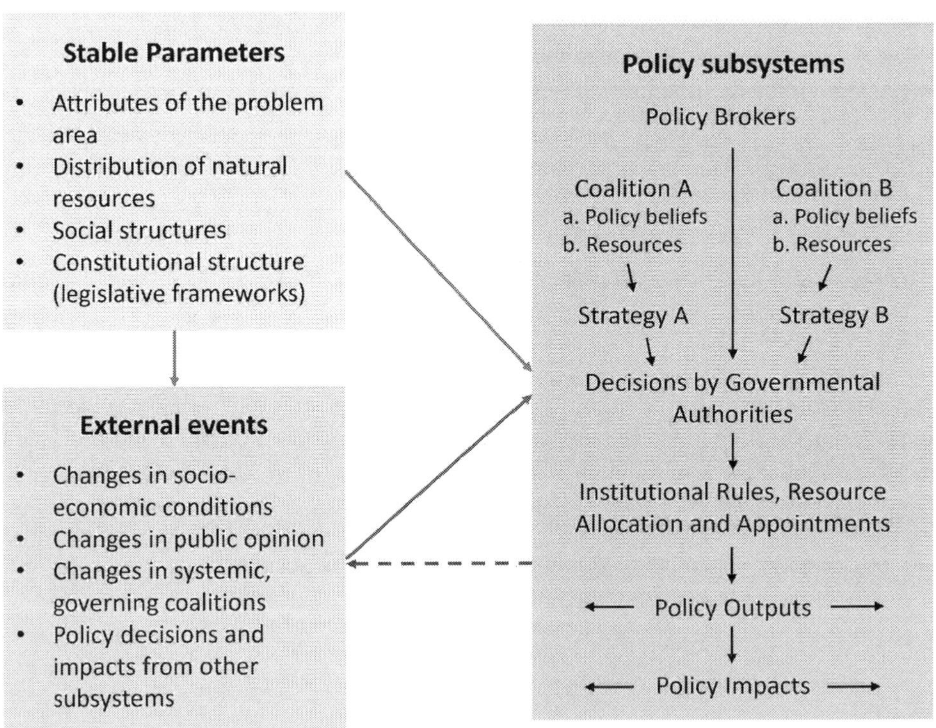

Figure 3.2 Example of the Advocacy Coalition Framework.

Source: adapted from Zhou et al. (2021)

characterised by their shared understanding of a policy problem, associated solutions, and goals. (3) **Policy windows** relate to opportunities for policy change, such as through interactions and conflicts between advocacy coalitions, shifts in public opinion, or changes in political leadership. (4) **Policy feedback** relating to external factors that facilitate or impede efforts of advocacy coalitions. This can include public opinion, changes in government, and socio-economic conditions as factors affecting policy outcomes (Figure 3.2). Taken together, focusing on the interplay between belief systems, coalition dynamics, and external factors, ACF has become an important tool for forest policy researchers, allowing for insights into how policy decisions are being formulated, contested, and transformed, over time. The framework is particularly useful in understanding how actors and interests interact in policymaking and how these interactions shape policy outcomes.

ACF has been applied extensively in forest policy research, such as to analyse the development of forest certification systems (Elliott and Schlaepfer, 2001), to review how and why forest policies change over time (Hysing and Olsson, 2008), to examine the role of stakeholder coalitions in forest politics (Harrinkari et al., 2016), and to study the mechanism of cross-sectoral coordination between nature protection and forestry (Sarvašová et al., 2013), to mention a few. For example, changes in forest management practices or the prioritisation of specific forest values (e.g., economic or ecological) can be better understood through ACF-based analyses. Understanding coalition dynamics is crucial for predicting policy outcomes and identifying potential areas of conflict or cooperation (Sabatier, 1987). Another illustrative example relates to policy learning, where ACF has been used to study how advocacy coalitions within a forest policy subsystem adapt and learn from new information and experiences, such as how policy actors incorporate new scientific findings into their policy positions (Moyson, 2017, Sabatier, 1988). ACF can, as such, be a valuable tool to analyse the dynamics of forest policy, mainly to shed light on the roles of interest groups, their beliefs and impacts on policy formulation and implementation.

Case example: Shifting coalitions and policy change

Research using ACF has generated thought-provoking results; however, it has also been argued that the framework can struggle with issues surrounding advocacy coalition identification, how to assess the dynamics of shifting coalitions, and the role of exploitive coalitions[6] in shaping policy change. In a study by Sotirov and Winkel (2016), efforts were made to address these concerns through the integration of Cultural Theory[7] and ACF. It set out to provide fresh perspectives on forest sector reforms and forest certification in Germany and Bulgaria. The paper presents four policy actors' cultural biases that guide and constrain policy core beliefs in policy subsystems. Moreover, the study offers insights into how actors seize opportunities presented by external and internal disruptions (through exploitative coalitions). It is argued that strategic issue reframing and changing coalitions and ideological congruence (related to veto and institutional players) enabled policy change. All-in-all, the paper demonstrates the explanatory power of ACF and the added value of theoretical integration in providing a more comprehensive picture of the policymaking process and associated outcomes.

2.5. Critical policy analysis

Critical policy analysis (CPA) represents a multidisciplinary approach that draws on various theories in social sciences, such as neo-Marxism and social constructivism (Arts, 2012, Kangas et al., 2006, Krott, 2012). This family of theories distinguishes themselves from 'traditional' policy analysis in that they emphasise the importance of power relationships and actors involved in the policymaking process. Rather than viewing policies (or knowledge) as being neutral or objective, CPA views policies as being shaped by the interests and values of dominant groups in society, where facts, values, norms, and meanings are intermingled. These theories move away from postpositivist and realist arguments in saying that reality does not necessarily exist independently from knowledge, essentially arguing that facts cannot be separated from norms and values. In other words, the development of scientific knowledge is determined by social forces, where knowledge is a set of beliefs or mental models people use to interpret actions and events in the world. For example, social constructivism has been widely used as a learning or knowledge theory, showing that people do not find or discover knowledge so much as construct or make it. The implications are that people invent concepts, models, and schemes to make sense of experiences, which are then continuously tested and modified based on new experiences and validated by use in society (Schwandt, 1994). On the far end of the spectrum, proponents of social constructivism argue that knowledge is a social process of consensus-building within communities or socially negotiated knowledge creation (Rannikmäe et al., 2020, Stapleton and Stefaniak, 2019).

One illustrative example is social media, which allows people to engage, share information, and collectively shape their worldviews. Users co-create meaning through discussions and interactions. To put it differently, social media aligns with the principles of social constructivism by providing a platform for knowledge co-creation, dialogue and collaboration, including exposure to diverse perspectives. As such, social media should ideally enable interactions among individuals with diverse beliefs to challenge established knowledge and help reshape how people understand the world (Schrader, 2015). However, social media can also reinforce existing viewpoints. For instance, politically conservative individuals often follow like-minded accounts and consume reinforcing content. This is a cognitive bias (or confirmation bias) whereby individuals tend to seek out and give more weight to information that confirms their existing beliefs while dismissing or ignoring knowledge that contradicts them (Workman, 2018).

In forest policy research, discourse theory, a subset of critical policy analysis, is frequently utilised. It centres around language and how it influences identities, ideas, and interests in policymaking (Macdonell, 1991, Torfing, 2005). More specifically, discourse theory is concerned with in what way language, communication, and discourse (ways of talking and thinking) are used to create, reinforce, or challenge social norms, hierarchies, and ideologies. Discourse theory is rooted in the idea that language and knowledge are not neutral but are shaped by power structures and social contexts. For instance, different discourses can create different versions of reality, and the dominant discourse commonly shapes how people perceive and understand the world. Researchers have uncovered underlying ideologies and beliefs that influence policymaking by analysing discourses, such as hidden assumptions, biases, and power structures (Figure 3.3). Examples of the application of discourse analysis in forest policy research include studies on the development of global forest politics and governance (Arts and Buizer, 2009, van Heeswijk and Turnhout, 2013), EU forest policymaking (Pülzl et al., 2018), the sustainability debate on forest biomass for energy (Mather-Gratton et al., 2021) and effects of climate change discourses on forest conservation (De Koning et al., 2014).

Another example can be taken from frame theory, which focuses on how policies are framed and represented to the public and policymakers. Frames are generally considered as a way by which people interpret (or 'frame') their immediate social and natural environment, providing a system or process whereby a person can classify and manage information. While there are several definitions of 'frames' and 'framing', it is the information filtering process that represents

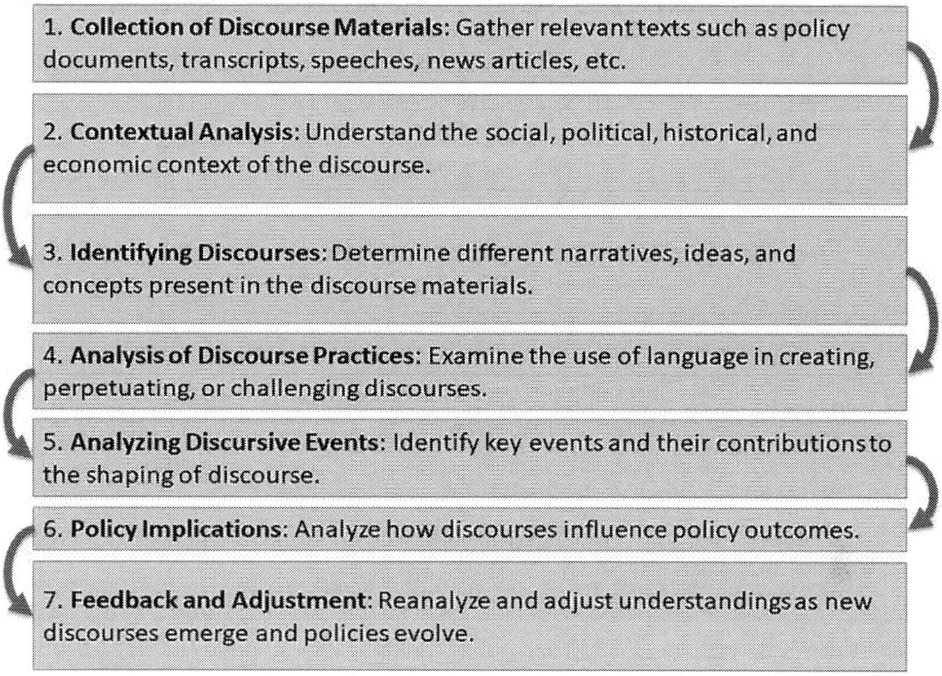

Figure 3.3 Example of steps taken for a discourse analysis.

Source: own figure

one of the shared assumptions underlying frame theory (Goffman, 1974, Gamson and Modigliani, 1989, Entman, 1993, Pan and Kosicki, 1993, Reese et al., 2001). It is commonly argued that frames affect how humans act and think in everyday life (Rein and Schön, 1996, Daviter, 2007). Framing is, in turn, the active application of a frame, such as when a person emphasises specific aspects of reality (Snow and Benford, 1988, Pan and Kosicki, 1993, Reese et al., 2001) or when people evaluate information which affects how they interpret experiences, facts, and events (Stern et al., 1999, Kaltenborn and Bjerke, 2002, Vugteveen et al., 2010). Forest policy researchers have regularly employed frame analysis to investigate EU forest policy fragmentation (Elomina and Pülzl, 2021), understanding policy conflicts associated with bioenergy and forestry (Söderberg and Eckerberg, 2013) and forest management practices (Heder Brandt et al.).

Case example: Forest biomass for energy in Europe

A recent study by Mather-Gratton et al. (2021) evaluated the ongoing debate regarding the use of forest biomass for renewable energy in Europe. The primary objective was to pinpoint the narratives embraced by different coalitions as they engage in discourse and vie for influence in policymaking. The paper's central concern was to unravel the narratives constructed by these coalitions for their respective audiences and assess their impact on other narratives and stakeholders. The study's findings reveal contrasting perspectives on sustainability and the natural world. These perspectives are encapsulated in three competing storylines regarding the use of forest biomass in Europe. Furthermore, the study sheds light on instances where these narratives intersect, offering valuable insights for fostering a more constructive and informed debate on the sustainability of forest biomass.

2.6. Other theories and approaches

In the preceding sections, the focus has been on a selection of theories and approaches used in forest policy research. However, these represent only a small portion of the theories and analytical tools available to researchers engaged in forest policy analysis. Beyond the theories introduced, a diverse array of additional frameworks, perspectives, and methodologies have contributed significantly to understanding forest policy dynamics. In the following section, three additional theories and approaches will be briefly outlined, recognising their importance in offering insights and comprehensive analyses of forest policy.

2.6.1. Pathways to sustainability

The pathways to sustainability approach, introduced by Leach et al. (2007), tackles the governance challenges stemming from complex, interconnected, multi-scale systems. It is a framework geared towards achieving sustainability in forest management by embracing a holistic and integrative perspective. As it underscores, sustainability is intricately shaped by dynamics, complexities, uncertainties, diverse narratives, and values-driven objectives. Moreover, it stresses the importance of recognising dominant but ineffective approaches and forging new, successful pathways to address challenges (Leach et al., 2010).

This approach reframes sustainability challenges within complex socio-ecological systems that constantly evolve along multiple pathways. It strongly emphasises contextual analysis, pathway dynamics, interactions, and the necessity of learning and adapting, all in pursuit of adaptive governance (Leach et al., 2018, Ostrom, 2002, Scharlemann et al., 2020). For instance, it sheds light on the persistence of dominant yet unsuccessful approaches. The pathways approach also underscores the significance of developing effective alternative pathways towards sustainability (Leach), acknowledging the diverse means and strategies, each with its synergies and trade-offs concerning various dimensions of inequality.

2.6.2. Common pool resources

The common pool resource theory, derived from Hardin (1968), describes shared and non-excludable resources with a finite supply. These resources are often subject to the 'tragedy of the commons', where individuals acting in their self-interest deplete the resource, leading to overconsumption (Ostrom, 2008b, Ostrom, 2008a, Yoder et al., 2022). The theory focuses on the ability of people to collaborate to overcome management dilemmas inherent to common-pool resources. It categorises goods into four specific categories: private goods, common goods, club goods, and public goods, and recognises that if managed carefully, common-pool resources can be extended by forming a negative feedback loop between the stock variable and the fringe variable. For example, consider a forest ecosystem where the stock variable is the population of a particular tree species. If the population starts to decline (e.g., due to disease or habitat loss), this triggers a negative feedback loop with fringe variables, such as conservation efforts or reforestation initiatives.

The general idea is that by recognising the challenges posed by common-pool resources and their vulnerability to overuse, policymakers can develop effective policies and regulations to ensure equitable and sustainable forest management (Fleischman et al., 2014). Methodologically, the theory has been applied using varied approaches such as game theory models to simulate interactions among individuals or groups sharing forest resources (Bednarik et al., 2019), case studies of forest commons to analyse how communities or user groups manage and govern shared forests (Yoder et al., 2022), and varied policy analysis methods to assess the effectiveness of different policies regimes in preventing or addressing the tragedy of the commons (Bouriaud and Schmithüsen, 2005, Ostrom, 2008a). These methodological approaches have helped forest policy researchers gain insights into the dynamics of managing common resource pools and the tragedy of the commons.

2.6.3. Political ecology

Political ecology, an interdisciplinary field, focuses on the interplay between politics, economics, and the environment, including forest policy and management. Rooted in critical social theory, it emphasises the role of power dynamics, societal structures, and institutional frameworks in shaping the governance of natural resources, including forests (Smith, 2007, Forsyth, 2005). Recognising that forests are not only physical ecosystems, extending into social systems integrated with political and economic processes, political ecology underlines the importance of understanding social, economic, and political factors affecting forest management. This includes stakeholder interests, power dynamics, and equitable distribution of benefits and costs (Blaikie and Brookfield, 1987, Muthee et al., 2022).

In the realm of political ecology, a combination of methods and approaches, such as case studies, political economic analysis, and the political ecology of knowledge,[8] have been employed to provide a more holistic picture. Political ecology has been applied extensively in forest policy research to examine diverse aspects of forest management. It delves into issues like deforestation, forest governance, indigenous land rights, and the influence of global markets on forests. Examples of such inquiries include investigations into the global timber trade and its repercussions on forests (Humphreys, 2006), assessments of the politics and ramifications of forest certification programs in promoting sustainable forestry practices (Cashore, 2002), and analyses of the political and environmental factors that contribute to forest fires (Fairhead et al., 2012). Through the lens of political ecology, these studies have sought to explain how power dynamics affect policymaking processes and resource allocation.

3. Conclusions

This chapter introduced public policy analysis as applied by the forest-based sector, introducing five distinct yet interconnected theoretical and methodological approaches: institutional policy analysis, policy network analysis, the advocacy coalition framework, critical policy analysis, and rational policy analysis. Each brings its perspectives and toolkits to the research process, trying to explain different dynamics that shape the development, implementation, and evaluation of public policies. For instance, institutional policy analysis directs attention to the role of formal and informal institutions in the policymaking process. Policy network analysis offers insights into the relationships between actors and organisations that influence policy outcomes. The advocacy coalition framework demonstrates the effects of ideas and values among stakeholder coalitions trying to shape policy objectives. Critical policy analysis draws attention to prevailing power structures, ideologies, and normative assumptions in the policy discourse. In contrast, rational policy analysis illustrates the utility of systematically evaluating the cost and benefits of efficient and effective policy solutions. These diverse theoretical and methodological approaches act as lenses, each offering a unique perspective on forest policies and practices.

Importantly, these approaches have not remained confined to theory but have affected forest policy research and practice. Consider, for instance, ACF in the context of forest management. Researchers applying ACF have analysed the dynamics of stakeholder coalitions involved in forest policy, revealing how belief systems and interests have influenced the design of forest policy. This has improved our understanding of the negotiation processes surrounding forest resource management, ultimately contributing to more informed policymaking (Hysing and Olsson, 2008, Sotirov et al., 2021). Likewise, policy network analysis has exposed the power structures and information flows that shape forest governance (Baulenas et al., 2021, Borg et al., 2015). Insights gained from such analyses have informed strategies for enhancing the effectiveness of forest conservation efforts and ensuring equitable access to forest resources.

As you delve deeper into the application of different theoretical and methodological approaches, it will become evident that these approaches often converge to offer more meaningful results. Considering this, one of the significant benefits of learning and applying different theoretical and methodological practices in research, particularly in the context of shared natural resources, such as forests, resides in the potential for more holistic insights. As complex socio-ecological systems, forests arguably require integrative approaches to understand their management and governance fully. For instance, while institutional policy analysis might highlight the formal rules and regulations governing forest resource use, critical policy analysis can reveal the power structures and inequities embedded within these institutions. At the same time, policy network analysis can illuminate the relationships that underpin forest governance, highlighting key actors and their roles. In other words, integrating different perspectives and methodologies allows researchers to look beyond disciplinary boundaries to explore the synergies and trade-offs inherent in forest management. This underscores the importance of interdisciplinary work and the importance of having access to a multidisciplinary analytical toolbox (see Chapter 10) in addressing complex and interconnected policy issues.

Want to find out more?

The scope of this section has been to provide a brief snapshot of some theoretical and methodological approaches within the broad field of public policy analysis. It mainly serves as an introduction, offering a taste of the diverse tools and frameworks available for understanding and researching policymaking processes. For those seeking more knowledge, resources, and guidance on policy analysis, see the references provided below, or check the many references provided throughout the chapter:

- *The Policy Analyst's Handbook* (Irwin, 2004).
- 'A guide to policy analysis as a research method' (Browne et al., 2019).
- *A Practical Guide for Policy Analysis* (Bardach and Patashnik, 2019).
- *Forest Policy Analysis* (Krott, 2005).

The following set of questions may also help guide further inquiry into specific areas of interest:

1. How do institutions, such as government agencies and regulatory bodies, shape the policymaking process and outcomes?

2. What role does organisational culture play in influencing policy decisions within institutions?
3. What are the main components of a policy network, and how do they interact in the policymaking process?
4. How can the advocacy coalition framework help us understand policy stability and change in complex policy areas?
5. What challenges might researchers face when applying critical policy analysis in practice?
6. How can rational policy analysis techniques be integrated with other frameworks to provide more comprehensive insights?

4. Assignment II: Analysing forest policy using different theoretical frameworks

Description

This assignment is based on Chapter 3, and it is meant to deepen the student's understanding of the role of theoretical frameworks in forest policy analysis, focusing on applying theoretical knowledge to real-world policy problems, including the development of policy recommendations.

Purpose

Students are assigned to select one of the theoretical frameworks considered in Chapter 3 or Annex I, such as institutional policy analysis, policy network analysis, advocacy coalition framework or critical policy analysis (or another theory, if they wish). They will then apply the selected theoretical framework to a forest policy problem that they consider relevant or find interesting.

Outline of task

- Read Chapter 3 and Annex I for an overview of different theories and approaches used in forest policy research.
- *Literature review*: Briefly review the selected theoretical framework, summarising its key components and relevance to forest policy research.
- *Case study selection*: Choose and outline a forest policy problem, such as deforestation, biodiversity conservation or climate change mitigation, that interests you.
- *Analysis*: Apply the theoretical framework to the selected case study. This can include identifying the main actors, their interests, relevant governance structures, conflicts and coalitions. Consider how the framework can help to explain the policy development process, implementation challenges, and potential policy outcomes.
- *Draft policy recommendations*: Based on your analysis, propose a few recommendations to address the identified challenges and enhance policy implementation.

Deliverable

- Prepare a report including a literature review, case study analysis (including a methodological section), and policy recommendations (length: 2500–3000 words).
- Based on your case study, prepare a 1-page proposal for a class discussion (including some open questions).

5. Notes

1 Advocacy coalitions are groups of political actors who share ideas and coordinate effectively to influence government decisions.
2 See www.sciencedirect.com/journal/forest-policy-and-economics.
3 See www.mdpi.com/journal/forests.
4 See www.tandfonline.com/journals/tjfr20.
5 **Advocacy coalitions/networks** refer to the organised groups and individuals that work together to promote specific policy goals and objectives within a particular policy subsystem (Figure 3.2). These networks play a central role in shaping policy outcomes.
6 Coalitions that take advantage of issues triggered by external and internal disruptive events to push for policy change.
7 Cultural theory argues that individual's attitudes and behaviours regarding risk are influenced by their cultural backgrounds, values, and worldviews. The theory suggests that individuals can be classified into different cultural groups or worldviews, each with a distinct risk perception and management approach.
8 This approach explores how different forms of knowledge (indigenous, scientific, local) shape environmental discourses and policies.

6. References

AGGESTAM, F. & GIURCA, A. 2021. The art of the 'green' deal: Policy pathways for the EU Forest Strategy. *Forest Policy and Economics*, 128, 102456.

AGGESTAM, F., KONCZAL, A., SOTIROV, M., WALLIN, I., PAILLET, Y., SPINELLI, R., LINDNER, M., DERKS, J., HANEWINKEL, M. & WINKEL, G. 2020. Can nature conservation and wood production be reconciled in managed forests? A review of driving factors for integrated forest management in Europe. *Journal of Environmental Management*, 268, 110670.

AGGESTAM, F. & VOGELPOHL, T. 2009. Incorporating the institutional dimension into sustainability impact assessment for an advanced sustainability analysis of the forest-wood chain (Poster). Uppsala, Sweden.: EFORWOOD final conference, 23–24 September 2009: Shape your sustainability tools.

ARROW, J. K., CROPPER, L. M., GOLLIER, C., GROOM, B., HEAL, M. G., NEWELL, G. R., NORDHAUS, D. W., PINDYCK, S. R., PIZER, A. W., PORTNEY, R. P., STERNER, T., TOL, J. S. R. & WEITZMAN, K. M. 2014. Should governments use a declining discount rate in project analysis? *Review of Environmental Economics and Policy*, 8, 145–163.

ARTS, B. 2012. Forests policy analysis and theory use: Overview and trends. *Forest Policy and Economics*, 16, 7–13.

ARTS, B. & BUIZER, M. 2009. Forests, discourses, institutions: A discursive-institutional analysis of global forest governance. *Forest Policy and Economics*, 11, 340–347.

ARTS, B. & VAN DE GRAAF, R. 2009. Theories for forest policy: An overview. *CIAS discussion paper No.8.*

BANERJEE, O. & ALAVALAPATI, J. 2014. Forest policy modelling in an economy-wide framework. *In:* KANT, S. & ALAVALAPATI, J. (eds.) *Handbook of Forest Resource Economics.* Routledge.

BARDACH, E. & PATASHNIK, E. M. 2019. *A Practical Guide for Policy Analysis: The Eightfold Path to More Effective Problem Solving (6th edition)*, SAGE.

BAULENAS, E., KRUSE, S. & SOTIROV, M. 2021. Forest and water policy integration: A process and output-oriented policy network analysis. *Environmental Policy and Governance*, 31, 432–450.

BAYCHEVA-MERGER, T., SOTIROV, M., HOLMGREN, S. & SELTER, A. 2018. Institutional and actor-oriented factors constraining expert-based forest information exchange in Europe: A policy analysis from an actor-centred institutionalist approach. *Forests* 9, 129.

BEDNARIK, P., LINNEROOTH-BAYER, J., MAGNUSZEWSKI, P. & DIECKMANN, U. 2019. A game of common-pool resource management: Effects of communication, risky environment and worldviews. *Ecological Economics*, 156, 287–292.

BELAND LINDAHL, K., BAKER, S., RIST, L. & ZACHRISSON, A. 2016. Theorising pathways to sustainability. *International Journal of Sustainable Development & World Ecology*, 23, 399–411.

BESUSSI, E. 2006. Policy networks: conceptual developments and their European applications. *CASA Working Papers*. London: Centre for Advanced Spatial Analysis (UCL).

BLAIKIE, P. & BROOKFIELD, H. 1987. *Land and Degradation and Society (1st ed.)*, Routledge.

BOARDMAN, A. E., GREENBERG, D. H., VINING, A. R. & WEIMER, D. L. 2018. *Cost–benefit Analysis: Concepts and Practice*. Cambridge: Cambridge University Press.

BONAZOUNTAS, M., KALLIDROMITOU, D., KASSOMENOS, P. A. & PASSAS, N. 2005. Forest fire risk analysis. *Human and Ecological Risk Assessment: An International Journal*, 11, 617–626.

BORG, R., TOIKKA, A. & PRIMMER, E. 2015. Social capital and governance: a social network analysis of forest biodiversity collaboration in Central Finland. *Forest Policy and Economics*, 50, 90–97.

BORGATTI, S. P. & FOSTER, P. C. 2003. The network paradigm in organizational research: A review and typology. *Journal of Management*, 29, 991–1013.

BORRASS, L., KLEINSCHMIT, D. & WINKEL, G. 2017. The 'German model' of integrative multifunctional forest management – Analysing the emergence and political evolution of a forest management concept. *Forest Policy Econ.*, 77, 16–23. https://doi.org/10.1016/j.forpol.2016.06.028.

BOURIAUD, L. & SCHMITHÜSEN, F. 2005. Allocation of property rights on forests through ownership reform and forest policies in central and eastern European Countries. *Schweizerische Zeitschrift fur Forstwesen*, 156, 297–305.

BRANDER, L. M., VAN BEUKERING, P., NIJSTEN, L., MCVITTIE, A., BAULCOMB, C., EPPINK, F. V. & CADO VAN DER LELIJ, J. A. 2020. The global costs and benefits of expanding Marine Protected Areas. *Marine Policy*, 116, 103953.

BRESSERS, H. T. A. & O'TOOLE, L. J. 1998. The selection of policy instruments: a network-based perspective. *Journal of Public Policy*, 18, 213–239.

BRETON, E. & DE LEEUW, E. 2011. Theories of the policy process in health promotion research: a review. *Health Promot Int*, 26, 82–90.

BROUSSEAU, E. & GLACHANT, J.-M. 2008. *New Institutional Economics*. Cambridge University Press.

BROWNE, J., COFFEY, B., COOK, K., MEIKLEJOHN, S. & PALERMO, C. 2019. A guide to policy analysis as a research method. *Health Promot Int*, 34, 1032–1044.

CASHORE, B. 2002. Legitimacy and the privatisation of environmental governance: How non-state market-driven (NSMD) governance systems gain rule-making authority. *Governance*, 15, 503–529.

CASHORE, B., AULD, G. & NEWSOM, D. 2004. *Governing Through Markets: Forest Certification and the Emergence of Non-State Authority*. Yale University Press.

CHAUDHARY, A., BURIVALOVA, Z., KOH, L. P. & HELLWEG, S. 2016. Impact of forest management on species richness: Global Meta-analysis and economic trade-offs. *Scientific Reports*, 6, 23954.

CHRISTINE, M. & LUUK, D. 2021. Cost–benefit analysis as a basis for risk-based rockfall protection forest management. *In:* MICHAELA, T., CRISTIAN, A., FRANK, P. & KARL, K. (eds.) *Protective Forests as Ecosystem-based Solution for Disaster Risk Reduction (Eco-DRR)*. Rijeka: IntechOpen.

CLARKE, B., SWINBURN, B. & SACKS, G. 2016. The application of theories of the policy process to obesity prevention: a systematic review and meta-synthesis. *BMC Public Health*, 16, 1084.

COMPSTON, H. 2009. *Policy Networks and Policy Change: Putting Policy Network Theory to the Test*. London: Palgrave Macmillan.

COVENEY, J. 2010. Analysing public health policy: Three approaches. *Health Promotion Practice*, 11, 515–521.

CREWETT, W. 2015. Street-level bureaucrats at work: A municipality-level institutional analysis of community-based natural resource management implementation practice in the pasture sector of Kyrgyzstan. *Sustainability*, 7.

CROITORU, L. 2007. Valuing the non-timber forest products in the Mediterranean region. *Ecological Economics*, 63, 768–775.

CULLERTON, K., DONNET, T., LEE, A. & GALLEGOS, D. 2017. Joining the dots: The role of brokers in nutrition policy in Australia. *BMC Public Health*, 17.

DAVITER, F. 2007. Policy framing in the European Union. *Journal of European Public Policy*, 14, 654–666.

DE KONING, J., WINKEL, G., SOTIROV, M., BLONDET, M., BORRAS, L., FERRANTI, F. & GEITZENAUER, M. 2014. Natura 2000 and climate change - polarisation, uncertainty and pragmatism in discourses on forest conservation and management in Europe. *Environmental Science and Policy*, 39, 129–138.

DEWULF, A., CRAPS, M. & DERCON, G. 2004. How issues get framed and reframed when different communities meet: A multi-level analysis of a collaborative soil conservation initiative in the Ecuadorian Andes. *Journal of Community and Applied Social Psychology*, 14, 177–192.

DIMAGGIO, P. J. & POWELL, W. W. 1991. *The New Institutionalism and Organizational Analysis*. Chicago: University of Chicago Press.

ELLIOTT, C. & SCHLAEPFER, R. 2001. Understanding forest certification using the Advocacy Coalition Framework. *Forest Policy and Economics*, 2, 257–266.

ELOMINA, J. & PÜLZL, H. 2021. How are forests framed? An analysis of EU forest policy. *Forest Policy and Economics*, 127, 102448.

ENTMAN, R. M. 1993. Framing: Toward clarification of a fractured paradigm. *Journal of Communication*, 43, 51–58.

FAIRHEAD, J., LEACH, M. & SCOONES, I. 2012. Green grabbing: A new appropriation of nature? *The Journal of Peasant Studies*, 39, 237–261.

FLEISCHMAN, F. D., LOKEN, B., GARCIA-LOPEZ, G. A. & VILLAMAYOR-TOMAS, S. 2014. Evaluating the utility of common-pool resource theory for understanding forest governance and outcomes in Indonesia between 1965 and 2012. *International Journal of the Commons*, 8, 304–336.

FORSYTH, T. 2005. The political ecology of the ecosystem approach for forests. *In:* SAYER, J. & MAGINNIS, S. (eds.) *Forests in Landscapes: Ecosystem Approaches for Sustainability.* London: Earthscan.

GALE, F. & CADMAN, T. 2014. Whose norms prevail? Policy networks, international organizations and 'sustainable forest management'. *Society & Natural Resources*, 27, 170–184.

GAMSON, W. A. & MODIGLIANI, A. 1989. Media discourse and public opinion on nuclear power: A constructionist approach. *American Journal of Sociology*, 95, 1–37.

GILLIAM, F. D. & BALES, S. N. 2001. Strategic frame analysis: Reframing America's youth. *Social Policy Report*, XV, 3–14.

GLÜCK, P. 2000. Theoretical perspectives for enhancing biological diversity in forest ecosystems in Europe. *Forest Policy and Economics*, 1, 195–207.

GOFFMAN, E. 1974. *Frame Analysis: An Essay on the Organisation of Experience*. New York: Harper & Row.

GORMLEY, W. T. 1987. Institutional policy analysis: A critical review. *Journal of Policy Analysis and Management*, 6, 153–169.

HANEWINKEL, M., HUMMEL, S. & ALBRECHT, A. 2011. Assessing natural hazards in forestry for risk management: a review. *European Journal of Forest Research*, 130, 329–351.

HANLEY, N. 2001. Cost–benefit analysis and environmental policymaking. *Environment and Planning C: Government and Policy*, 19, 103–118.

HARDIN, G. 1968. The Tragedy of the Commons. *Science*, 162, 1243–1248.

HARRINKARI, T., KATILA, P. & KARPPINEN, H. 2016. Stakeholder coalitions in forest politics: revision of Finnish Forest Act. *Forest Policy and Economics*, 67, 30–37.

HAYS, S. P. 1981. The Environmental Movement. *Journal of Forest History*, 25, 219–221.

HEDER BRANDT, P., OLSSON, A., DAHLQUIST, K. & INAL, T. 'Profitability is sustainability:' framing of forest management practices by the Swedish forest industry. *Scandinavian Journal of Forest Research*, 1–13.

HEILBRON, J. L. 1990. Introductory essay. *In:* FRANGSMYR, T., HEILBRON, J. L. & RIDER, R. E. (eds.) *The quantifying spirit of the Eighteenth Century.* Los Angeles: University of California Press.

HOEFER, R. 2021. Theory in Social Policy Research: Rationality and Its Discontents. *Journal of Policy Practice and Research*, 2, 233–237.

HOOD, C. 1990. J. G. March and J. P. Olsen Rediscovering Institutions: The Organisational Basis of Politics, New York, Free Press, 1989. *Journal of Public Policy*, 10, 349–351.

HUMPHREYS, D. 2006. *Logjam: Deforestation and the Crisis of Global Governance* (1st ed.), Routledge.

HYSING, E. & OLSSON, J. 2008. Contextualising the Advocacy Coalition Framework: theorising change in Swedish forest policy. *Environmental Politics*, 17, 730–748.

IRWIN, G. L. 2004. *The Policy Analyst's Handbook: Rational Problem Solving in a Political World*, Routledge.

JENKINS-SMITH, H. C. & SABATIER, P. A. 1994. Evaluating the Advocacy Coalition Framework. *Journal of Public Policy*, 14, 175–203.

KALTENBORN, B. P. & BJERKE, T. 2002. Associations between environmental value orientations and landscape preferences. *Landscape and Urban Planning*, 59, 1–11.

KANGAS, A., LAUKKANEN, S. & KANGAS, J. 2006. Social choice theory and its applications in sustainable forest management – a review. *Forest Policy and Economics*, 9, 77–92.

KNIIVILÄ, M., OVASKAINEN, V., SAASTAMOINEN, O. & KNIIVILÄ, M. 2002. Costs and benefits of forest conservation: regional and local comparisons in Eastern Finland. *Journal of Forest Economics*, 8, 131–150.

KNOKE, D. & YANG, S. 2008. *Social Network Analysis*. Thousand Oaks, CA: Sage.

KOTECKÝ, V. 2015. Contribution of afforestation subsidies policy to climate change adaptation in the Czech Republic. *Land Use Policy*, 47, 112–120.

KROTT, M. 2005. *Forest Policy Analysis*. Dordrecht, The Netherlands: Springer.

KROTT, M. 2012. Value and risks of the use of analytical theory in science for forest policy. *Forest Policy and Economics*, 16, 35–42.

KUNREUTHER, H. 2002. Risk analysis and risk management in an uncertain world1. *Risk Analysis*, 22, 655–664.

LAZARIDOU, D. C., MICHAILIDIS, A. & TRIGKAS, M. 2021. Exploring environmental and economic costs and benefits of a forest-based circular economy: A literature review. *Forests*, 12, 436.

LEACH, M., REYERS, B., BAI, X., BRONDIZIO, E. S., COOK, C., DÍAZ, S., ESPINDOLA, G., SCOBIE, M., STAFFORD-SMITH, M. & SUBRAMANIAN, S. M. 2018. Equity and sustainability in the Anthropocene: a social–ecological systems perspective on their intertwined futures. *Global Sustainability*, 1, e13.

LEACH, M., SCOONES, I. & STIRLING, A. 2007. Pathways to sustainability: An overview of the STEPS Centre approach. *STEPS Approach Paper.* Brighton: STEPS Centre.

LEACH, M., SCOONES, I. & STIRLING, A. 2010. *Dynamic Sustainabilities: Technology, Environment, Social Justice*. London: Earthscan.

LEBAN, V., PEZDEVŠEK MALOVRH, S., ZADNIK STIRN, L. & KRČ, J. 2016. Forest biomass for energy in multi-functional forest management: Insight into the perceptions of forest-related professionals. *Forest Policy Econ.*, 71, 87–93. https://doi.org/10.1016/j.forpol.2015.07.005.

LOVRIĆ, M., DA RE, R., VIDALE, E., PETTENELLA, D. & MAVSAR, R. 2018. Social network analysis as a tool for the analysis of international trade of wood and non-wood forest products. *Forest Policy and Economics*, 86, 45–66.

LOWOOD, H. E. 1990. The calculating forester: quantification, cameral science, and the emergence of scientific forestry management in Germany. *In:* FRANGSMYR, T., HEILBRON, J. L. & RIDER, R. E. (eds.) *The quantifying spirit of the Eighteenth Century.* Los Angeles: University of California Press.

MACDONELL, D. 1991. *Theories of Discourse: An Introduction*. Wiley-Blackwell.

MARIN, B. & MAYNTZ, R. 1991. *Policy Networks: Empirical Evidence and Theoretical Considerations*. Boulder: Westview Press.

MATHER-GRATTON, Z. J., LARSEN, S. & BENTSEN, N. S. 2021. Understanding the sustainability debate on forest biomass for energy in Europe: A discourse analysis. *PLOS ONE*, 16, e0246873.

MEYER, J. W. & ROWAN, B. 1977. Institutionalised organisations: Formal structure as myth and ceremony. *American Journal of Sociology*, 83.

MONTGOMERY, D. J. 1987. Rational policy analysis: Normative foundations. *Public Productivity Review*, 10, 47–54.

MOYSON, S. 2017. Cognition and policy change: the consistency of policy learning in the advocacy coalition framework. *Policy and Society*, 36, 320–344.

MUTHEE, K., DUGUMA, L., WAINAINA, P., MINANG, P. & NZYOKA, J. 2022. A review of global policy mechanisms designed for tropical forests conservation and climate risks management. *Frontiers in Forests and Global Change*, 4.

NAGEL, T. A., FIRM, D., PISEK, R., MIHELIC, T., HLADNIK, D., DE GROOT, M. & ROZENBERGAR, D. 2017. Evaluating the influence of integrative forest management on old-growth habitat structures in a temperate forest region. *Biological Conservation*, 216, 101–107.

NAIDOO, R. & RICKETTS, T. H. 2006. Mapping the economic costs and benefits of conservation. *PLOS Biology*, 4, e360.

NEPSTAD, D., MCGRATH, D., STICKLER, C., ALENCAR, A., AZEVEDO, A., SWETTE, B., BEZERRA, T., DIGIANO, M., SHIMADA, J., SEROA DA MOTTA, R., ARMIJO, E., CASTELLO, L., BRANDO, P.,

HANSEN, M. C., MCGRATH-HORN, M., CARVALHO, O. & HESS, L. 2014. Slowing Amazon deforestation through public policy and interventions in beef and soy supply chains. *Science*, 344, 1118–1123.
NINAN, K. N. & INOUE, M. 2013. Valuing forest ecosystem services: What we know and what we don't. *Ecological Economics*, 93, 137–149.
NOCENTINI, S., CIANCIO, O., PORTOGHESI, L. & CORONA, P. 2021. Historical roots and the evolving science of forest management under a systemic perspective. *Canadian Journal of Forest Research*, 51, 163–171.
NORTH, D. C. 1990. *Institutions, Institutional Change and Economic Performance*. Cambridge: Cambridge University Press.
OSTROM, E. 1990. *Governing the Commons: The Evolution of Institutions for Collective Action*. New York: Cambridge University Press.
OSTROM, E. 2002. Chapter 24 Common-pool resources and institutions: Toward a revised theory. *Handbook of Agricultural Economics*. Elsevier.
OSTROM, E. 2008a. The challenge of common-pool resources. *Environment: Science and Policy for Sustainable Development*, 50, 8–21.
OSTROM, E. 2008b. Tragedy of the commons. *In:* DURLAUF, S. N. & BLUME, L. E. (eds.) *The New Palgrave Dictionary of Economics: Volume 1–8*. London: Palgrave Macmillan.
PALMER, J. & BULKAN, J. 2010. Legitimacy of public domain forest taxation, and combatting corruption in forestry. *International Forestry Review*, 12, 150–164.
PAN, Z. & KOSICKI, G. M. 1993. Framing analysis: An approach to news discourse. *Political Communication*, 10, 55–75.
PASCUAL, U., BALVANERA, P., DÍAZ, S., PATAKI, G., ROTH, E., STENSEKE, M., WATSON, R. T., BAŞAK DESSANE, E., ISLAR, M., KELEMEN, E., MARIS, V., QUAAS, M., SUBRAMANIAN, S. M., WITTMER, H., ADLAN, A., AHN, S., AL-HAFEDH, Y. S., AMANKWAH, E., ASAH, S. T., BERRY, P., BILGIN, A., BRESLOW, S. J., BULLOCK, C., CÁCERES, D., DALY-HASSEN, H., FIGUEROA, E., GOLDEN, C. D., GÓMEZ-BAGGETHUN, E., GONZÁLEZ-JIMÉNEZ, D., HOUDET, J., KEUNE, H., KUMAR, R., MA, K., MAY, P. H., MEAD, A., O'FARRELL, P., PANDIT, R., PENGUE, W., PICHIS-MADRUGA, R., POPA, F., PRESTON, S., PACHECO-BALANZA, D., SAARIKOSKI, H., STRASSBURG, B. B., VAN DEN BELT, M., VERMA, M., WICKSON, F. & YAGI, N. 2017. Valuing nature's contributions to people: the IPBES approach. *Current Opinion in Environmental Sustainability*, 26–27, 7–16.
POLSKI, M. M. & OSTROM, E. 1999. An institutional framework for policy analysis and design. *In:* COLE, H. D. & MCGINNIS, M. D. (eds.) *Elinor Ostrom and the Bloomington School of Political Economy: A Framework for Policy Analysis*. Lanham: Lexington Books.
POTEETE, A. R. & OSTROM, E. 2004. An institutional approach to the study of forest resources. *In:* POULSEN, J. (ed.) *Human impacts on tropical forest biodiversity and genetic resources*.
PÜLZL, H., WYDRA, D. & HOGL, K. 2018. Piecemeal integration: Explaining and understanding 60 years of European Union forest policymaking. *Forests*, 9, 719.
RAKATAMA, A., PANDIT, R., MA, C. & IFTEKHAR, S. 2017. The costs and benefits of REDD+: A review of the literature. *Forest Policy and Economics*, 75, 103–111.
RAMETSTEINER, E. & SIMULA, M. 2003. Forest certification – an instrument to promote sustainable forest management? *Journal of Environmental Management*, 67, 87–98.
RÄMÖ, J., TUPEK, B., LEHTONEN, H. & MÄKIPÄÄ, R. 2023. Towards climate targets with cropland afforestation – effect of subsidies on profitability. *Land Use Policy*, 124, 106433.
RANNIKMÄE, M., HOLBROOK, J. & SOOBARD, R. 2020. Social constructivism – Jerome Bruner. *In:* AKPAN, B. & KENNEDY, T. J. (eds.) *Science Education in Theory and Practice: An Introductory Guide to Learning Theory*. Cham: Springer International Publishing.
REESE, S. D., GANDY, J., O.H., & GRANT, A. E. 2001. *Framing Public Life: Perspectives on Media and Our Understanding of the Social World*. Abingdon: Routledge.
REIN, M. & SCHÖN, D. A. 1996. Frame-critical policy analysis and frame-reflective policy practice. *Knowledge and Policy*, 9, 85–104.
RHODES, R. A. W. 1997. *Understanding Governance: Policy Networks, Governance, Reflexivity and Accountability*. Open University Press.
SABATIER, P. A. 1987. Knowledge, policy-oriented learning, and policy change: An advocacy coalition framework. *Knowledge*, 8, 649–692.

SABATIER, P. A. 1988. An advocacy coalition framework of policy change and the role of policy-oriented learning therein. *Policy Sciences*, 21, 129–168.

SABATIER, P. A. & JENKINS-SMITH, H. 1996. *Policy Change and Learning: An Advocacy Coalition Approach*. Westview Press.

SARVAŠOVÁ, Z., ŠÁLKA, J. & DOBŠINSKÁ, Z. 2013. Mechanism of cross-sectoral coordination between nature protection and forestry in the Natura 2000 formulation process in Slovakia. *Journal of Environmental Management*, 127, S65–S72.

SCHARLEMANN, J. P. W., BROCK, R. C., BALFOUR, N., CABERNARD, L., BURGESS, N. D., GUTH, M. K., INGRAM, D. J., LANE, R., MARTIN, J. G. C., WICANDER, S. & KAPOS, V. 2020. Towards understanding interactions between Sustainable Development Goals: the role of environment–human linkages. *Sustainability Science*, 15, 1573–1584.

SCHRADER, D. E. 2015. Constructivism and learning in the age of social media: Changing minds and learning communities. *New Directions for Teaching and Learning*, 2015, 23–35.

SCHWANDT, T. A. 1994. Constructivist, interpretivist approaches to human inquiry. *Handbook of qualitative research*. Thousand Oaks, CA, US: Sage Publications.

SCOTT, J. 2017. *Social Network Analysis*, 4th ed. London: Sage.

SCOTT, W. R. 1995. *Institutions and Organisations*. Thousand Oaks, CA: Sage.

SHREVE, C. M. & KELMAN, I. 2014. Does mitigation save? Reviewing cost–benefit analyses of disaster risk reduction. *International Journal of Disaster Risk Reduction*, 10, 213–235.

SMITH, N. 2007. *Nature as Accumulation Strategy*. Socialist Register.

SNOW, D. A. & BENFORD, R. D. 1988. Ideology, Frame Resonance, and Participant Mobilization. *In:* KLANDERMANS, B., KRIESI, H. & TARROW, S. (eds.) *From Structure to Action: Social Movement Participation Across Cultures*. Greenwich: JAI Press.

SÖDERBERG, C. & ECKERBERG, K. 2013. Rising policy conflicts in Europe over bioenergy and forestry. *Forest Policy and Economics*, 33, 112–119.

SOLBERG, B. 1986. Forest sector simulation models as methodological tools in forest policy analysis. *Silva Fennica*, 20, 420–427.

SOTIROV, M. & WINKEL, G. 2016. Toward a cognitive theory of shifting coalitions and policy change: linking the advocacy coalition framework and cultural theory. *Policy Sciences*, 49, 125–154.

SOTIROV, M., WINKEL, G. & ECKERBERG, K. 2021. The coalitional politics of the European Union's environmental forest policy: Biodiversity conservation, timber legality, and climate protection. *Ambio*, 50, 2153–2167.

STAPLETON, L. & STEFANIAK, J. 2019. Cognitive Constructivism: Revisiting Jerome Bruner's Influence on Instructional Design Practices. *TechTrends*, 63, 4–5.

STERN, C. P., DIETZ, T., ABEL, T., GUAGNANO, A. G. & KALOF, L. 1999. A Value-Belief-Norm Theory of Support for Social Movements: The Case of Environmentalism. *Human Ecology Review*, 6, 81–97.

TOPPINEN, A. & KUULUVAINEN, J. 2010. Forest sector modelling in Europe – the state of the art and future research directions. *Forest Policy and Economics*, 12, 2–8.

TORFING, J. 2005. Discourse Theory: Achievements, Arguments, and Challenges. *In:* HOWARTH, D. & TORFING, J. (eds.) *Discourse Theory in European Politics: Identity, Policy and Governance*. London: Palgrave Macmillan UK.

US FOREST SERVICE. 1968. *Highlights in the History of Forest Conservation*. Washington, DC: Forest Service, US Dept. of Agriculture.

VAN HEESWIJK, L. & TURNHOUT, E. 2013. The discursive structure of FLEGT (Forest Law Enforcement, Governance and Trade): The negotiation and interpretation of legality in the EU and Indonesia. *Forest Policy and Economics*, 32, 6–13.

VUGTEVEEN, P., LENDERS, H. J. R., DEVILEE, L. A. J., LEUVEN, S. E. W. R., VAN DER VEEREN, J. H. M. R., WIERING, A. M. & HENDRIKS, A. J. 2010. Stakeholder value orientations in water management. *Society & Natural Resources: An International Journal of the Commons*, 23, 805–821.

WASSERMAN, S. & FAUST, K. 1994. *Social Network Analysis: Methods and Applications*. Cambridge: Cambridge University Press.

WEIBLE, M. C. & SABATIER, P. A. 2018. *Theories of the Policy Process*. New York: Routledge.

WILLIAMSON, O. E. 1991. Comparative economic organization: The analysis of discrete structural alternatives. *Administrative Science Quarterly*, 36, 269–296.

WOLFF, S. & SCHWEINLE, J. 2022. Effectiveness and economic viability of forest certification: A systematic review. *Forests*, 13, 798.

WORKMAN, M. 2018. An empirical study of social media exchanges about a controversial topic: Confirmation bias and participant characteristics. *The Journal of Social Media in Society*, 7, 381–400.

YODER, L., WAGNER, C. H., SULLIVAN-WILEY, K. & SMITH, G. 2022. The promise of collective action for large-scale commons dilemmas: reflections on common-pool-resource theory. *International Journal of the Commons*.

YOUSEFPOUR, R., JACOBSEN, J. B., THORSEN, B. J., MEILBY, H., HANEWINKEL, M. & OEHLER, K. 2012. A review of decision-making approaches to handle uncertainty and risk in adaptive forest management under climate change. *Annals of Forest Science*, 69, 1–15.

ZHOU, X., LI, X., SONG, W., KONG, X. & LU, X. 2021. Farmland transitions in China: An advocacy coalition approach. *Land* [Online], 10.

ZITTOUN, P. & PETERS, B. 2016. *Contemporary Approaches on Public Policy: Theories, Controversies, Perspectives*. Basingstoke: Palgrave Macmillan.

4 Policy pathways for the European Green Deal and the EU Forest Strategy

1. Setting the stage

Chapter 2 highlights the significant incoherence in EU forest-focused and forest-related policies. This incoherence suggests that the design of different EU policies affecting forests often fails to consider relevant policy interactions adequately. There needs to be coordination and alignment in achieving a coherent policy mix that can effectively balance necessary trade-offs. For instance, a critical question arises when considering such trade-offs: What should be prioritised, conserving biodiversity or meeting climate change targets? This challenge is particularly relevant in the context of forest management. For the forest-based sector, this lack of coordination is attributable to the wide range of policy domains and instruments that influence forest management. This situation has resulted in policy fragmentation and the emergence of conflicting policy targets (Howlett et al., 2010, Vogelpohl and Aggestam, 2011, Edwards and Kleinschmit, 2013, Winkel, 2017). Policy fragmentation[1] appear in the spread of forest-focused and forest-related policies across various EU policy domains, including agriculture, energy, water, climate, and rural development (Elomina and Pülzl, 2021). This complex policy landscape is further complicated by the absence of an appropriate steering instrument that can coordinate a clear policy pathway for EU forests (Lazdinis et al., 2009, 2019, Kleinschmit et al., 2018, Wolfslehner et al., 2019, 2020).

While a certain degree of fragmentation might be inevitable given the diverse interests and use of forest-based products, the main concern arises from the conflicting and contradictory policy targets and objectives stemming from this fragmentation. These inconsistencies create challenges for the management of forests and forest-based industries. This includes conflicting target setting in policy frameworks such as the European Climate Law (Regulation, 2021/1119), land use and forestry in relation to the 2030 climate and energy framework(Regulation, 2018/841), trade and the EU timber regulation (Regulation, 995/2010), renewable energy targets (Directive, 2018/2001), and conservation objectives set out via the habitats and birds directive (Directive, 92/43/EEC, 2009/147/EC). The key point here is that these policy instruments set objectives and targets that must be more consistent and coordinated. For example, the European Climate Law includes a legal objective for the EU to reach climate neutrality by 2050, while the land use and forestry (LULUCF) regulation set a climate neutrality objective for the land use, forestry, and agriculture sectors already for 2035. While the overarching target may be the same, the varied timelines have significant implications in terms of implementation. Arguably, the climate neutrality target may also contradict conservation efforts. For instance, the EU 2030 Biodiversity Strategy (EC, 2020a) sets out to legally protect a minimum of 30 per cent of the EU's land and sea area, of which 10 per cent should be strictly Protected Areas (EC, 2022a). This implies that

DOI: 10.4324/9781003198987-4

initiatives to protect certain habitats strictly might conflict with objectives to increase biomass production as part of the EU's climate action.

The conflict between different EU policy goals and targets underscores a deeper systemic issue in EU policymaking. The absence of a coherent and coordinated approach to forest policy means that environmental and socio-economic objectives are not addressed in tandem. This hampers individual policies' effectiveness and undermines the overall approach to sustainable forest management (SFM). More importantly, the horizontal incoherence and fragmentation of EU forest policy highlights a critical governance challenge associated with managing these contradictory objectives, even more so as the number of EU policy instruments affecting forests continuously increases. In addition, EU objectives and targets are set against a complex backdrop of vertical[2] (national, regional, and international) policy instruments (Brukas and Sallnäs, 2012, Sotirov et al., 2015, Kern et al., 2017, Aggestam and Pülzl, 2018, 2020), which can lead to additional confusion, inefficiency, and contradictory goals.

1.1. A green deal for Europe

In recent years, the EU has adopted several pivotal policy instruments and regulatory proposals, most notably the European Green Deal (EC, 2019). This ambitious package of measures is designed to guide the EU's economy and society towards increased sustainability. Crucially for this chapter, the Green Deal also has implications for forests, highlighting the interconnected nature of environmental policies. For example, one central theme in the Green Deal is the protection and restoration of biodiversity, a goal that includes the increased protection and conservation of forests. Additionally, the Green Deal advocates for sustainable land use, including SFM practices. Another critical aspect of the Green Deal concerns the EU's capacity for carbon absorption, storage, and use, which implies forestry measures focused on carbon sequestration. Moreover, as part of the actions planned under the Green Deal, there was also the introduction of a new EU Forest Strategy, a non-binding (or soft) policy instrument. In 2021, the European Commission proposed this Strategy, which was developed against the backdrop of several new initiatives, such as the updated EU Biodiversity Strategy (EC, 2020a), an EU taxonomy for sustainable activities (Regulation, 2020/852), the 2050 Climate Change Mitigation Strategy (EC, 2018a), the Adaptation Strategy (EC, 2021b) and Circular Economy Action Plan (EC, 2020c). All these instruments collectively underpin the Green Deal.

Dealing with the varied goals under the Green Deal, all of which are critically important, represents a significant challenge. The key challenge resides in paying attention to all goals and integrating them in a cohesive and coordinated manner. However, historical precedents suggest that such integration often falls short during implementation, as demonstrated by earlier iterations of the forest strategy (Pelli et al., 2012, Wolfslehner et al., 2019). To address this, several approaches need to be considered. For example, there would be a need for enhanced coordination between different policy domains to ensure alignment of objectives and actions, thereby avoiding siloed approaches. There is also a need for adaptable and flexible implementation strategies tailored to specific contexts, ensuring the effectiveness of integrated policy goals.

This chapter will examine the influence of the European Green Deal on the formulation of the new EU Forest Strategy. To achieve this, the focus will be on a review of academic literature relating to implementing previous forest strategies, analysing the Green Deal's outlook on forestry and forests, and examining the perspectives of key stakeholders in the forest-based sector. The methodology will apply the 'pathways to sustainability' approach introduced by Leach et al. (2010), coupled with qualitative content (Berg, 2003, Elo et al., 2014) and frame analysis (Entman, 1993, Schön and Rein, 1994). These approaches will be applied to scientific publications and policy documents. The aim will be to critically evaluate the policies that have historically influenced the forest-based sector. By examining the first and second EU Forest Strategies, the chapter hopes to outline insights and lessons that could inform the effective implementation of the new Strategy. Furthermore, the European Green Deal, along with its related strategies and instruments, may offer new pathways for directing the development of the new Forest Strategy.

> The European Green Deal aims to enhance sustainability, emphasising biodiversity protection, sustainable land use, and carbon sequestration, directly impacting forest management. However, integrating these various objectives under the Green Deal presents challenges, particularly in implementation, suggesting the need for improved policy coordination and adaptable implementation strategies to ensure cohesive policy action. Considering the growing number of relevant EU policy instruments and their interactions with national and international policies, effective integration and flexible strategies are essential for SFM in the context of the Green Deal.

2. Context and conceptual approach

2.1. The EU Forest Strategy – a recap

The first Strategy was adopted in 1998 and provided general guidelines for coordinating other EU forest-related policies (EC, 1998). It emphasised key principles of SFM and addressed topics such as competitiveness, job creation, forest protection, and the delivery of forest ecosystem services. In addition, the Strategy identified areas of EU competence and relevant processes and platforms for coordination, including the Standing Forestry Committee (SFC), the Civil Dialogue Group on Forestry and Cork, and the Expert Group on Forest-based Industries. The tool for implementing the Strategy was the EU Forest Action Plan, which covered the period from 2007 to 2011, focusing on voluntary cooperation between EU Member States with some coordination actions being implemented by the European Commission (EC, 2006, Pelli et al., 2012).

The European Commission adopted the second Forest Strategy to respond to new challenges facing forests and the forest-based sector in 2013 (EC, 2013). The Strategy served as an updated framework to address increasing demands on forests while also addressing changes in societal and policy priorities. The primary tool for implementation was the voluntary Forest Multi-Annual Implementation Plan (Forest MAP), which outlined relevant activities (EC, 2015). Unlike the first Strategy, the second Strategy acknowledged the complex and fragmented forest policy environment; however, it still covered a wide range of conflicting policy activities and goals. These goals included applying the ‘cascading use of wood’ principle, using forest biomass for energy and carbon sequestration, and halting global deforestation (Wolfslehner et al., 2019, EC, 2015, 2018b).

As outlined by the European Commission in 2015, the Forest MAP highlights a fundamental challenge in EU forest policy (EC, 2015), stressing the importance of enhancing policy coherence and consistency (EC, 2015). However, due to the principle of subsidiarity, which limits direct EU intervention in matters better handled at the national level, the European Commission finds itself constrained regarding direct action on forests. Consequently, the Commission’s approach has emphasised the importance of policy coordination to achieve coherence. From the EU’s standpoint, effective coordination is crucial as numerous forest-related functions interact with different policy domains over which the EU has competence, such as climate change, agriculture, rural development, and nature protection. This cross-sectoral aspect emphasises the need for a harmonised approach, where the EU can leverage its competencies in different areas to advance a more integrated and effective forest policy framework.

Section 3.1.1. provides more information about the current and previous versions of the EU Forest Strategy.

2.2. Policy pathways to sustainability

The coordination challenges facing EU forest-related policy can be attributed to the combination of different policy domains, such as different sectoral perspectives, policy goals, and implementation tools, that affect forests directly or indirectly (Vogelpohl and Aggestam, 2011, Sotirov et al., 2015, Winkel, 2017, Kern et al., 2017, Aggestam and Wolfslehner, 2018). These policy domains create different models for policymaking and management with defined objectives and targets. However, the different policy pathways to achieve these objectives and targets may be framed by different policy problems (Leach et al., 2010, Beland Lindahl et al., 2017). For example, forests play an essential role in protecting and improving the quality of Europe's water bodies by acting as natural filters for water and helping to regulate water cycles. However, while the EU's Water Framework Directive includes several provisions related to forests and their role in protecting and improving the quality of Europe's water resources (e.g., Article 14 on River Basin Management Plans), forests are only directly noted once as a possible pressure point. This is in an Annex to the EU Water Framework Directive, where forestry is not mentioned, despite its relevance to water security (Directive, 2000/60/EC). So, even though forests help protect water resources, they still need to be adequately considered by the water sector despite their significance to water security.

'Policy pathways' refer to how a policy system can change over time, depending on different issues, such as changes in forest cover. Understanding these pathways involves looking at historical patterns, current trends, and potential future scenarios (Leach et al., 2007, 2010). Different scales may be necessary for this analysis, and these may overlap or act simultaneously. The focus of this chapter is on emerging frames. Frames can be compared to perspectives that shape how people, communities, or societies perceive, organise, and communicate about reality (Schön and Rein, 1994). They can be thought of as narratives that define what counts as 'fact' and what arguments are considered relevant and compelling by the actors advocating them. These frames can limit the number of perceived actions by ignoring specific options and promoting others. For instance, Rein and Schön (1994) describe frames as narratives that 'underlie the particular problem-setting stories one finds in any particular policy controversy' (p. 88), whereby frames function as a filter that limits the number of actions available. What actors perceive as political problems or solutions largely depends on the frames used in the political discourse (Hajer, 1993, 1997). A policy pathway is thus a theme that emerges when combining different frames (Beland Lindahl et al., 2016, 2017). This means a policy pathway can be understood as the enacted action strategy based on different goals, values, and functions associated with forests, as perceived by different actors or sectors.

This chapter builds on the 'pathways to sustainability' approach introduced by Leach et al. (2007). It focuses on how relevant policy problems are framed in the scientific literature and through statements made by forest-relevant representative organisations (Leach et al., 2007, 2010). The approach is based on the idea that sustainability is a dynamic and ongoing process that involves multiple stakeholders, including governments, businesses, civil society organisations, and individuals. Each of these stakeholders may have different values, priorities, and perspectives on sustainability, which can lead to different pathways to achieving a sustainable future (Beland Lindahl et al., 2017, Aggestam and Giurca, 2021, Elomina and Pülzl, 2021). The pathways to sustainability approach comprises three elements: (1) historical patterns, (2) current trajectories, and (3) future scenarios. The goal here is to identify critical policy problems for the new EU Forest Strategy as characterised by the scientific community and selected actors. The 'pathways to sustainability' approach will applied to identify potential obstacles and future trajectories for the new Strategy. However, this analysis will not discuss the normative values associated with the policy problems. The desired policy pathway is one that effectively achieves the key objectives of the new EU Forest Strategy.

> The 'pathways to sustainability' approach focuses on understanding how policy systems evolve, considering historical patterns, current trends, and future scenarios. It emphasises the importance of stakeholder perspectives, including governments, businesses, civil society, and individuals, each bringing unique values and priorities to forest management. The approach involves analysing historical patterns of forest policy, current trajectories, and future scenarios for sustainability.

2.3. Policy pathways in the context of the European Green Deal and EU Forest Strategy

Policy pathways in the context of the European Green Deal are understood as the routes through which forest-relevant policy objectives can be pursued and implemented over time. These pathways are influenced by different policy domains, each with its distinct perspectives, objectives, and targets. Moreover, these domains create models for policymaking and management, implying that the pathways to achieving policy objectives differ, shaped by the varied policy problems and sectoral frames. Having this in mind, the concept of policy pathways emphasises the dynamic nature of policy development. In other words, policy pathways are not static. They must adapt to emerging environmental challenges, new scientific findings, and changes in public opinion or political priorities. This adaptability is crucial to ensure that policies remain effective and relevant. Understanding these pathways involves considering the narratives and frames that shape stakeholders' perceptions.

For the current chapter, policy pathway refers to the specific forest-relevant trajectories or approaches that can be taken to achieve policy objectives set out in the European Green Deal. It represents the actions, decisions, and strategies policymakers can adopt to navigate the current state to a desired future outcome. This implies that each policy pathway is distinct, not just a linear route, from problem to solution:

- *Multiple pathways for different objectives*: The Green Deal covers various environmental objectives, from climate mitigation to biodiversity conservation. Different policy pathways can be explored for each of these areas. For example, one pathway might focus on enhancing carbon sequestration through forest growth, while another might prioritise preserving biodiversity in forest ecosystems.

- *Pathways shaped by stakeholder frames*: Stakeholders, including government agencies, environmental groups, and industry, bring their perspectives or frames to the policy discourse. These frames influence the choice of pathways by highlighting certain objectives, such as economic benefits or environmental protection, over others. Understanding these frames helps to explain why certain policy pathways may be favoured over others.
- *Cross-sectoral considerations*: Due to the lack of a common policy on forests, the pathways in EU forest policy are at the intersection of different policy domains, such as agriculture, energy, and water. This cross-sectoral aspect implies the need for pathways that can be coordinated and harmonised across different sectors to avoid contradictory outcomes.

3. Methodological approach

This chapter builds on an earlier publication – The Art of the 'Green' Deal: Policy Pathways for the EU Forest Strategy (Aggestam and Giurca, 2021). The present text has been updated and expanded to reflect policy developments since its publication. For example, the new EU Forest Strategy was not yet adopted when the original paper was published.

3.1. Data collection and analysis

The study employs a two-pronged approach to explore policy pathways within the context of the European Green Deal, combining an integrative literature review with qualitative content analysis:

1. *Integrative literature review*: This step involves reviewing relevant academic literature to identify and compile key themes and issues relevant to policy pathways in forest-related EU policies, guided by the principles outlined in Torraco (2005) and Snyder (2019).
2. *Qualitative Content Analysis*: A detailed examination of different types of policy documents, including regulations, reports, and strategy documents. The goal is to identify patterns, themes, and narratives to help explain existing policy pathways. The approach for this analysis is based on the framework suggested by Elo et al. (2014).

The data was collected from three main sources, as detailed in the study design (Figure 4.1). The figure includes a visual representation of the data sources, which clarifies the research process and the relationship between the different approaches and data sources. The study design aims to provide an overview of possible policy pathways by combining different research methods and data sources, which is crucial for this analysis (Johnson et al., 2007, Muskat et al., 2012).

3.1.1. Reviewing policy documents

The qualitative content analysis focused on the European Commission's (EC) communication regarding the Green Deal, issued on December 11, 2019 (EC, 2019). Various policy documents were considered as relevant for the Green Deal, especially its implications for forest-related issues. These included the EU Biodiversity Strategy (EC, 2020a), EU taxonomy for sustainable activities (EC, 2018c), 2050 Climate Change Mitigation Strategy (EC, 2018a), Farm to Fork Strategy (EC, 2020b), revised EU Bioeconomy strategy (EC, 2018f) and EU Circular Economy Strategy (EC, 2020c). The study also incorporated relevant EC-funded studies and evaluations,

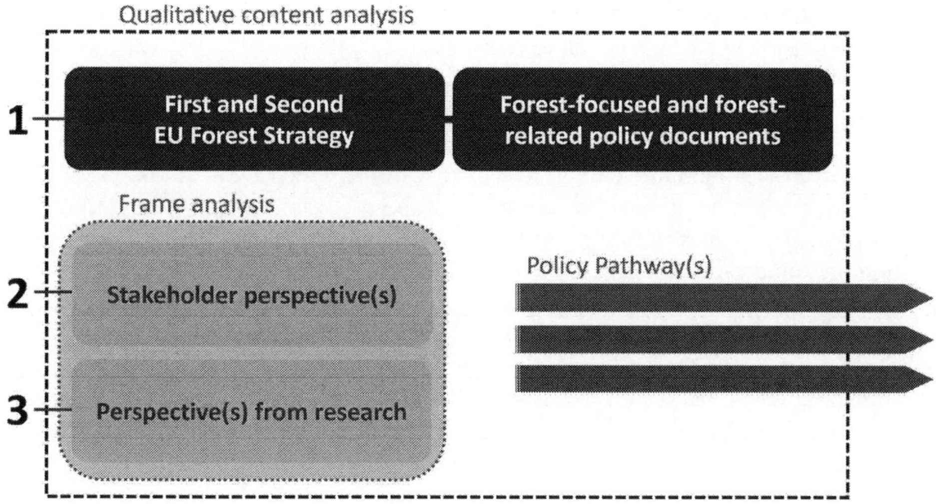

Figure 4.1 Study design.

Source: adapted from Aggestam and Giurca (2021)

focusing on the first and second Strategies and their implementation tools (EC, 1998, 2006, 2013, 2015). All documents were accessed from the EU's online Publications Office and EUR-LEX, the official repository for EU law and public documents.[3]

The analysis applied an iterative approach involving multiple readings of the documents to identify key statements and terms pertinent to the Strategy's implementation (Srivastava and Hopwood, 2009). This approach was chosen because it allows for a more thorough examination of the documents, enabling the identification of themes and patterns that emerge through a structured search process.

Data annotations were performed using the redaction tool in Adobe PDF Reader. The analysis and codification of the data were conducted in Microsoft Excel. This codification entailed pinpointing each marked section's thematic focus, relevance to forest-related concerns, and any associated strategies or initiatives in the text. This approach allowed for a more detailed understanding of the Strategies' implementation context and how different policy areas intersect with forest-related issues.

Subsequently, the codified data served as a basis for identifying key themes and patterns within the policy documents. These findings are crucial, offering valuable insights into implementing the Green Deal, particularly its connections to forest-related issues.

3.1.2. Analysing stakeholder opinions

In this study, 'stakeholder organisations' refer to any entity, including representative organisations and institutions, that are involved in forest-related issues. These organisations are important in shaping forest-related policies and strategies, and they have a significant interest in the outcomes of these policies. The range of these entities is broad, covering industry associations like the Confederation of European Forest Owners, non-governmental organisations (NGOs) such as FERN and the International Union for Conservation of Nature (IUCN), and research institutions like the European Forest Institute. Each organisation contributes its distinct perspective and expertise to the discourse around forest-related matters.

For inclusion in the study sample, an organisation was required to have published a statement, report, or policy brief on the Green Deal that is publicly available. This criterion ensured that the study included opinions and perspectives from organisations that have a formal position on the Green Deal and are actively engaged with the Green Deal. An online search was conducted using a combination of keywords: 'European', 'EU', 'Green Deal', 'forests', and 'forest-based sector'. This search yielded 18 publicly available opinions from 21 organisations and institutions, all detailed in Aggestam and Giurca (2021).

The stakeholder opinions, often classified as grey literature, were subjected to the same review process as the policy documents. This method ensured a comprehensive analysis of the available information on the Green Deal. Such an approach is essential to fully understand various organisations' diverse perspectives and insights, each with its unique background and expertise. This information is crucial for understanding the complexities of forest-related policy and offers insights for implementing the Green Deal.

3.1.3. Reviewing the scientific literature

The Web of Science, an academic citation search engine, was utilised to identify scientific literature on the EU Forest Strategy. Recognised as a leading tool in research, this search engine is extensively used by researchers for identifying and accessing scholarly literature (Li et al., 2018). A comprehensive search strategy was implemented to ensure relevant literature coverage. This involved using a combination of search terms like 'EU', 'European', 'European Union', 'forest policy', 'forest-related policy', and 'forest strategy'. The timeframe for the search was set from 1998 to 2020, aligning with the period from the adoption of the first Strategy. This process resulted in the identification of 33 publications/reports.

Each publication was evaluated to determine its relevance to the study objectives. To be considered important, the papers needed to address the EU Forest Strategy directly. This criterion was applied to ensure the analysis included the most relevant literature. Additionally, the bibliographies of these selected publications were reviewed for other significant works related to EU forest policy. This supplementary step expanded the total count of relevant publications/reports to 35 (Jalali and Wohlin, 2012).

The complete compilation of these documents is available in Aggestam and Giurca (2021). Following the approach outlined in section 3.1.1, these documents underwent a detailed review, enabling a thorough review and analysis of the literature on the EU Forest Strategy.

3.2. Analytical framework

The data collection and annotation process were crucial in this research. After annotating and coding the data, it underwent an extensive review to ensure alignment with the study's guiding questions and objectives, as detailed in the analytical framework (Figure 4.2). This framework, adopted from the methodologies of Beland Lindahl et al. (2017) and Kleinschmit et al. (2017), operates on the assumption that both opinion statements and academic literature are instrumental in improving our understanding of the targeted issues.

However, the scope of the analysis was somewhat constrained by the limitations of the available data. Additionally, the analysis was conducted before the new Strategy was adopted, which restricted the possibility of conducting a more in-depth analysis that could have traced the

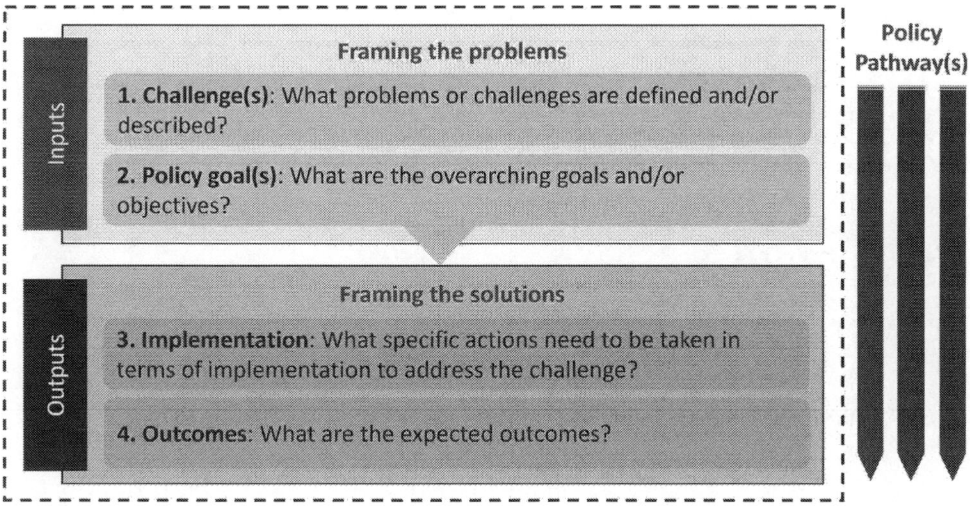

Figure 4.2 Analytical framework.

Source: adapted from Aggestam and Giurca (2021)

impact of different frames on the policy pathways being examined. Despite these constraints, the analytical framework provided a robust foundation for the study. It facilitated an in-depth examination of the key policy pathways and the factors influencing them, ensuring a comprehensive overview of the subject matter.

The collected and coded data was organised and grouped according to the questions outlined in Figure 4.2. This step involved systematically describing the content, focusing on challenges, policy goals, implementation aspects (where applicable), and outcomes of the policies. This approach aimed to clarify the policy issues (inputs) linked to implementing the first and second EU Forest Strategy and to identify viable solutions (outputs) for the new Strategy. This process was instrumental in identifying key challenges that the new policy framework should address. For example, a notable challenge identified was 'policy incoherence', characterised by existing policies having conflicting objectives and targets, leading to confusion among policymakers and stakeholders. The corresponding policy goal in this scenario might be to 'reduce fragmentation and improve coordination' between different policies and actors. The implementation phase could involve strategies for resolving policy trade-offs and crafting a more cohesive policy framework. The desired outcome would be 'enhanced coordination in forest policy', benefitting all involved parties.

Throughout the analysis, comments and additional references were added to the Excel file as part of the iterative and reflective process, supplementing the data with further context and insight. To synthesise the findings, themes and related subcategories were identified for each question. For instance, 'fragmentation' emerged as a theme, encompassing issues such as 'conflicting policy objectives' and 'trade-offs' in relation to inputs and outputs. These summaries were then used to develop the policy pathways outlined in Table 4.1. This provided a clear representation of the key policy issues and potential solutions.

3.3. Methodological considerations

Reflecting on the methodology, a notable challenge was differentiating between policy goals and outcomes, given their consistent overlap. For instance, a policy goal like 'reducing incoherence'

Table 4.1 Policy pathways for the new Strategy.

Meta pathway 1#	*Policy integration and coherence*				
1. Challenge(s)* 2. Policy goal(s)	↓ ***Fragmentation*** → Reducing policy incoherence & addressing trade-offs	← ***Gaps*** ↓ Expanded & delimited definition of EU forest policy	***Relevance*** → Increased relevance of the EU Forest Strategy at the Member State level	← ***Legal status*** → EU regulation or directive on forests	← ***Cost impacts*** Financial and human cost impact of forest-relevant policy
3. Implementation	Mechanism for managing conflicting policy objectives & targets	Cross-sectoral perspective/definition of EU forest policy	Renewed EU Forest Strategy	International (e.g., Forest Europe) or EU-driven process to negotiate a binding agreement	Mechanism to streamline selected areas of EU legislation (e.g., REFIT 2.0)
4. Outcome	Coherent EU forest policy	Inclusive and integrative EU forest policy	Improved implementation of the third Strategy	EU competence in forests	Reduced direct & indirect cost impacts of EU policy
Action(s)					
	Clearly defined activities, as part of the third Strategy or the green deal, for resolving conflicting policy objectives & targets	Consultative process with EU Member States, Directorates-General (DGs), and relevant stakeholders on a definition of EU forest policy	Member State commitment to the Action Plan for the GREEN DEAL on implementing the EU Forest Strategy	A deliberative process with EU Member State on a legally binding agreement	Propose a set of activities, as part of the Action Plan for the third Strategy, on reducing the cost impact of legislation for the forest sector
Meta pathway 2#	*Governance and coordination*				
1. Challenge(s)* 2. Policy goal(s)	↑ ***Coordination*** → Reconcile complex conflicts in EU forest policy processes	← ***Institutional setup*** → Participatory governance to operationalise shared policy goals	← ***Forest ownership*** → Improved data availability on forest ownership	←***Forest information*** → Harmonised information on Europe's forests	← ***Forest communication*** Increase public interest in forest-related issues at EU level

(Continued)

Table 4.1 (Continued)

Meta pathway 2#	*Governance and coordination*				
3. Implementation	Mechanisms of coordination in forests, both vertically and horizontally	Integrated mechanisms for steering coordinated actions	Moderate the interface between forest ownership and forest ecosystem services	Intersectoral communication process on data collection & analysis methods	Tailoring future forest communication for specific stakeholder groups
4. Outcome	Agree on a common set of priorities in EU forest policy	Regionally and locally adapted governance approaches	Harmonised definitions & data on forest ownership	New modes of knowledge production & learning	Changing public perspectives on forest use & management
Action(s)					
	Pilot a multi-level coordination approach to synchronise existing policies & national strategies on forests	Development of adaptable & place-based institutional solutions for EU forest governance	Review how land ownership structures, interests, and values influence forest management through regulation, incentivisation, and advice.	Engage EU Member States and relevant stakeholders in a policy dialogue on harmonising further and standardising information on forests.	Develop diversified communication strategies to convey the added value of the third Strategy.

*The challenges characterise the sub-pathways under the respective meta pathways.

Source: Adopted from Aggestam and Giurca (2021).

would naturally lead to an outcome such as 'improved policy coherence'. However, treating these elements as distinct categories was beneficial in clarifying ambiguities present in the documents. In other words, this separation enabled a clearer understanding of policymakers' intentions versus the policies' outcomes.

Moreover, when analysing scientific literature, the categorisation of text into policy goals and outcomes presented a challenge, as these categories do not align well with the primary focus of a scientific paper. Recognising this distinction highlights the need for adaptability in the analytical framework, especially in future applications. It may be essential to modify the categories used in the framework to better fit the nature of the documents under review, whether it is a policy document or a scientific paper. This adjustment would enhance the framework's effectiveness in accurately interpreting and categorising the information across various documents.

It can also be noted that European and forest-relevant policies are dynamic and evolving fields. The perspectives on forests and the range of actors involved in the policymaking process are constantly changing. Consequently, future studies would greatly benefit from incorporating in-depth interviews as an additional data collection method. This approach would facilitate a deeper understanding of the policy landscape, including how various actors frame the problems and the proposed solutions. Incorporating interviews would also enhance the analysis of the frames used by these actors and their impact on the policymaking process. Interviews could also yield additional insights into the translation of policy goals into outcomes and how different stakeholders perceive these.

4. Results and analysis

This section will analyse the alignment between the European Green Deal and the new EU Forest Strategy in achieving their respective goals. The analysis will zoom in on the key elements of both policies, including SFM, climate change, biodiversity, and agriculture. The analysis will consider how (or if) these elements are integrated and harmonised within the implementation processes of the European Green Deal and the new EU Forest Strategy.

4.1. Framing the EU Forest Strategy

The new EU Forest Strategy sets out targeted goals and actions to promote SFM and leverage forests in addressing environmental challenges. These objectives include:

- *Enhancing carbon sequestration*: Improve the capability of forests to absorb carbon dioxide as part of climate change mitigation efforts.
- *Sustainable wood use*: Encourage using wood as a renewable resource for energy and materials, supporting a circular economy.
- *Biodiversity conservation and restoration*: Protect biodiversity within forest ecosystems.
- *Rural development support*: Support the growth of rural economies through sustainable forest resource use.
- *Inclusive stakeholder engagement*: Facilitate participation from various groups, including forest owners, industry, and civil society, in policymaking and management processes.

The Green Deal represents a broader policy framework, incorporating various policies and actions aimed at environmental sustainability and ecological resilience. Key components of the Green Deal include:

- *Renewable energy transition*: Shift towards sustainable sources of energy.
- *Energy efficiency*: Enhance energy use efficiency across sectors.
- *Sustainable transport*: Support eco-friendly transportation models.
- *Sustainable agricultural and forestry*: Encourage practices that preserve natural resources and ecosystems.
- *Biodiversity protection*: Ensure the protection of diverse ecosystems.
- *Circular economy*: Promote recycling and sustainable consumption to minimise waste and resource depletion.

4.1.1. Forests in the context of the Green Deal

The European Green Deal references forests 20 times, primarily in relation to the new EU Forest Strategy. Within the Green Deal, forests and forestry are notably linked to three areas:

1. Emissions trading and reduction: The regulation of land use, land use change, and forestry (LULUCF) aims to manage and reduce emissions effectively.
2. Agro-forestry and the CAP: The integration of agro-forestry practices within the framework of the CAP highlights the intersection of agriculture and forest management.
3. Biodiversity and climate change: Forest management plays a crucial role in conserving biodiversity and addressing climate change.

The Green Deal frames the policy objectives of the new Forest Strategy, suggesting that it is aligned with existing and emerging policy targets, including the EU climate law and the 2050 target for net-zero GHG emissions. Nonetheless, the Green Deal's focus on forests remains limited, primarily emphasising the service-related aspects of the bioeconomy.

4.1.2. Challenges in EU forest policy implementation

Reflecting on the first Strategy's implementation reveals a tendency among EU Member States to attribute minimal significance to non-binding, voluntary instruments like the Forest Strategy (Pelli et al., 2012, Wolfslehner et al., 2019). This raises questions about the potential impact of the new Strategy, even if it aligns more closely with policy priorities and domains where the EU has competencies (Aggestam and Pülzl, 2020, Lier et al., 2021, Gordeeva et al., 2022). Moreover, a notable discrepancy exists between the Green Deal and the new Forest Strategy regarding governance. The Green Deal proposes maintaining EU forest governance mechanisms for the current programming period (2021–2027). In contrast, the new Strategy indicates a shift, with plans to review the governance structure of the Standing Forestry Committee (SFC) and a proposal for an updated governance approach. This new approach aims to merge the SFC with the Working Group on Forest and Nature (EC, 2021a, 2021c). Such changes potentially imply significant modifications in the Strategy's implementation by the European Commission and EU Member States.

The European Green Deal and the new EU Forest Strategy share several closely aligned objectives. For example, The Green Deal targets transforming the EU's economy to achieve zero net GHG emissions by 2050. It emphasises increasing carbon storage in forests and other vegetation as a crucial element of this transformation. The Strategy focuses on enhancing SFM practices, thereby increasing forests' role in addressing climate change, biodiversity loss, and effective water management. Furthermore, the Green Deal proposes that the Strategy should encompass the entire forest value chain, highlighting the ecological and socio-economic services that forests provide. Specific measures under consideration include imported forest products and value chains. Aligning with the European Parliament resolution on the Strategy, there is a focus on supporting 'legal, sustainable, and deforestation-free production and supply chains' (para. 45). Moreover, the Commission's target to plant 3 billion trees by 2030 is a key component of the Strategy, underscoring a commitment to biodiversity conservation (EC, 2020a).

Despite these alignments, there are challenges in creating a more integrative EU framework for forests. As a non-binding (or soft) policy instrument, the strategy's effectiveness in tackling cross-sectoral challenges remains uncertain (Pelli et al., 2012, Wolfslehner et al., 2019). Key areas requiring further attention include trade and forest products. For example, the Green Deal needs to emphasize trade in connection with forests, indicating potential gaps in policy coverage where the Commission has competencies. In addition, the European Parliament resolution calls for incorporating 'illegally acquired timber in trade agreements' (Parliament Resolution, 2020, para. 47). However, the Strategy focuses primarily on 'enhancing monitoring and enforcement by Member States' (p. 26), suggesting a need for broader action (EC, 2021c).

4.1.3. Forests in the context of agriculture and rural development

The CAP is a framework of policies and regulations designed to support the agricultural sector and rural areas. Recent initiatives have aimed at reforming the CAP, attempting to align it more with sustainability goals and make it 'fit for the future' (e.g., EC, 2018d, EC, 2018e). A notable feature of these reforms is the increased freedom granted to EU Member States in promoting SFM. However, the European Parliament has suggested tying financial assistance to adopting SFM practices, a significant step towards integrating sustainability into forest management. While historical financing measures have focused predominantly on afforestation, it is uncertain if the new reforms will push national RDPs to cover a broader range of SFM practices in practice. For instance, there has also been a recent push for extensive tree planting, which might amplify afforestation efforts at the expense of other measures that could contribute to SFM. Even more, despite innovative financing models in previous programming periods, afforestation has remained the predominant form of financing (Alliance Environnement et al., 2018). This trend highlights the need to evaluate how new measures and initiatives influence the broader spectrum of SFM objectives. In addition, it will be crucial to balance forest-relevant initiatives with other strategies that contribute equally to SFM (e.g., afforestation, while integral to SFM, should be complemented with other practices, such as addressing climate change). In other words, the challenge for the future will be to develop policies that not only incentivise afforestation but also integrate a wider range of SFM practices, ensuring an integrative approach to forest management.

4.2. Stakeholder Perspectives on the Strategy

4.2.1. Forest-based industries

Representative organisations from forest-based industries have raised concerns that the Green Deal should more explicitly acknowledge the important role of forests and forest-based industries in achieving sustainability objectives. Key points raised include the need for a comprehensive industrial forest strategy. For example, organisations like the Swedish Forest Industries have advocated for a strategy that prioritises greater utilisation of wood, emphasising the role of the bioeconomy. They have further noted that the European Commission needs to better appreciate the contributions of forests and forest-based products to environmental and economic sustainability (Swedish Forest Industries, 2020), such as reducing carbon emissions and promoting a circular economy. In line with this argument, the World BioEconomy Forum and other stakeholders have noted that the Green Deal does not focus enough on the bioeconomy,[4] despite its potential in substituting non-renewable materials and mitigating climate change. Collectively, the forest-based industries express concern that the economic importance of forests is yet to be fully recognised in the Strategy, as highlighted in a joint statement by the Confederation of European Forest Owners (CEPF), which points out the oversight of the forestry and wood economy in the current Strategy (CEPF, 2019, p. 2).

Despite the Green Deal's perceived shortcomings in addressing the forest-based sector, the European Parliament and the Council have made explicit links between the circular bioeconomy and the forest-based industries in recent resolutions and conclusions (Parliament Resolution, 2020, para. 29). This development suggests a growing recognition of the sector's role, yet the practical implementation of these considerations remains to be seen. Moreover, from an industry perspective, the European Commission needs to better integrate the forest-based industry's role in achieving sustainability goals.

4.2.2. Forest owners

Organisations representing forest owners have highlighted the need for greater stakeholder engagement in shaping forest-related policies under the European Green Deal. They stress the importance of early involvement to ensure the effectiveness and suitability of these policies (CEPF et al., 2019, CEPF, 2019, EUSTAFOR, 2021). These organisations underscore the diversity of the forest-based sector across the EU, highlighting differing regional characteristics such as forest types, ownership structures, and socio-economic contexts. They argue against a generic, 'one-size-fits-all' policy model, advocating for region-specific strategies considering the unique requirements and challenges of areas with varying ownership patterns, from small-scale private forests to large-scale industries. A bottom-up approach is proposed for policymaking, covering participatory planning, local implementation, and tailored financing. The approach that is being advocated emphasises the importance of engaging stakeholders directly in the policymaking process to reflect varied regional perspectives and specificities, localised implementation in terms of allowing for adaptation to specific regional and community needs, enhancing the responsiveness and relevance of policies, and adaptive financing regarding the provision of resources and flexibility at national and local levels to support policy implementation and encourage adoption of SFM practices (CEPF, 2019, CEPF et al., 2019).

It is also argued that a more inclusive approach can yield multiple benefits. For example, it is suggested that policies shaped by stakeholder inputs are likely to be more applicable and effective, and tailoring policies to local conditions ensures they are better suited to address specific challenges and opportunities, such as supporting forest owners and managers in adapting to changing climate conditions and implementing SFM practices. The vision set out by organisations that represent forest owners implies that increased involvement of stakeholders in the policymaking process, coupled with a bottom-up approach, is crucial for the success of the Green Deal. The critique of the policymaking process suggests that the EU is not doing enough to listen to the perspectives of different stakeholder groups.

4.2.3. Research organisations

Research organisations have likewise voiced concerns regarding how the Green Deal considers the forest-based sector. For example, while there is a broad consensus on the goal of achieving EU carbon neutrality by 2050, as Hetemäki (2020) underscored, opinions vary on the approaches and methodologies to achieve this goal. This divergence indicates varying perspectives and interests regarding the role of forests in the Green Deal and more broadly. Another interesting observation comes from a recent ThinkForest webinar, during which a significant majority (88% of participants) noted that the forest bioeconomy was underrepresented in the Green Deal.[5] This suggests a need for a greater emphasis on the forest-based sector's potential to contribute to the EU's carbon neutrality objective. Similarly, in a recent publication, the European Forest Institute argued for clearer forest-related policy objectives within the Green Deal. The institute stresses the need for governance mechanisms to cover all forest ecosystem services and highlights the importance of considering the global dimensions of EU forest policy. This includes acknowledging how EU policies might shift environmental impacts to other regions (Hetemäki, 2020).

Research organisations have also proposed using plantation forests to enhance carbon sequestration in the EU, although with a balanced view regarding environmental and socio-economic implications. It is, for example, argued that plantation forests can increase the carbon sequestration capacity of EU forests (Freer-Smith et al., 2019). Furthermore, there are calls for innovative bio-based solutions and the development of new integrative models of forest management that can fully account for the diverse services forests provide and align with the Green Deal's environmental and socio-economic objectives. As expected, the emphasis on scientifically informed solutions is also a cornerstone of all these perspectives. It is argued that all solutions and management models must be grounded in the latest scientific knowledge, which covers the complex interplay of forest ecosystems across different forest types and conditions, impacts of climate change, and socio-economic factors.

4.2.4. Non-governmental organisations

Environmental non-governmental organisations (NGOs) have similarly raised concerns, emphasising the need for more sustainable practices and integrative perspectives. For instance, the International Union for Conservation of Nature (IUCN) has urged a shift in focus towards restoring forest ecosystems and soils (IUCN, 2019, 2020). They argue against establishing monoculture plantations, noting their ecological unsustainability and impacts on biodiversity (e.g., monoculture plantations, often characterised by the cultivation of a single species, can lead to decreased genetic diversity and increased vulnerabilities to pests and diseases). Similarly, FERN has challenged the European Commission's emphasis on bioenergy subsidies. They have argued against the myth that burning wood for energy is carbon neutral (FERN, 2019a, 2019b), pointing out that this practice can contribute to the destruction of natural habitats and elevate carbon emissions. Like nearly all the other stakeholder organisations, NGOs advocate for a more holistic approach to forest management that integrates social-economic and environmental aspects of SFM. Expanding on this, the consistent call for more integration suggests that the policymaking process underlying the Green Deal needs to be more open. Furthermore, NGOs highlight an important perspective in the discourse on forest management, namely, the need for policies that promote diversity in forest species and structures, ensuring that forest ecosystems become more resilient to climate change, diseases, and pests.

4.2.5. Enhancing forest integration in the Green Deal

As noted from the preceding analysis, the Green Deal's policy goals concerning forests have been subject to significant debate, with various actors highlighting uncertainties and calling for clearer implementation strategies. On one side, stakeholders from forest-based industries advocate for enhanced coordination among forest-related strategies. They propose a robust EU forest strategy post-2020 that emphasises effective forest management while setting clear and realistic targets for biodiversity conservation (CEPF et al., 2019, EUSTAFOR, 2021, Swedish Forest Industries, 2020). These stakeholders, including private and public forest owners and managers, stressed the need for their involvement in policymaking, such as leveraging their understanding of local and regional forest conditions. On the other side, these differing viewpoints underscore a significant challenge for the EU, namely, how to reconcile the diverse objectives and trade-offs between policies and sectoral goals impacting forests (Pelli et al., 2012, Wolfslehner et al., 2019, Aggestam and Wolfslehner, 2018, Aggestam and Pülzl, 2018, 2020). This challenge is recognised in the Council's conclusion on the Strategy (Council Conclusion, 2020), which acknowledges the 'increasing fragmentation of EU forest-related policies' (para. d). Such fragmentation arises from the EU's siloed and technocratic policymaking process and the varied objectives of its forest-relevant policies, including economic growth, environmental protection, and social welfare. This complexity extends beyond widespread conflicts and trade-offs, such as biodiversity versus biomass energy production, to less obvious sectors like packaging, construction, transport, and trade, each with distinct regulatory instruments influencing the forest-based sector. Consequently, achieving a coherent and consistent EU forest policy that addresses all aspects of SFM while addressing the trade-offs between different policies and sectors remains a formidable challenge. Moreover, the complex web of perspectives, interests and objectives that comprise the EU's forest policy landscape suggests the need for innovative and new solutions.

> Stakeholder organisations have expressed that the Green Deal must more explicitly recognise the role of forests and forest-based industries. Industry representatives advocate for an industrial forest strategy that prioritises wood use and emphasises the bioeconomy's role. Forest owners call for greater engagement in policymaking, stressing the need for tailored, region-specific strategies and a bottom-up approach to policy design and implementation. Research organisations highlight the need for clearer forest-related policy objectives within the Green Deal. NGOs focus on the need for SFM practices, advising against monoculture plantations and the overemphasis on bioenergy subsidies. Overall, there is a call for a more integrative and inclusive approach to forest policy under the Green Deal, addressing different stakeholder groups' varied and sometimes conflicting interests.

4.3. Academic literature on the EU Forest Strategy

Based on the problems identified in academic literature, three areas of research have been identified for EU forest-related policy (Figure 4.3). These areas include:

1. *Implementation:* This strand of research examines the different types of policy instruments that affect EU forests and the new Strategy. This includes laws, regulations, and other measures to implement forest-related policy goals. The research in this area can provide insight into how the new Strategy can effectively use policy instruments to achieve its goals.
2. *Coherence*: This strand of research looks at what constitutes EU forest policy and the level of integration between different policy areas. This includes analysing how EU forest policy interacts with other policies, such as climate change, biodiversity conservation, and land use.

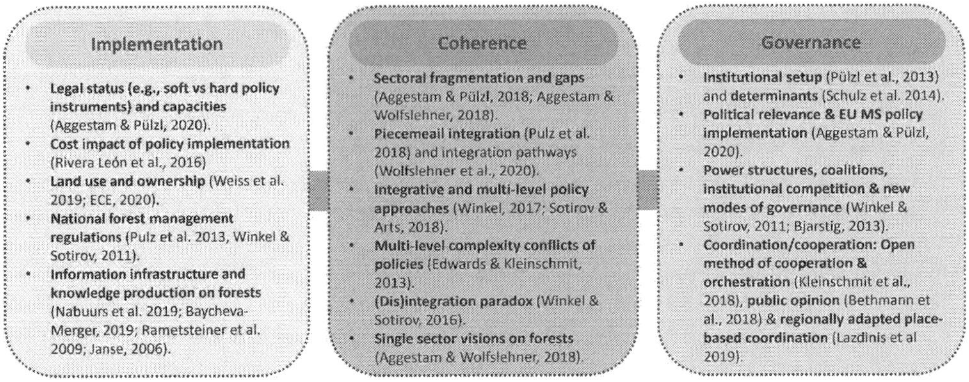

Figure 4.3 Three areas of research that are relevant to the new Strategy.

Source: adapted from Aggestam and Giurca (2021)

The research in this area can provide insight into how the new Strategy can ensure consistency and coordination across different policy areas.

3. *Governance*: This strand of research examines the institutional and socio-economic framework that underpins the implementation of the new Strategy. This includes analysing the roles and responsibilities of different actors in implementing EU forest-related policy, such as governments, the private sector, and civil society. The research in this area can provide insight into how the new Strategy can ensure effective and efficient governance of EU forests.

These strands of research can be seen as 'meta pathways', referring to overarching policy pathways or roadmaps that could be used to guide or influence the broader trajectory of policy development and implementation, such as directing how specific policies could be shaped to achieve certain objectives or as a guide for more detailed policy formulation. In other words, while these meta-pathways principally represent domains of knowledge, they can provide a high-level view of the intended policy pathway, highlighting major steps and strategies required to reach goals under the Green Deal and the Strategy. For example, the Green Deal aims for a transformative approach that can integrate socio-economic and environmental factors; however, this requires coherent policies across policy domains. By analysing the effectiveness of different policy instruments, these meta-pathways could help in understanding how new regulations should be designed, ensuring they are practical, enforceable, and aligned with the objectives of the Green Deal.

4.4. Synergies and trade-offs between the Green Deal and Strategy

The analysis highlights that the European Green Deal and the new EU Forest Strategy demonstrate significant synergies, primarily in their emphasis on sustainable and climate-resilient forests. This shared objective is crucial for realising the EU's ambitious climate change mitigation and conservation targets. Central to the Green Deal is the vision of a climate-neutral Europe by 2050, a target that the new Forest Strategy actively supports by advocating for enhanced carbon sequestration in EU forests. This Strategy contributes to climate neutrality and merges with efforts to conserve biodiversity, underscoring the protection, enhancement, and restoration of forests and wooded areas. Furthermore, both initiatives prioritise using wood as a sustainable, bio-based material within the circular economy. The Green Deal's emphasis on increasing renewable energy and reducing fossil fuel dependence also intersects with the Strategy's focus on expanding the use of wood in construction, thus reinforcing the bioeconomy.

However, it is noteworthy that the Forest Strategy emphasises the environmental domain, with a strong focus on biodiversity conservation. This emphasis on environmental aspects could lead to trade-offs when balancing the objectives of climate mitigation and conservation. For instance, while increased forest coverage and stricter conservation measures benefit biodiversity, they may limit the potential to use forests for carbon sequestration and biomass production, which are key components in achieving climate neutrality. This underscores a dynamic where prioritising one goal (biodiversity conservation) might impact the efficiency or outcomes of another (climate change mitigation). Thus, while the alignment of the Strategy with the Green Deal's climate neutrality target is evident, possible trade-offs stress the need to harmonise these dual objectives, ensuring that efforts to protect and restore biodiversity do not affect progress towards climate neutrality and vice versa.

There are several trade-offs and conflicting objectives that would need to be carefully balanced:

1. *Bioenergy*: The Green Deal's focus on renewable energy, including bioenergy, could conflict with conservation efforts, especially if biomass production affects natural forests, which would impact biodiversity.
2. *Plantation forests*: The Strategy's objective to increase carbon sequestration through plantation forests may conflict with the Green Deal's focus on restoring natural forests, given the ecological limitations of monoculture plantations.
3. *Construction*: Promoting wood as a construction material must be balanced against sustainable logging practices to prevent the depletion of natural forests.
4. *Trade*: The EU's trade policies could harm natural forests, particularly if they encourage importing products from unsustainably harvested wood in Europe and globally.
5. *Transport*: The EU's transport policy, if not aligned with environmental goals, might contribute to climate change and, consequently, harm natural forests.
6. *Agriculture*: The EU's agriculture policy may have unintended negative impacts on forests if it promotes the expansion of agriculture at the expense of natural forests.

Addressing these trade-offs would require an approach or mechanism that can actively integrate the varied objectives of the Green Deal and the Strategy. This includes a careful assessment of policy impacts and creating a balanced policy framework that not only recognises the multiple policy objectives affecting forests but is also able to manage these trade-offs directly.

4.5. Policy pathways for the EU Forest Strategy

The policy pathways outlined in Table 4.1 include recommended actions for achieving the desired outcomes of the new Strategy. It is important to note that the listed pathways are not exhaustive and are based on the analysis of previous Strategies, EU forest-related policy and relevant academic work. The alternatives provided are limited and mainly intended to outline possible actions to improve the implementation and outcomes of the new Strategy. The table outlines two 'meta-pathways' subdivided into several sub-pathways characterised by the challenge they address. In line with the analytical framework (Figure 4.2), the table is structured into five cells, each representing different stages and aspects of the policy development process:

1. *Challenges*: Identifies key obstacles affecting the effectiveness of EU forest policy. Challenges such as policy fragmentation, gaps in policy coverage, issues of relevance, legal status, and cost impacts are highlighted. These challenges represent the starting point for the sub-pathways.
2. *Policy goals*: Outlines the objectives that the pathways wish to address. Goals like reducing policy incoherence, addressing trade-offs, expanding and delimiting the definition of EU forest policy, and considering financial and human costs are key to creating a cohesive approach.
3. *Implementation*: Focuses on the mechanisms and processes needed to achieve the outlined policy goals. For example, it suggests utilising mechanisms for managing conflicting policy objectives, adopting a cross-sectoral perspective on EU forest policy and streamlining EU legislation, etc.
4. *Outcomes*: The desired results of implementing the policy pathways are listed here. The outcomes include a coherent and inclusive EU forest policy, improved implementation of the third Strategy, reduced cost impacts of EU policy, etc.
5. *Actions*: This additional cell (not part of the original analytical framework) sets out specific activities to address the challenges and achieve the policy goals as part of implementing the Strategy. These actions include defining activities within the Green Deal, consulting with EU Member States, proposing activities to reduce legislation's costs, etc.

It can also be noted that the arrows in the table indicate the relationship between the respective sub-pathways. For example, addressing policy fragmentation goes hand in hand with addressing policy gaps and coordination. It should also be noted that the policy pathways outlined in Table 4.1 are not mutually exclusive, and they are interconnected. With this in mind, the arrows indicate interlinkages between the respective meta and sub pathways.

4.5.1. Analysing the sub-pathways

The sub-pathways 'fragmentation' and 'gaps' (Table 4.1) are based on research that has identified conflicting EU policy objectives within the forest value chain. These conflicting objectives can stem from policy goals such as renewable energy and conservation. The research further suggests that many of these policy instruments were not included in the second Strategy, such as competition and transport policy, indicating that the European Commission took a narrow perspective on what is considered as 'forest-related policy' in this book. This narrow perspective only accounts for parts of the forest value chain and not the increasing number of policy objectives that affect the forest-based sector. These sub-pathways raise questions as to whether the new Strategy, despite its broadened scope, can be expected to improve policy coherence meaningfully. In line with research findings on previous iterations of the Strategy, it is expected that the new Strategy will have limited success in improving coordination (Pelli et al., 2012, Wolfslehner et al., 2019, Aggestam and Pülzl, 2020, Aggestam and Giurca, 2021, Gordeeva et al., 2022). This suggests that new and innovative coordination modes and mechanisms are needed to move forward, such as establishing new institutional mechanisms, coordination structures and assessment mechanisms. More specifically, even though the Green Deal specifies that 'careful attention will have to be paid when there are potential trade-offs' (p. 4) and the Council's conclusion on the Strategy highlights the 'need to strategically address increasing fragmentation of EU forest-related policies and improve their coordination and coherence' (para. d), one fundamental question underlying these pathways is not whether there are trade-offs, but how the European Commission proposes to tackle the coherence and fragmentation issue. Ultimately, the new Strategy does not propose concrete measures or mechanisms to tackle these issues.

The 'relevance' and 'legal status' pathways (Table 4.1) are critical for EU forest policies. Reviews of the EU Forest Action Plan have shown that when EU and national policy goals align well, countries are more likely to take the Strategy seriously (Aggestam and Pülzl, 2020). This research shows that the actions taken by EU Member States correlate with the relevance each country attaches to past strategies. The EU Forest Action Plan is an example to demonstrate that EU forest policymaking can impact the national level. Even though there is no common forest policy across EU Member States, the Forest Action Plan set relevant forest topics for 2007–2011 on one policy agenda. It operationalised the principles of the first Strategy for shared implementation by EU Member States. However, as the Forest Action Plan was voluntary, no EU Member State chose to implement activities contrary to their national interests or agendas. The analysis would, as such, suggest that the high priority assigned to the new Strategy may contribute to making it more relevant at the national level. If the new Strategy is not considered relevant, it will likely not be taken seriously, as previous versions. Taken together, the 'relevance' and 'legal status' pathways are crucial for the success of EU forest-relevant policies. Achieving alignment between EU and national policy goals and making the Strategy relevant for the EU Member States would increase the chances of the Strategy being implemented and followed through.

Acting along the 'coordination' and 'institutional setup' sub-pathways (Table 4.1) would be particularly important for the new Strategy as it remains a non-binding (soft) policy instrument. Particularly, as forest governance in the EU takes place across multiple levels of policymaking, there needs to be a common set of priorities for European forests agreed upon by countries, interest groups, and the industry. Furthermore, the European Commission needs to identify institutional solutions that can adequately address the peculiarities of EU forest policy, such as improving vertical and horizontal coherence and addressing key priorities for the forest-based sector. In other words, the coordination and institutional setup pathways are crucial for the new Strategy's success. For example, the outcomes of the Forest Action Plan and the Multi-annual Implementation Plan for the EU Forest Strategy demonstrate that more progress needs to be made in this domain (Pelli et al., 2012, Wolfslehner et al., 2019). Another example is the omission of recommended links to SFM and the list of specific SFM requirements and practices in the recently adopted EU taxonomy for sustainable activities. This emphasises that coordination remains a significant challenge and that the institutional mechanisms for coordination and cooperation at the Community level need to be improved (Lazdinis et al., 2009, 2019, Bjärstig and Keskitalo, 2013, Kleinschmit et al., 2018). For example, the Strategy must address the large number of legislative instruments affecting forests (Aggestam and Pülzl, 2018, 2020). This implies that the new Strategy will need to adapt better to the multi-level governance arrangements underlying European forest management. If not, it faces the risk of addressing forests purely from a strategic perspective.

The 'forest ownership', 'forest information', and 'forest communication' sub-pathways are all closely related and have a significant impact on the management and sustainability of Europe's forests. The 'forest ownership' pathway deals with understanding and managing the various forms of land ownership and control over forest land, including private, public, and communal ownership. Research has revealed a lack of consistent and accessible information on land ownership in Europe due to variations in ownership rights and structures across the pan-European region (ECE, 2020, Weiss et al., 2019). The 'forest information' pathway focuses on collecting, managing, and disseminating data and information concerning forests, including forest fires, forest cover, biomass, and species diversity. This information is crucial for understanding the state of Europe's forests, identifying potential threats, and implementing effective management strategies. For example, studies have shown that forest information can be used to maintain political power or expand control over forest resources (Baycheva-Merger and Sotirov, 2020). The inter-related 'forest communication' pathway deals with how information and knowledge about forests is shared with stakeholders, including policymakers, forest managers, and the public. This includes creating opportunities for dialogue and collaboration, providing education and training, and promoting public awareness and engagement. Effective communication is essential for building trust and support for forest management policies and practices and addressing public concerns and misconceptions about forests. Improving communication about forests is also crucial to increase public understanding and acceptance of the forest-based sector (Lazdinis et al., 2019). Furthermore, having access to information about the EU's system and decision-making processes could decrease resistance to a formalised EU forest policy (Bjärstig, 2013, Bjärstig and Keskitalo, 2013). While the three pathways are set out as interdependent, they are also mutually reinforcing. For example, the lack of consistent and accessible information on forest ownership can impede the effective management and conservation of forests. At the same time, poor communication can lead to misunderstandings and mistrust among stakeholders. Additionally, improving the production, management, and dissemination of forest information can support the development of more effective SFM policies and improve public understanding and support.

4.5.2. Interconnectivity of the policy pathways

As previously noted, the different meta and sub-pathways are interrelated and can affect each other. For instance, the 'coordination' pathway is a foundation for the 'fragmentation' pathway. Also, the 'gaps' pathway is fundamental for the 'fragmentation' pathway. For example, to have coherent EU forest-relevant policies, it is necessary to define what these policies should include clearly. Moreover, a clear definition makes it easier to assess and determine whether the policies are coherent. Similarly, there must be a shared understanding of what EU forest policy should include to identify policy gaps. With this understanding, it would be easier to determine where the policy gaps are. As indicated, this interconnectedness is illustrated by the lines and arrows in Table 4.1, which show how the different sub-pathways are related. This inconsistency is also evident in the academic literature, which presents varied perspectives on policy fragmentation and gaps. While the academic literature demonstrates that EU forest-relevant policy is complex and affected by multiple factors, the varied perspectives make it hard to define actions to tackle the fragmentation and policy gaps.

The analysis further suggests that specific policy pathways are mutually reinforcing. One such pathway is the 'cost impacts' pathway, which refers to the financial and human costs of implementing forest-related policy. Recent studies have shown that these costs can be a significant barrier to implementing forest-related policy, mainly when the policy instrument is voluntary. For instance, research has demonstrated the need for streamlining and reducing the financial and human cost impact of forest-related policy on forest-based industries (Rivera León et al., 2016). This cost can be a barrier to implementation, even more so if it is a voluntary instrument. This implies that the new Strategy needs to better align with national priorities to avoid imposing high costs at the national level. This way, it is more likely that the EU Member States will effectively implement and take the Strategy seriously.

Lastly, the policy options listed in Table 4.1 are intended to be a starting point for further reflection and discussion rather than a definitive list of possible solutions. The 'pathways to sustainability' approach provides one way to examine the most likely outcomes based on previous policy decisions and the current political and economic context. It also highlights the interdependency of policy decisions and the potential for unintended consequences. The frame analysis provides a method for understanding how stakeholders, the media, and the public perceive and discuss the different pathways. It helps to identify the dominant narratives and to understand how the different policy options are perceived in relation to each other. This is important as it can help identify potential roadblocks or areas of resistance towards specific options and indicate the potential for public support. Together, the pathway approach and frame analysis provide a mixed approach to understanding potential policy options and the context in which they are being discussed (Beland Lindahl et al., 2016, 2017, Kleinschmit et al., 2017, Leach et al., 2007, 2010, Aggestam and Giurca, 2021, Elomina and Pülzl, 2021). This information can help inform the policymaking process and ensure the selected solutions are effective, efficient, and socially acceptable.

> The new EU Forest Strategy, closely aligned with the European Green Deal, highlights the vital role of forests in carbon sequestration, biodiversity conservation, and sustainable land use, aiming for a climate-neutral and circular economy. The Strategy requires a dynamic and adaptable approach to EU forest policy. Critical to its success is the understanding of interconnected policy pathways, where enhanced coordination and a clear definition of policy goals and gaps are essential to address policy fragmentation and incoherence issues. Such efforts are crucial in ensuring the effective implementation and success of the Strategy.

5. What can we learn from analysing the European Green Deal and the new EU Forest Strategy?

The European Green Deal, introduced in 2019, is a comprehensive package of measures for a climate-neutral, circular, and sustainable economy in the EU (EC, 2019). The Green Deal sets out a series of ambitious targets and policies to reduce GHG emissions, increase the use of renewable energy, and protect biodiversity, among other goals. Forests play an important role in the Green Deal as forests are significant carbon sinks that are crucial in mitigating climate change. On the other hand, the EU Forest Strategy is a soft policy document, which means it sets out general objectives and principles. However, it is up to the EU Member States to implement the Strategy through their national policies and programs. This means that EU Member States have a significant role in shaping how the Strategy is implemented (if at all). Moreover, EU forest-focused and forest-related policy is dynamic and constantly evolving. This is due to the changing socio-economic and environmental conditions affecting forest management and the services they provide, including how countries respond to these changes. For this reason, it is essential to monitor the progress and effectiveness of the new EU Forest Strategy and identify ways to adapt implementation accordingly. This implies that more than having a mid-term and ex-post analysis is required.

This chapter aimed to consider policy pathways for the new Strategy by reviewing the European Green Deal, assessing what representative organisations say about forests and the Green

Deal, and reviewing what the scientific work on the first and second Strategy can teach us about the future. The goal has been to provide insights that can inform the implementation of a comprehensive and effective EU Forest Strategy that also supports the objectives of the European Green Deal. For example, the primary policy connections between forests and the European Green Deal concern the CAP, Land Use, Land Use Change, Forestry (LULUCF), biodiversity, and climate regulations and directives. These policy instruments fall clearly under the EU's jurisdiction. The CAP is the EU's framework for supporting farmers and rural communities and includes sustainable land use and forest management provisions. LULUCF, on the other hand, is a set of regulations aiming to reduce emissions from land use, land-use change and forestry.

Another important aspect of the European Green Deal is the protection of biodiversity. Forests are home to many plant and animal species, and the EU has committed to protecting and restoring biodiversity as part of its Green Deal. The EU also has climate regulations and directives that aim to reduce GHG emissions and increase the use of renewable energy sources. Forests are highlighted as a means of reducing carbon emissions and are included in carbon accounting and land use plans. The Green Deal also addresses deforestation and threats to forests and biodiversity. However, the Green Deal does not consider the multiple benefits that forests provide to society, such as water regulation and soil protection. Also, it does not address the potential contributions of the forest-based industries to the Green Deal, such as the forest bioeconomy, which includes activities such as wood production, bioenergy, and non-timber forest products. The main links between forests and the Green Deal are found in the development of the new Strategy, whereby key objectives and functions for the Strategy are outlined.

Coordination and fragmentation in forest-related policies have been long-standing issues on the policy agenda. Despite ongoing efforts to address these issues, they remain a priority, as illustrated by statements from the Council of Europe and the European Commission (Council Conclusion, 2020, Parliament Resolution, 2020). For instance, it is presumed that a lack of coordination has resulted in different approaches to SFM in the EU taxonomy for sustainable activities (Regulation, 2020/852). There are also concerns about conflicting policy objectives and targets, as raised by representative organisations, environmental NGOs, and the scientific community. These concerns include focusing on carbon sequestration at the expense of other ecosystem services, such as recreation and cultural values. Additionally, there is a concern that climate is prioritised over conservation, which could negatively impact biodiversity. The first and second Strategies have not successfully addressed these issues, partly because they only considered a limited set of policy objectives and did not address trade-offs between relevant policy instruments (Pelli et al., 2012, Wolfslehner et al., 2019). For example, the EU Forest Action Plan and the Forest MAP only focused on a limited set of policy objectives. They did not consider the trade-offs generated between relevant policy instruments (Aggestam and Pülzl, 2018, 2020). To address these issues, a streamlined EU forest-related policy framework is needed to coordinate forest policy objectives and reduce conflicts. Additionally, it would need to directly address the trade-offs between relevant policy instruments so that EU policy goals can coexist in a logical and non-contradictory fashion (Howlett et al., 2010, Howlett and Rayner, 2013). However, the question remains whether the new Strategy can provide the necessary tools to achieve the policy objectives of the Green Deal and promote synergies within and across sectors. These coordination tools would require the ability to align EU policy goals so that they reinforce rather than undermine each other (Cejudo and Michel, 2017, Wolfslehner et al., 2020).

5.1. What role can the policy pathways play?

The new Strategy could achieve genuine coordination if it addressed the core challenges under the respective meta pathways in Table 4.1. This includes sub-pathways such as the legal status, institutional setup, fragmentation, and coordination. For example, it would not be possible to address conflicting policy objectives and targets by only changing the legal status and institutional setup underlying the policy domains affecting forests, such as energy, water, and agriculture. This implies that the new Strategy would need to consider integrating forest-related policies with other policy domains and aligning legal frameworks and institutional setup to support this integration. Another example is the 'fragmentation' pathway, which is inherently interconnected with the 'coordination' pathway (as indicated by the arrows in Table 4.1). Fragmentation refers to the need for coordination and consistency between different policies and regulations that affect forests. This can lead to conflicting policy objectives and targets and a lack of coherence in implementing forest-related policies. To address this, the new Strategy would need to consider ways to integrate and coordinate different policies and regulations affecting forests, such as by creating a common framework for SFM. However, this has yet to happen. Furthermore, these pathways challenge the new Strategy because the solutions available may be outside the European Commission's jurisdiction. For example, integrated mechanisms for steering coordinated actions may require changes in the legal status and institutional setup that fall under the jurisdiction of EU Member States. Nevertheless, despite these limitations, the new Strategy would need to move beyond describing challenges and policy goals to proposing clear mechanisms and processes to address issues identified by, for example, the Council of Europe and the European Parliament (Council Conclusion, 2020, Parliament Resolution, 2020). This call for concrete action is also supported by the scientific community (Winkel, 2017, Kleinschmit et al., 2018, Aggestam and Pülzl, 2018, Wolfslehner et al., 2020, Gordeeva et al., 2022) and representative organisations (CEPF et al., 2019).

The EU Member States have been giving little weight to previous iterations of the EU Forest Strategy, mainly because it is a soft policy instrument lacking enforcement mechanisms (Pelli et al., 2012, Wolfslehner et al., 2019). This implies that the same argument will likely apply to the new Strategy. However, it should be pointed out that the Green Deal requires that all EU actions and policies contribute to the objectives of the Green Deal (EC, 2019). This may increase the relevance of the new Strategy at both the EU and national level. It is also an open question whether this means that the new Strategy is more binding, even if it is a voluntary instrument. In terms of recent developments, there is also the Glasgow Leaders' Declaration on Forests and Land Use,[6] which aims to mobilise action to protect and restore forests and other ecosystems, reduce deforestation and forest degradation, and promote sustainable land use practices (Uhlich and Burck, 2022). The Glasgow Declaration may help to improve the importance of the EU Forest Strategy by providing a clear international commitment to protect and restore forests, which aligns with the goals of the EU Forest Strategy and can serve as a framework for the EU to strengthen its efforts in the forest policy domain. Nevertheless, there are still many decisions to be made. For example, legally binding nature restoration targets were not outlined in detail until 2022 (EC, 2022b), and the nature restoration regulation was only adopted in 2024. Additionally, while most of the actions called for in the Green Deal fall within the jurisdiction of the European Commission, such as biodiversity, trade, and climate, it is unclear how the rest of the policy domains will be addressed. It is, furthermore, unlikely that the European Commission can push for significant change in the forest-based sector due to the absence of enforcement capabilities (Brukas and Sallnäs, 2012, Aggestam and Pülzl, 2018, Kleinschmit et al., 2018). This means that the new Strategy should consider ways to enforce the policies and regulations related to forest management. Moreover, the absence of a definition for EU forest-related policy at the community level makes it difficult to delimit or focus the Strategy on a few concrete measures or provide a better strategic vision for forests. This indicates that the new Strategy need to be more specific and clearer in its objectives, mechanisms, and processes.

5.2. Final reflections on the EU Forest Strategy

Many challenges that policy researchers have identified over the past 20 to 30 years in forest policy, such as policy incoherence and fragmentation, continue to be prevalent today. This is a common theme in forest policy research and is acknowledged by the Council of Europe and the European Commission (Council Conclusion, 2020, Parliament Resolution, 2020). The conflicting policy objectives that characterise EU forest-related policy correspond to a more profound governance challenge that can only be addressed by changing the tools and mechanisms available to implement the new Strategy. This is a recurring issue in forest policy research and is also acknowledged by experts in the field (Bjärstig and Keskitalo, 2013, Winkel, 2017, Kleinschmit et al., 2018, Wolfslehner et al., 2020). A future redesign of the new (or next) Strategy would be required to make it more adaptive to these challenges.

The complexity of EU forest-related policy is further exacerbated by the need to understand how EU Member States and the EU's political priorities intersect at the national level. Determining whether vertical coherence between policy domains and instruments can be achieved at all is crucial. This is an ongoing research topic, and several studies have been conducted on this subject (Kern et al., 2017, Sotirov et al., 2015). Higher political priorities, such as the transition towards a carbon-free economy, the circular bioeconomy, or the preservation of biodiversity, may contribute towards more coherent forest-related priorities across the Union. These priorities may have the potential to align conflicting policy objectives and be used to guide a possible redesign of the new Strategy. In conclusion, it is important to note that having a shared agenda on forests is not enough to address the complex challenges facing EU forest-related policy. A comprehensive and adaptive approach is required, considering the legal constraints, the EU's political priorities, and a better understanding of how EU Member States and the EU intersect at the national level.

The 'pathways to sustainability' approach has provided valuable information about how different actors frame the European Green Deal and its forest narratives. As outlined in Table 4.1, the pathways have also highlighted potential policy options for the new Strategy. Given the back-casting nature of the analysis, many of these options may already be familiar to some readers. However, to fully understand these pathways, it is crucial to consider historical precedents and the context in which they have developed (Leach et al., 2007, 2010). Overall, the results make it clear that more progress has yet to be achieved in implementing these pathways. Another purpose of this chapter has been to demonstrate how the 'pathways to sustainability' approach can be adapted to an ongoing policy process in which actors have been actively pushing for different narratives as part of the policy design. By doing so, this discussion has proved beneficial in providing some clues for how the new Strategy can move forward. Furthermore, it allows us to consider diverse perspectives and interests concerning forest-related issues.

Finally, lessons learned from this exercise demonstrate that further development and redesign of the methodology could increase its usability. For example, collecting feedback on the pathways (e.g., preferred options) from the actors involved could provide valuable insights into improving the Strategy and making it more effective. Additionally, providing recommendations on the appropriate strategic level at which the pathways could be used may also be useful as it helps to ensure that the Strategy is implemented at the appropriate level and that the pathways align with associated goals and objectives. It is also clear from the academic literature analysis that the analytical framework was not perfect for these types of documents. This would suggest the need for methodological adjustments in applying the 'pathways to sustainability' approach, at least if the work looks beyond policy documents.

Want to find out more?

For those seeking more knowledge on the European Green Deal or EU forest policy more generally, please see the references provided below or some of the many references throughout the text.

- Sustainability in the European Union: Analysing the Discourse of the European Green Deal (Eckert and Kovalevska, 2021).
- The European Green Deal: Assessing its current state and future implementation (Siddi, 2020).
- Rethinking the European Green Deal: An Industrial Policy for a Just Transition in Europe (Pianta and Lucchese, 2020).

The set of questions listed below may also help guide further inquiry into specific areas of interest:

1. How does the European Green Deal address the challenge of policy coherence in forest management, considering the diverse policy domains influencing forest-related decisions?
2. What are some significant trade-offs and conflicts between different objectives (like bioenergy vs. conservation) under the Green Deal and the new Forest Strategy?
3. How do different stakeholders, such as forest-based industries, forest owners, research organisations, and NGOs, view the role of forests in the Green Deal?
4. Considering the past strategies and the dynamic nature of EU forest policy, what are the main challenges in implementing the new Forest Strategy?
5. What are the potential implications of the new EU Forest Strategy and the European Green Deal for the future of forest policy in the EU?

6. Assignment IV: Developing policy pathways

Description

This assignment is based on Chapter 4 and focuses on enabling students to critically analyse and develop possible policy pathways to implement the third EU Forest Strategy, focusing on recent policy developments, the role of stakeholder engagement, and the strategic planning required to address forests at the EU level.

Purpose

Students are asked to conduct an in-depth analysis of the EU Forest Strategy, emphasising the challenges and opportunities associated with policy integration, coherence, governance, and coordination in achieving forest-related EU policy objectives.

Outline of task

- Review Chapter 4 (see also Chapter 2) to better understand the new EU Forest Strategy and the '*pathways to sustainability*' framework, considering elements such as policy integration, coherence, and associated EU forest governance challenges.

- *Policy pathway design*: Design a policy pathway that addresses one of the challenges identified in Chapter 4 (or elsewhere), such as policy incoherence, fragmentation, or lack of coordination. The pathway should include specific actions, policy goals, implementation mechanisms, and expected outcomes.
- *Stakeholder analysis*: Identify and analyse the role of different stakeholders in the forest-based sector, such as forest owners, industry representatives, NGOs, and research organisations. Consider how the stakeholders' perspectives and interests could be integrated into the policy pathway.
- *Presentation*: Prepare a presentation outlining the proposed policy pathway, including a rationale for the chosen approach, an analysis of stakeholder roles and how the policy pathway addresses the identified challenge within the context of the EU.

Suggested additional reading

- EU Forest Strategy and related policy papers (available on EUR-Lex[1] and other EU websites).
- Aggestam, F.& Pülzl, H. (2018) Coordinating the uncoordinated: The EU Forest Strategy. *Forests*, 9, 125.
- Aggestam, F. & Giurca, A. (2021) The art of the 'green' deal: Policy pathways for the EU Forest Strategy. *Forest Policy and Economics*, 128: 102456.
- Lier, M., Köhl, M., Korhonen, K.T., Linser, S., Prins, K. & Talarczyk, A. (2022) The new EU Forest Strategy for 2030: A new understanding of sustainable forest management? *Forests*, 13(2): 245.

Deliverable

- Prepare a 10-minute presentation (10 to 15 slides) summarising the findings and recommendations for effective EU Forest Strategy policy implementation.

7. Notes

1 Horizontal incoherence occurs when policies from different departments (or sectors) at the same level of governance (e.g., horizontally across government ministries) are not well-aligned or are contradictory, while fragmentation refers to the existence of many separate policies (e.g., within a government) that lack a unified or coordinated approach.

2 Vertical policy incoherence refers to when policies at higher levels of governance (e.g., international) are not in sync with those at lower levels (e.g., national or regional) or vice versa, leading to conflicting implementation processes across levels.

3 See http://eurlex.europa.eu.

4 See www.paperfirst.info/a-forthcoming-european-green-deal-is-good-but-what-about-the-bioeconomy-and-the-role-of-forest-industries (accessed 24 January 2023).

5 See https://efi.int/policysupport/thinkforest/science_greendeal (accessed 24 January 2023).

6 Signed during COP26 of the UN Framework Convention on Climate Change.

8. References

AGGESTAM, F. & GIURCA, A. 2021. The art of the 'green' deal: Policy pathways for the EU Forest Strategy. *Forest Policy and Economics*, 128, 102456.

AGGESTAM, F. & PÜLZL, H. 2018. Coordinating the uncoordinated: the EU Forest Strategy. *Forests*, 9, 125, DOI: https://doi.org/10.3390/f9030125.

AGGESTAM, F. & PÜLZL, H. 2020. Downloading Europe: A regional comparison in the uptake of the EU Forest Action Plan. *Sustainability*, 12.

AGGESTAM, F. & WOLFSLEHNER, B. 2018. Deconstructing a complex future: Scenario development and implications for the forest-based sector. *Forest Policy and Economics*, 94, 21–26. https://doi.org/10.1016/j.forpol.2018.06.004.

ALLIANCE ENVIRONNEMENT, ORÉADE-BRÈCHE SARL, IEEP & EFI. 2018. Evaluation study of the forestry measures under Rural Development. Luxembourg: Publications Office of the European Union.

BAYCHEVA-MERGER, T. & SOTIROV, M. 2020. The politics of an EU forest information system: Unpacking distributive conflicts associated with the use of forest information. *Forest Policy and Economics*, 112, 102110.

BELAND LINDAHL, K., BAKER, S., RIST, L. & ZACHRISSON, A. 2016. Theorising pathways to sustainability. *International Journal of Sustainable Development & World Ecology*, 23, 399–411.

BELAND LINDAHL, K., STÉNS, A., SANDSTRÖM, C., JOHANSSON, J., LIDSKOG, R., RANIUS, T. & ROBERGE, J.-M. 2017. The Swedish forestry model: More of everything? *Forest Policy and Economics*, 77, 44–55.

BERG, B. 2003. *Qualitative Research Methods for the Social Sciences*, 5th edition. Boston, MA: Allyn & Bacon.

BJÄRSTIG, T. 2013. The Swedish forest sector's approach to a formalized forest policy within the EU. *Forest Policy and Economics*, 26, 131–137.

BJÄRSTIG, T. & KESKITALO, C. H. E. 2013. How to influence forest-related issues in the European Union? Preferred strategies among Swedish forest industry. *Forests*, 4, 693–709. https://doi.org/10.3390/f4030693.

BRUKAS, V. & SALLNÄS, O. 2012. Forest management plan as a policy instrument: Carrot, stick or sermon? *Land Use Policy*, 29, 605–613.

CEJUDO, G. M. & MICHEL, L. C. 2017. Addressing fragmented government action: coordination, coherence, and integration. *Policy Science*, 50, 745–767.

CEPF. 2019. *EU forests in the European Green Deal* [Online]. www.cepf-eu.org/news/eu-forests-european-green-deal: CEPF.

CEPF, EUSTAFOR, COPACOGECA, CEI-BOIS, CEPI, EOS, ELO, FECOF, CEETTAR, UEF & EUROPE., B. 2019. Multifunctional European forests are crucial to reach the objectives of a European Green Deal. Brussels: http://www.cepf-eu.org/sites/default/files/Multifunctional%20European%20forests%20are%20crucial%20to%20reach%20the%20objectives%20of%20a%20European%20Green%20Deal_20191010.pdf (accessed 16 June 2020).

COUNCIL CONCLUSION 2020. Perspectives for the EU forest-related policies and EU forest strategy post 2020. *12695/1/20.* Brussels: General Secretariat of the Council.

DIRECTIVE 92/43/EEC. The conservation of natural habitats and wild fauna and flora. *L 206/7.* Official Journal of the European Union: European Council.

DIRECTIVE 2000/60/EC. Establishing a framework for Community action in the field of water policy. *L 327/1.* Brussels: Official Journal of the European Communities.

DIRECTIVE 2009/147/EC. Conservation of wild birds. Brussels: Official Journal of the European Union.

DIRECTIVE 2018/2001. Promotion of the use of energy from renewable sources. *L 328/82* Brussels: Official Journal of the European Union

EC 1998. A forestry strategy for the European Union. *COM 1998/649 final.* Brussels: European Commission.

EC 2006. EU Forest Action Plan. *COM(2006) 302 final.* Brussels: European Commission.

EC 2013. A new EU Forest Strategy: for forests and the forest-based sector. *COM(2013) 659 final.* Brussels: European Commission.

EC 2015. Multi-annual Implementation Plan of the new EU Forest Strategy. *SWD(2015) 164.* Brussels: European Commission.

EC 2018a. A Clean Planet for all: A European strategic long-term vision for a prosperous, modern, competitive and climate neutral economy. *COM(2018) 773 final.* Brussels: European Commission.

EC 2018b. Progress in the implementation of the EU Forest Strategy 'A new EU Forest Strategy: for forests and the forest sector'. *COM(2018) 811 final.* Brussels: European Commission.

EC 2018c. Proposal for a regulation amending Regulation (EU) 2016/1011 on low carbon benchmarks and positive carbon impact benchmarks. *COM(2018) 355 final.*

EC 2018d. Proposal for a regulation establishing rules on support for strategic plans to be drawn up by Member States under the Common agricultural policy (CAP Strategic Plans) and financed by the European Agricultural Guarantee Fund (EAGF) and by the European Agricultural Fund for Rural

Development (EAFRD) and repealing Regulation (EU) No 1305/2013 of the European Parliament and of the Council and Regulation (EU) No 1307/2013 of the European Parliament and of the Council. *COM(2018) 392 final.* Brussels: European Commission.
EC 2018e. Proposal for a regulation on the financing, management and monitoring of the common agricultural policy and repealing Regulation (EU) No 1306/2013. *COM(2018) 393 final.* Brussels: European Commission.
EC 2018f. A sustainable Bioeconomy for Europe: Strengthening the connection between economy, society and the environment. *COM(2018) 673 final.* Brussels: European Commission.
EC 2019. The European Green Deal. *COM(2019) 640 final.* Brussels: European Commission.
EC 2020a. EU Biodiversity Strategy for 2030: Bringing nature back into our lives. *COM(2020) 380 final.* Brussels: European Commission.
EC 2020b. A Farm to Fork Strategy for a fair, healthy and environmentally-friendly food system. *COM(2020) 381 final.* Brussels: European Commission.
EC 2020c. A new Circular Economy Action Plan: For a cleaner and more competitive Europe. *COM(2020) 98 final.*
EC 2021a. Annex to the New EU Forest Strategy for 2030: Roadmap of the Commission's action to implement the pledge to plant 3 billion additional trees by 2030 in the EU. *COM(2021) 572 final.* Brussels: European Commission.
EC 2021b. Forging a climate-resilient Europe: the new EU Strategy on Adaptation to Climate Change. *COM(2021) 82 final.* Brussels: European Commission.
EC 2021c. New EU Forest Strategy for 2030. *COM(2021) 572 final.* Brussels: European Commission.
EC 2022a. Criteria and guidance for protected areas designations. *SWD(2022) 23 final.* Brussels: European Commission.
EC 2022b. Proposal for a regulation on nature restoration. *COM(2022) 304 final.* Brussels: European Commission.
ECE 2020. Who owns our forests? Forest ownership in the ECE region. Geneva, Switzerland: United Nations Economic Commission for Europe.
ECKERT, E., & KOVALEVSKA, O. 2021. Sustainability in the European Union: Analyzing the Discourse of the European Green Deal. *Journal of Risk and Financial Management*, 14(2), 80. https://doi.org/10.3390/jrfm14020080
EDWARDS, P. & KLEINSCHMIT, D. 2013. Towards a European forest policy: Conflicting goals. *Forest Policy and Economics*, 33, 87–91.
ELO, S., KÄÄRIÄINEN, M., KANSTE, O., PÖLKKI, T., UTRIAINEN, K. & KYNGÄS, H. 2014. Qualitative content analysis: A focus on trustworthiness. *SAGE Open*, 4, 2158244014522633.
ELOMINA, J. & PÜLZL, H. 2021. How are forests framed? An analysis of EU forest policy. *Forest Policy and Economics*, 127, 102448.
ENTMAN, R. M. 1993. Framing: Toward clarification of a fractured paradigm. *Journal of Communication*, 43, 51–58.
EUSTAFOR. 2021. *The New EU Forest Strategy for 2030: Position of European Forest Owners and Managers* [Online]. https://eustafor.eu/uploads/20211004_The-New-EU-Forest-Strategy-for-2030-Position-of-European-Forest-Owners-and-Managers.pdf: EUSTAFOR.
FERN 2019a. European Green Deal offers hope for the future of forests. www.fern.org/news-resources/european-green-deal-offers-hope-for-the-future-of-forests-2063.
FERN 2019b. Von der Leyen's first 100 days: Are Europe's forests – and the climate – in good hands? www.euractiv.com/section/biomass/opinion/von-der-leyens-first-100-days-are-europes-forests-and-the-climate-in-good-hands/: FERN.
FREER-SMITH, P., MUYS, B., BOZZANO, M., DRÖSSLER, L., FARRELLY, N., JACTEL, H., KORHONEN, J., MINOTTA, G., NIJNIK, M. & ORAZIO, C. 2019. Plantation forests in Europe: challenges and opportunities. *From Science to Policy 9.* https://doi.org/10.36333/fs09: European Forest Institute.
GORDEEVA, E., WEBER, N. & WOLFSLEHNER, B. 2022. The New EU Forest Strategy for 2030: An Analysis of Major Interests. *Forests*, 13, 1503.
HAJER, M. 1993. Discourse coalitions and the institutionalization of practice: The case of acid rain in Great Britain. *In:* FISCHER, F. & FORESTER, J. (eds.) *The Argumentative Turn in Policy Analysis and Planning.* Duke University Press.
HAJER, M. 1997. *The Politics of Environmental Discourse: Ecological Modernization and the Policy Process*, New York, Oxford University Press.

HETEMÄKI, L. 2020. The Green Deal and the EU Forest Sector. *Background document for the Dinner Discussion on the EU forest related polices at the Embassy of Sweden, Helsinki, 12 February 2020.* Helsinki, Finland: European Forest Institute.

HOWLETT, M. & RAYNER, J. 2013. Patching vs packaging in policy formulation: Assessing policy portfolio design. *Politics and Governance*, 1, 170–182.

HOWLETT, M., RAYNER, J., GOEHLER, D., HEIDBREDER, E., PERRON-WELCH, F., RUKUNDO, O., VERKOOIJEN, P. & C., W. 2010. Overcoming the challenges to integration: embracing complexity in forest policy design through multi-level governance. *In:* RAYNER, J., BUCK, A. & KATILA, P. (eds.) *EMBRACING COMPLEXITY – MEETING THE CHALLENGES OF INTERNATIONAL FOREST GOVERNANCE.* Vienna: International Union of Forest Research Organisations (IUFRO).

IUCN 2019. Will the European Green Deal meet its ambitious expectations? www.iucn.org/news/europe/201912/will-european-green-deal-meet-its-ambitious-expectations: International Union for Conservation of Nature and Natural Resources.

IUCN 2020. The future of European forest biodiversity and ecosystem services – what can we expect from the Green Deal? www.iucn.org/news/europe/202003/future-european-forest-biodiversity-and-ecosystem-services-what-can-we-expect-green-deal: International Union for Conservation of Nature.

JALALI, S. & WOHLIN, C. Systematic literature studies: Database searches vs. backward snowballing. Proceedings of the 2012 ACM-IEEE International Symposium on Empirical Software Engineering and Measurement, 20–21 Sept. 2012, pp. 29–38.

JOHNSON, R. B., ONWUEGBUZIE, A. J. & TURNER, L. A. 2007. Toward a definition of mixed methods research. *Journal of Mixed Methods Research*, 1, 112–133.

KERN, F., KIVIMAA, P. & MARTISKAINEN, M. 2017. Policy packaging or policy patching? The development of complex energy efficiency policy mixes. *Energy Research & Social Science*, 23, 11–25.

KLEINSCHMIT, D., ARTS, B., GIURCA, A., MUSTALAHTI, I., SERGENT, A. & PÜLZL, H. 2017. Environmental concerns in political bioeconomy discourses. *International Forestry Review*, 19, 41–55.

KLEINSCHMIT, D., PÜLZL, H., SECCO, L., SERGENT, A. & WALLIN, I. 2018. Orchestration in political processes: Involvement of experts, citizens, and participatory professionals in forest policy making. *Forest Policy Econ.*, 89, 4–15. https://doi.org/10.1016/j.forpol.2017.12.011.

LAZDINIS, M., ANGELSTAM, P. & LAZDINIS, I. 2009. Governing forests of the European Union: Institutional framework for interest representation at the European Community Level. *Environmental Policy and Governance*, 19, 44–56.

LAZDINIS, M., ANGELSTAM, P. & PÜLZL, H. 2019. Towards sustainable forest management in the European Union through polycentric forest governance and an integrated landscape approach. *Landscape Ecology*, 34, 1737–1749.

LEACH, M., SCOONES, I. & STIRLING, A. 2007. Pathways to Sustainability: an overview of the STEPS Centre approach. *STEPS Approach Paper.* Brighton: STEPS Centre.

LEACH, M., SCOONES, I. & STIRLING, A. 2010. *Dynamic Sustainabilities: Technology, Environment, Social Justice*. London: Earthscan.

LI, K., ROLLINS, J. & YAN, E. 2018. Web of Science use in published research and review papers 1997–2017: a selective, dynamic, cross-domain, content-based analysis. *Scientometrics*, 115, 1–20.

LIER, M., KÖHL, M., KORHONEN, K. T., LINSER, S. & PRINS, K. 2021. Forest relevant targets in EU policy instruments: can progress be measured by the pan-European criteria and indicators for sustainable forest management? *Forest Policy and Economics*, 128, 102481.

MUSKAT, M., BLACKMAN, D. & MUSKAT, B. 2012. Mixed methods: Combining expert interviews, cross-impact analysis and scenario development. *Electronic Journal of Business Research Methods*, 10, 9–21.

PARLIAMENT RESOLUTION 2020. On the European Forest Strategy: The way forward. *P9_TA(2020)0257.* Brussels: European Parliament.

PELLI, P., AGGESTAM, F., WEISS, G., INHAIZER, H., KEENLEYSIDE, C., GANTIOLER, S., BOGLIO, D. & POLÁKOVÁ, J. 2012. Ex-post evaluation of the EU Forest Action Plan. http://ec.europa.eu/agriculture/evaluation/market-and-income-reports/forest-action-plan-2012_en.htm.

PIANTA, M., & LUCCHESE, M. 2020. Rethinking the European Green Deal: An Industrial Policy for a Just Transition in Europe. *Review of Radical Political Economics*, 52(4), 633–641. https://doi.org/10.1177/0486613420938207

REGULATION 995/2010. Laying down the obligations of operators who place timber and timber products on the market. *OJ L 295.* Brussels: Official Journal of the European Union.

REGULATION 2018/841. Inclusion of greenhouse gas emissions and removals from land use, land use change and forestry in the 2030 climate and energy framework, and amending Regulation (EU) No 525/2013 and Decision No 529/2013/EU. *L 156/1.* Brussels: Official Journal of the European Union.

REGULATION 2020/852. Establishment of a framework to facilitate sustainable investment, and amending Regulation (EU) 2019/208. *L 198/13.* Brussels: Official Journal of the European Union.

REGULATION 2021/1119. Establishing the framework for achieving climate neutrality and amending Regulations (EC) No 401/2009 and (EU) 2018/1999 ('European Climate Law'). *In:* EUROPEAN PARLIAMENT (ed.) *L 243/1.* Brussels: Official Journal of the European Union.

RIVERA LEÓN, L., BOUGAS, K., AGGESTAM, F., PÜLZL, H., ZOBOLI, E., RAVET, J., GRINIECE, E., VERMEER, J., MAROULIS, N., ETTWEIN, F., VAN BRUSSELEN J. & GREEN, T. 2016. An assessment of the cumulative cost impact of specified EU legislation and policies on the EU forest-based industries. Brussels: DG GROW.

SCHÖN, D. A. & REIN, M. 1994. *Frame Reflection: Toward the Resolution of Intractable Policy Controversies.* New York: Basic Books.

SIDDI, M. 2020. The European Green Deal: Assessing its current state and future implementation. FIIA Working Paper 114.

SNYDER, H. 2019. Literature review as a research methodology: An overview and guidelines. *Journal of Business Research*, 104, 333–339.

SOTIROV, M., LOVRIC, M. & WINKEL, G. 2015. Environmental governance between Europeanization and domestic politics: Analyzing the implementation of EU biodiversity conservation policy in Bulgaria and Croatia. *Environment and Planning C: Government and Policy 2014*, 33, 986–1004.

SRIVASTAVA, P. & HOPWOOD, N. 2009. A practical iterative framework for qualitative data analysis. *International Journal of Qualitative Methods*, 8, 76–84.

SWEDISH FOREST INDUSTRIES. 2020. Broader view of transition needed in European Green Deal [Online]. www.forestindustries.se/news/news/2020/02/Broader-view-of-transition-needed-in-European-Green-Deal.

TORRACO, R. J. 2005. Writing integrative literature reviews: Guidelines and examples. *Human Resource Development Review*, 4, 356–367.

UHLICH, T. & BURCK, J. 2022. *Fact Sheet: Glasgow Leaders' Declaration on Forests and Land Use.* German Federal Ministry for Economic Cooperation and Development, G7 | G20 Track 2 Dialogue.

VOGELPOHL, T. & AGGESTAM, F. 2011. Public policies as institutions for sustainability: potentials of the concept and findings from assessing sustainability in the European forest-based sector. *Eur. J. For. Res.*, 131, 57–71. https://doi.org/10.1007/s10342-011-0504-6.

WEISS, G., LAWRENCE, A., HUJALA, T., LIDESTAV, G., NICHIFOREL, L., NYBAKK, E., QUIROGA, S., SARVAŠOVÁ, Z., SUAREZ, C. & ŽIVOJINOVIĆ, I. 2019. Forest ownership changes in Europe: State of knowledge and conceptual foundations. *Forest Policy Econ.*, 99, 9–20. https://doi.org/10.1016/j.forpol.2018.03.003.

WINKEL, G. 2017. Towards a sustainable European forest-based bioeconomy- assessment and the way forward. What Science Can Tell Us 8. European Forest Institute. www.efi.int/publications-bank.

WOLFSLEHNER, B., AGGESTAM, F., PULZL, H., HENDRIKS, K., LINDNER, M., SOTIROV, M., KULIKOVA, E., SECCO, L., LOVRIC, M., NABUURS, G.-J., HURMEKOSKI, E., PETTENELLA, D., ARETS, E., BAULENAS SERRA, E., BOZZANO, M., DERKS, J., GATTO, P., LINSER, S., MASIERO, M., MAVSAR, R. & GIESSEN, L. 2019. Study on progress in implementing the EU Forest Strategy. Final report. Luxembourg: Publications Office of the European Union. DOI:10.2762/160685.

WOLFSLEHNER, B., PÜLZL, H., KLEINSCHMIT, D., AGGESTAM, F., WINKEL, G., CANDEL, J., ECKERBERG, K., FEINDT, P., MCDERMOTT, C., SECCO, L., SOTIROV, M., LACKNER, M. & LAZYA ROUX, J. 2020. European forest governance post-2020. *From Science to Policy 10.* Joensuu, Finland: European Forest Institute. https://doi.org/10.36333/fs10

5 Systemic change for the forest-based sector

Implementing the circular bioeconomy

1. The path to a sustainable future

Forests and the forest-based sector have the potential to play a significant role in the circular bioeconomy (Hetemäki, 2014, Hurmekoski et al., 2019, Jankovský et al., 2021), which is a concept that aims to create a sustainable system (Box 5.1) for producing and using biological resources. This can be achieved by minimising waste, pollution, and the depletion of natural resources while maximising the economic and social benefits of the bio-based economy. The forest-based sector is well-positioned to contribute to this concept. It is a significant source of renewable raw materials that can produce many products, from energy to construction materials. As the sector continues to evolve and diversify, new opportunities for contributing to the circular bioeconomy are emerging (Hetemäki et al., 2017, Winkel, 2017, Jankovský et al., 2021). For example, technological advances make it possible to extract more value from wood and other forest products, such as biofuels or bioplastics.

Additionally, there is growing interest in using wood as a building material, as it has a lower carbon footprint than traditional materials like concrete and steel. However, one of the challenges in this development is the need for a clear understanding of what the circular bioeconomy means (D'Amato et al., 2017, Carus and Dammer, 2018). The concept has mainly been discussed in academic and policy circles, with different interpretations and definitions used by different stakeholders. This can create confusion and prevent the development of practical solutions. While existing conceptualisations may bring important, broad, strategic overviews of how the circular bioeconomy could be implemented, they need to be more specific to provide hands-on and actionable examples of how value chains must be reconfigured and optimised to become circular. These detailed examples are necessary for companies and policymakers to understand what changes must be made to create a circular bioeconomy. It is, therefore, essential to develop a more concrete understanding of the circular bioeconomy, with clear examples of how the forest sector can contribute to this concept.

Box 5.1 What is sustainability?

Sustainability involves managing forest resources to meet present needs without compromising the needs of future generations. It covers economic, environmental, and social dimensions, focusing on forests' long-term health and productivity (Brundtland Commission, 1987, MCPFE, 2005, SFC, 2015). Sustainable systems in the forest-based sector refer to practices like Sustainable Forest Management (SFM), efforts to balance timber harvesting with

DOI: 10.4324/9781003198987-5

conservation, supporting biodiversity, and maintaining forest ecosystem services (Holvoet and Muys, 2004, Spathelf, 2009, Linser et al., 2018). This integrative approach is crucial for preserving forests as carbon sinks, reservoirs of biodiversity, and as an economic resource (Sotirov et al., 2015, Lazdinis et al., 2019, Lier et al., 2021).

The argument for transitioning to a circular bioeconomy is intrinsically interlinked with the current overuse of natural resources (Wiedmann et al., 2015, Giampietro, 2019). The prevailing linear economic model must be revised, which relies on raw materials extraction, production, and disposal. This linear approach is commonly referred to as the 'take, make, and dispose' model, which has had significant environmental and socio-economic impacts, positive and negative. For instance, human consumption of ecological resources has reached a point where the global community currently consumes resources as if it had 1.75 Earths (WWF, 2022). This is a staggering statistic, considering the limited resources available. Global material consumption has been increasing alarmingly over the last 50 years. It grew from 28.6 billion tonnes in 1972 to over 100 billion in 2022 (CGRI, 2022, 2023). In addition, approximately 90 billion tonnes of materials are discarded as waste yearly (WB, 2022).[1] This increase in consumption has resulted in the depletion of natural resources and has significantly impacted the environment. The 2023 Circularity Gap Report shows that despite a growing awareness concerning overconsumption, the global economy is only 7.2% circular (CGRI, 2022, 2023). This indicates a more than 90% circularity gap when considering the minerals, fossil fuels, metals, and biomass that enter the global economy yearly (EMF, 2020). This gap represents a major challenge. A circular bioeconomy model aims to close this gap by reusing and recycling resources, reducing the dependency on finite resources and minimising waste. This is a crucial step towards achieving a sustainable future for the planet.

1.1. Aim of the chapter

This chapter will examine the potential for circular bioeconomy principles to be implemented across different value chains in the forest-based sector. Forests provide an interesting case study to explore circularity as the use of timber and wood-based products has significant environmental impacts while being a renewable resource. The value chain approach is utilised in this chapter as it can help identify opportunities for closing production loops and raw material cycles that are closely linked to how the industry is structured (Kampelmann, 2020, Shabani et al., 2013, FAO, 2013). A value chain refers to the activities required to create and deliver a product or service to the customer. Analysing different value chains makes it possible to identify opportunities and develop strategies for implementing circular bioeconomy principles.

The chapter will start by clarifying some of the conceptual ambiguities surrounding the circular bioeconomy and define what circularity means in the context of this chapter. This will be done by providing an overview of historical developments and current debates on the circular bioeconomy, outlining the fundamental principles and concepts relevant to this chapter. This will be followed by a value chain analysis of forest-based sectors that use wood as a raw material. The value chain analysis will cover various stages, such as harvesting, processing, and manufacturing, and will examine the opportunities and challenges of implementing circular bioeconomy principles. Additionally, it will consider the governance mechanisms needed to support the implementation of circular bioeconomy principles and the role of different stakeholders in this process.

Finally, the chapter will examine the main strengths and weaknesses of the proposed conceptual approach and suggest a direction for future research on the topic. The strengths and weaknesses will be identified by assessing the proposed approach in terms of contributing to understanding circular bioeconomy principles, its limitations, and areas for improvement. The chapter will also suggest future research that can be carried out to explore further the implementation of a circular bioeconomy in the forest-based sector.

2. Understanding and defining the circular bioeconomy

Forests are a land-based renewable resource, providing many environmental and societal benefits (Hetemäki et al., 2017). One of the most notable features of forests is that they provide a natural resource, wood, that is reusable and recyclable. Wood is a versatile material used in various products and applications, such as construction, furniture, and paper. Wood also has the unique characteristic of sequestering carbon, which means it can absorb and store carbon dioxide from the atmosphere. This can help to mitigate the effects of climate change. Furthermore, wood is biodegradable, which means it can be returned to the biosphere as nutrients when it can no longer be reused or recycled (Jarre et al., 2020, Sherwood, 2020, EC, 2016). Wood is also an intriguing material due to its particular properties. For example, while wood fibres deteriorate over time, they can be recycled up to 25 times with minimal quality loss. On the one hand, this means wood fibres are interesting for circular systems (Eckhart, 2021); on the other hand, the limits to recycling imply that resource loops cannot be entirely closed.

These unique characteristics of wood make it an interesting case study for the circular bioeconomy, which aims to create sustainable systems that mimic the natural cycles of life. Previous studies have addressed the circular bioeconomy potential of the forest-based sector.[2] For example, Stegmann et al. (2020) identified strategies for bioeconomy clusters' feedstock and product focus by investigating the role played by biorefineries, circular solutions, recycling and cascading. Hassegawa et al. (2022) reviewed some of the most innovative circular bioeconomy products that are already on the market and could enter international markets soon. Similarly, Hurmekoski et al. (2018) examined how the position of the wood-based industry in a given value chain determines the respective production value in the circular bioeconomy. Despite these studies, a concise and universally adopted conceptualisation of the circular bioeconomy for value chains of the forest-based sector is still missing. Further research is needed to fully understand the potential of wood as a sustainable resource and develop strategies for integrating it into circular economic systems.

2.1. What is the circular economy?

The circular economy framework, which focuses on closing the loop on the use of resources, has roots that go back to the beginning of modern industrial practices. The initial focus of these practices was on linear production models, where resources were extracted, used to produce goods, and then discarded as waste. However, as early as the onset of industrialisation, efforts were made to repurpose objects and materials to optimise resource utilisation and improve economic efficiency (Reike et al., 2018, Potting et al., 2017, Ritzén and Sandström, 2017). Over time, these early efforts gave rise to a growing recognition that we need more sustainable production and consumption models. The concept of circularity emerged as a response to these concerns and represents a departure from traditional linear models of production and consumption. Instead of seeing resources as single-use inputs, the circular economy framework emphasises the importance of keeping resources in use for as long as possible and minimising waste. As a result, the circular economy framework has gained widespread recognition for promoting sustainability and increasing resource utilisation efficiency. It is a holistic approach that considers the entire lifecycle of resources, from extraction and production to use and disposal, intending to minimise waste and maximise value (Figure 5.1). The framework provides a roadmap for creating a more sustainable and resilient future by promoting the circular use of resources and reducing dependency on finite resources.

The circular economy framework is not anchored to a specific economic theory or model. However, it has strong roots in ecological and environmental economics and industrial ecology (Lieder and Rashid, 2016, Kapur and Graedel, 2004). Moreover, as noted above, the concept of circularity can be traced back to the early days of modern industrial practices and efforts to optimise resource use and improve economic efficiency – the idea of reusing materials dates to this time. From an academic standpoint, the 'spaceman economy' is often noted as being the first reference to a circular system. In this context, Boulding (1966) introduced a 'closed system' where all outputs from consumption are recycled back into inputs for production. Over time, other academic contributions have shaped the development of the circular economy, such as 'Limits to Growth' (Meadows et al., 1972), 'Overshoot' (Catton, 1980), 'Cradle-to-Cradle' design (Braungart and McDonough, 2002), and the 'Performance Economy' (Stahel and Clift, 2016). However, the formal introduction of the circular economy can be credited to environmental economists Pearce and Turner (1990) in their review of the linear economic system.

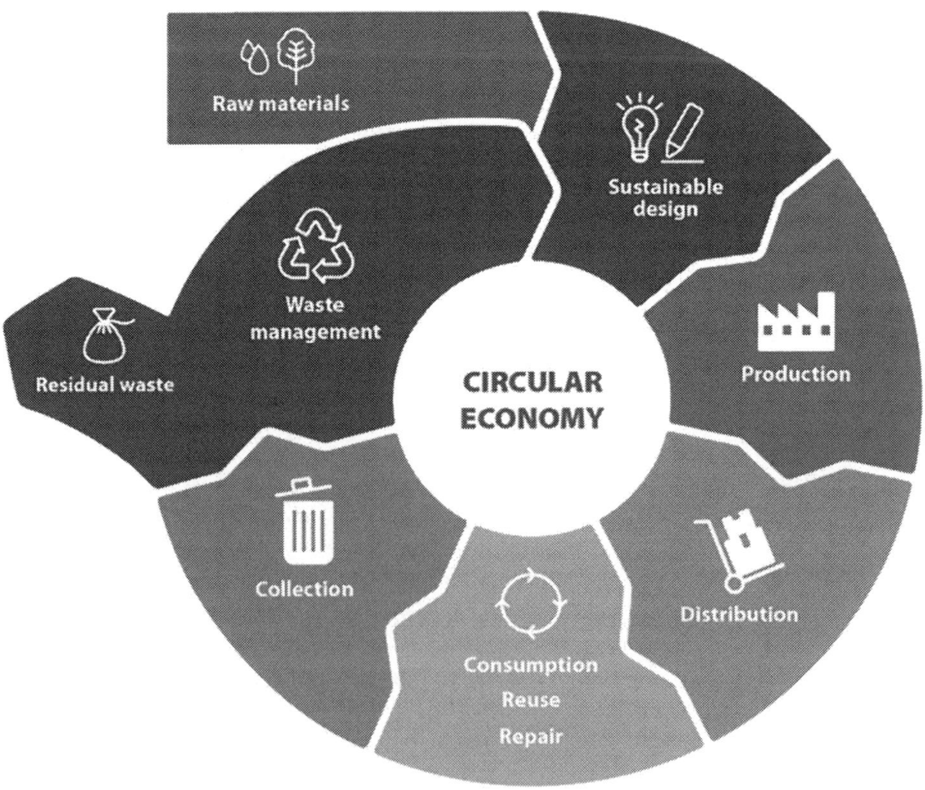

Figure 5.1 The circular economy model.

Source: European Parliament Research Service

They drew inspiration from Boulding's (1966) ideas and the principle that everything is an input to everything else when they set out to develop the concept of a circular economy.

Initially, the focus was on managing waste and improving recycling efforts, specifically for waste streams such as paper and plastics (Reike et al., 2018). The objective was to find technological solutions to turn waste into valuable inputs for other processes. However, despite these efforts, reusing materials and reducing material consumption has not been widely adopted (Ghisellini et al., 2016, Ritzén and Sandström, 2017). Over the last two decades, the circular economy has evolved into a more comprehensive approach encompassing socio-economic factors. This shift can be attributed, in part, to political developments, such as the Rio Declaration (UN, 1992), which takes a systemic view on natural resource use. Additionally, the 2030 Agenda for Sustainable Development (UN, 2015), which sets out a global plan to achieve sustainable development, highlights the significance of transitioning to a circular economy to meet the Sustainable Development Goals (Schroeder et al., 2019). The circular economy has thus come a long way from its initial focus on waste management and recycling to a broader approach that considers socio-economic and political factors in the pursuit of sustainability.

> The EU's Action Plan on the Circular Economy characterises circularity as an economy in which waste is minimised and the value of products, materials, and resources is maintained for as long as possible (EC, 2020). It is a system of economic activity designed to be regenerative and restorative by nature. Similarly, the Ellen MacArthur Foundation[3] defines a circular economy as a process whereby the economy is decoupled from the consumption of finite resources by eliminating waste and pollution, the circulation of products and materials, and the regeneration of nature. The foundation also proposes that a circular economy can achieve this by implementing strategies such as product design for longevity, sharing and leasing models, and closed-loop supply chains. These perspectives show that the circular economy concept varies across sectors and disciplines. However, they share common principles, such as responsible natural resource use and minimising waste and pollution (CGRI, 2022, 2023).

It can also be noted that non-academic actors have been active in promoting the circularity concept, including private sector organisations, governments, and international organisations (CGRI, 2022, 2023). These actors have recognised the social, economic, and environmental benefits of circular business models and policies and have advocated for their implementation. One prominent example is the Ellen MacArthur Foundation, which has been a leading voice in promoting circularity and has played a crucial role in shaping the discourse on the circular economy (EMF, 2012, 2015). The foundation has been working to raise awareness of circularity's benefits, build a coalition of organisations committed to circular practices, and provide practical tools and guidance for businesses to transition to a circular model. Another example is the European Union, which has been actively promoting circularity on the international stage. The EU has been instrumental in shaping the conceptual framework for the circular economy and has advocated for its implementation through various policy initiatives and programs. The EU's action plan for the circular economy, part of its new strategy for the European industry and the European Green Deal (see Chapter 4), is a clear example of its commitment to promoting circularity (EC, 2015, EC, 2020). In addition to these private sector and governmental actors, international organisations such as the United Nations Environment Programme have promoted the circular economy (UNEP, 2018). By adopting the EU's conceptual framework, they have been working to raise awareness of circularity's benefits and promote its implementation globally.

> The circular economy is often characterised as an approach that can reduce natural resource use by slowing, closing, or narrowing resource loops (Geissdoerfer et al., 2017, Mazur-Wierzbicka, 2021). For forests, the circular economy can reduce the need for virgin wood by closing resource loops. This is done by using resources already in the system. Understanding these principles and the ways and means by which they can be applied is the key to effectively implementing the circular economy. For example, this can be achieved by using wood products that can be recycled or repurposed or by developing new technologies that can extract more value from biomass. These initiatives need to be supported by policies and regulations encouraging the development of a circular bioeconomy, such as promoting the use of bio-based materials in construction and manufacturing.

2.2. *What does the circular bioeconomy mean?*

The father of the bioeconomy, Georgescu-Roegen, was one of the first to use the term 'bioeconomics' in response to conventional economists who favoured the pursuit of growth (Vivien et al., 2019). He advocated against waste and promoted 'sufficiency' among consumers through his 'minimal bioeconomic programme' (Georgescu-Roegen, 1975). Today, the bioeconomy, as defined by the European Commission, relates to the use of renewable biological resources from forests and agricultural sectors as inputs for industry, such as raw materials, chemicals, and energy (D'Amato et al., 2017, McCormick and Kautto, 2013). These resources include wood, crops, and other types of biomass that can be used to produce a wide range of products, from food to biofuels to bioplastics. The bioeconomy is essential for transitioning to a more sustainable and resilient economy. It can help reduce the dependence on fossil fuels, create new jobs and businesses, and promote rural development. However, the bioeconomy is not inherently 'circular'. This is why circular economy principles have been incorporated to minimise waste generation and extend the lifecycle of products, materials, and resources (Stegmann et al., 2020, D'Amato et al., 2017). A circular economy is restorative and regenerative by design. The aim is to keep resources in use for as long as possible, extract the maximum value from them while in use, and then recover and regenerate products and materials at the end of life. In other words, by combining the principles of the circular economy and the bioeconomy, it is possible to create a more sustainable and resilient system for producing and using forest resources. This system can minimise waste, pollution, and the depletion of natural resources.

The concept of a circular bioeconomy is surrounded by conflicting narratives in the academic literature. Some scholars have attempted to operationalise circularity within current economic systems (Fullerton, 2015, Rifkin, 2011), while others advocate for a transformation of the socio-economic order (Trainer and Alexander, 2019, Latouche, 2009). These differing perspectives reflect conflicting views on the societal capacity to overcome resource limitations and decouple economic growth from ecological degradation (Calisto Friant et al., 2020). Giampietro (2019) has noted that two distinct narratives on the circular bioeconomy have emerged. The dominant narratives associated with the circular bioeconomy are the growth-oriented and the limits-oriented narratives:

- The **growth-oriented narrative** views the circular bioeconomy as a new economic paradigm driven by technological progress and seeks to promote perpetual economic growth. This perspective focuses on the potential of the circular bioeconomy to generate economic benefits and create new business opportunities while also addressing environmental and resource challenges. Proponents of this narrative believe that the circular bioeconomy can provide a solution to the current linear economic model based on the extraction and disposal of natural resources.

- The **limits-oriented narrative** is based on an entropic view that acknowledges the limitations of economic growth. This perspective argues that the circular bioeconomy should not be seen as promoting perpetual economic growth but rather as creating a more sustainable and resilient economy that operates within the boundaries of available natural resources. Proponents of this narrative believe that the circular bioeconomy should be seen as a means of reducing the environmental impact of the economy and creating a more equitable distribution of resources.

These two narratives highlight the different perspectives on the circular bioeconomy and the challenges of balancing economic growth and sustainability. The varied narratives have also led to confusion and disagreement among researchers and practitioners, making it challenging to understand what circularity and the bioeconomy mean across applications and disciplines. For example, a systematic analysis conducted in 2017 found 114 different definitions of the circular economy (Kirchherr et al., 2017), which, in addition to problems with comparability, can lead to misinterpretation, greenwashing, and image depreciation (Nobre and Tavares, 2021). However, most definitions can be classified into two broad categories: those that focus on material use and those that concentrate on system change (Korhonen et al., 2018):

1. Definitions that focus on **material use**, such as the three Rs of sustainability (reduce, reuse, recycle) or the more comprehensive 9R approach (see below), emphasise the need to minimise the use of raw materials, maximise product reuse, and recycle raw materials to the highest possible quality. This approach is more geared towards a reuse economy, where materials are correctly recycled and products retain a stable quality.
2. Definitions focusing on **system change** stress the importance of closing production cycles, using renewable energy sources, and adopting a systems-thinking perspective. The 9R approach, which combines material use and system change, is a more comprehensive approach that covers both dimensions of circularity (Potting et al., 2017). The 9Rs comprise *refuse*, *rethink*, *reduce*, *reuse*, *repair*, *refurbish*, *remanufacture*, *repurpose*, *recycle*, and *recover* (Figure 5.2).

The 9Rs model, as illustrated in Figure 5.2, covers a range of strategies central to the circular bioeconomy, with each 'R' representing a distinct approach towards resource sustainability. For instance, 'Repair' focuses on fixing damaged items rather than discarding them. Within the forest-based sector, these strategies can be applied in different ways. For instance, designing wood products for straightforward disassembly, such as a modular wooden house, reflects the 'Rethink' principle. Additionally, offering services to repair existing wooden products, like damaged furniture, aligns with 'Reuse'. However, it is important to recognise that interpretations of the 'Rs' can vary across disciplines, each attaching its meaning to these principles (Khaw-ngern et al., 2021). This chapter will follow the framework presented in Figure 5.2 for a consistent reference to the 9Rs in the context of the forest-based sector.

Another relevant aspect is bioenergy production (falling under 'recover'). Compared to other resources, energy cycles cannot be recycled; therefore, they are only typically included in circular systems if the entire carbon cycle is considered. This effectively means that most definitions or models of circular systems do not consider energy production from biomass as circular (Giuntoli et al., 2022, Holmatov et al., 2019). For the wood-based industry, it is more common to adopt a cascading approach, which involves using wood-based materials sequentially (Besserer et al.,

0. **Refuse**: Make product redundant 1. **Rethink**: Make product use more intensive 2. **Reduce**: Use the least materials possible for production	Stage of the value chain when a product provides its primary function to the user, including efforts to optimise material efficiency, such as ensuring that products can be used for as long as possible.
3. **Reuse**: Use the product in different applications or, where possible, turn the product into a service to be used by different users 4. **Repair**: Amend the functionality of the product 5. **Refurbish:** Upgrade the product's functionality with the latest technologies and design 6. **Remanufacture:** Dismantle the existing product to use its parts in new products 7. **Repurpose:** Dismantle products to include them in new products with different functions	Parts of the value chain stage where end-users can interact with producers to update the functionality of the products, for example, to extend their lifespan.
8. **Recycle**: Recycle residues into secondary materials 9. **Recover**: Incineration with energy recovery	Value chain stages where specialised enterprises treat products at the end of life to turn them into secondary materials for other businesses.

Figure 5.2 Value retention loop and guiding principles (9Rs).

Source: adapted from Ellen MacArthur Foundation

2021). For example, by-products such as black liquor can be used to co-produce heat and power, where energy flows in a one-way direction and cannot be recycled. While studying circular systems and their impact on the forest-based sector is an ongoing research area, it is important to consider the role of energy production and the carbon cycle in these systems.

The forest-based sector relies heavily on bioenergy, and the unique properties of wood as a material make it a subject of ongoing research and debate. The general argument is that a cascading approach may be more appropriate for wood-based products as it considers the hierarchy of use, such as repurposing wood after its initial use (Mair and Stern, 2017, Stegmann et al., 2020, UBA, 2017). It also accounts for the natural limitations of wood, such as limits to recycling (Eckhart, 2021). However, while circular economy models have integrated cascading use (Box 5.1), one point of contention remains whether the absence of energy recovery means it can still be considered circular (Vis et al., 2016, Mair and Stern, 2017, UBA, 2017).

> The circular bioeconomy is characterised by diverging narratives: one growth-oriented, advocating for economic growth through technological advancements, and one limits-oriented, emphasising environmental constraints and equitable resource distribution. Consequently, the circular bioeconomy encompasses various interpretations, from material use (emphasising Reduce, Reuse, Recycle) to systemic change (focusing on closed production cycles and renewable energy). Central to this discussion is the 9Rs model, which integrates various strategies for resource sustainability.

2.3. Defining a circular and forest-based bioeconomy

This chapter applies the circular economy model proposed by Potting et al. (2017). This model views the circular (bio)economy as consisting of three interlocking value retention loops, shown in Figure 5.3. These loops encompass the entire life cycle of a product or material, from its

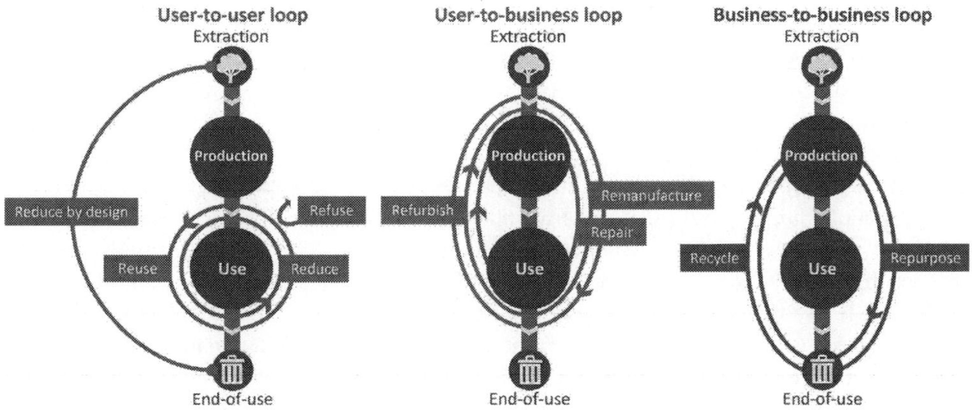

Figure 5.3 Guiding principles across the value retention loops.

Source: own figure

extraction to its production and use and, finally, to its end-of-life disposal. Integrating the concepts of user-to-user, user-to-business, and business-to-business loops into the circular bioeconomy model, together with the 9Rs, adds conceptual layers to help explain how circularity functions at different stages of a product life cycle:

- **User-to-user loop:** The direct exchange or reuse of products between end-users or consumers. It involves the circulation of goods within the consumer sphere. For example, in the forest-based sector, this could mean selling your second-hand furniture, thereby extending the product life cycle.
- **User-to-business loop:** Covering products or materials being returned from the user back to businesses. It can involve return schemes, where consumers return products to the manufacturer for refurbishment, remanufacturing, or recycling. This loop promotes resource efficiency and allows businesses to control the product life cycle.
- **Business-to-business loop:** Encompass the exchange or recycling of materials between different businesses. In the context of the forest-based sector, it might involve offcuts or wood waste from one manufacturing process being used as raw materials in another. This loop can foster industrial symbiosis, where one industry's waste becomes the input for another, increasing the efficient use of resources and reducing waste.

Incorporating design is considered a fundamental principle in this framework. Products designed with circularity in mind can move more easily between the loops. For instance, a wood-based product designed to be easily disassembled allows for effective refurbishing or recycling, facilitating its movement through and across the loops. This approach reduces waste and encourages sustainable practices throughout the life cycle of a product (Potting et al., 2017).

3. Methodological approach

The present chapter builds on an earlier publication – Implementing Circular-Bioeconomy Principles across Two Value Chains of the Wood-Based Sector: A Conceptual Approach (Aggestam and Giurca, 2022). It is also, in part, based on work carried out for the United Nations Economic Commission for Europe (Aggestam et al., 2021) and the European Commission (Rivera León et al., 2016).

3.1. Focus and objective

This chapter explains the flow of materials within two key value chains in the wood-based industry: woodworking and pulp, paper, and paperboard. The chapter uses a value chain analysis framework, which includes the 9R approach, to examine the entire life cycle of a product, from extraction to production to end-of-life. The 9R approach helps ensure the product's life cycle is carefully considered in the analysis. The five specific sectors within the forest-based industry that the chapter maps and analyses are sawnwood, bioenergy, construction, pulp and paper, and cellulose manufacturing. To further guide the analysis, the study applies the three value retention loops that Potting et al. (2017) characterised. This framework considers the entire life cycle of a product, from extraction to production to end-of-life, which helps to ensure that the analysis considers the full scope of the product's life cycle and the flow of materials within the value chains.

The objective of this chapter is to provide a comprehensive understanding of the material flow across the two key value chains in the forest-based industry and to highlight the potential for circularity within the industry. However, it is also important to note that the analysis primarily focuses on qualitative aspects of the different forest-based value chains. This is due to the wide variability of the sectors in question, both horizontally and vertically, and the macro-level scope of the value chain mapping. Therefore, the primary objective of this analysis is to highlight the requirement for diverse governance approaches across the forest-based industry rather than providing a comprehensive quantitative analysis.

3.2. Analytical framework

Value chain analysis (VCA) is a tool to evaluate the potential for circularity in forest-based industries. By mapping the various activities involved in producing and using different wood-based products, the VCA provides a clear picture of how each value chain can adopt circular bioeconomy principles (FAO, 2007, 2013). The VCA highlights the sequence of events from the extraction of raw materials to the product's end-of-life and identifies key segments crucial for ensuring circularity. The VCA effectively examines the feasibility of circularity by considering all product life cycle stages (Fearne et al., 2012, FAO, 2013). The VCA aims to facilitate a deeper understanding of how a circular bioeconomy can be implemented in the forest-based sector by illustrating the interconnectedness of different stages in producing and using wood-based products.

The value chain graphs in Figures 5.4 and 5.5 provide an overview of the different activities involved in producing and using wood-based products. These graphs are based on the Statistical Classification of Economic Activities in the European Community (NACE) Revision 2 codes (EUROSTAT, 2008, 2017, 2019, Directive, 1893/2006), which provides a standardised classification system for different economic activities. However, it is acknowledged that NACE may only capture some products and production processes considered wood-based. Therefore, additional product classifications, such as the FAO classification, were considered. The value chain graphs demonstrate the products belonging to each product group and the extent of processing they undergo (primary, secondary, or tertiary). The graphs also explain the interplay between product groups and their relationships within the forest-based sector. It should be noted that waste streams have yet to be indicated in the value chain graphs, as the focus is on the production and use of wood-based products rather than their end-of-life disposal.

Modular value chains were used to study EU forest-based industries to evaluate the aggregate financial effect of specific EU regulations and policies on these industries (Rivera León et al., 2016). Two key factors drive the choice to utilise these value chains: (1) these value chains were created with the involvement of industry experts from relevant sectors, a total of 12 industry organisations[4] participated in their development, (2) with some modifications, these value

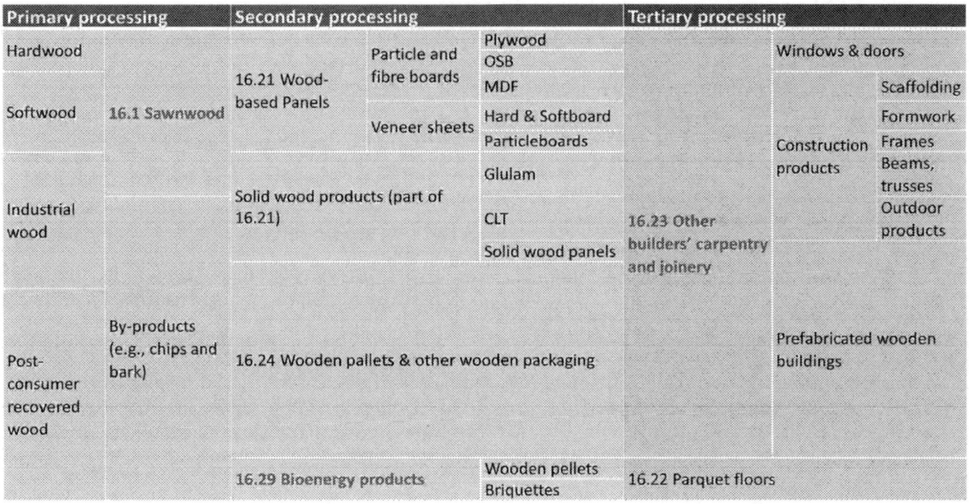

Figure 5.4 Wood processing and manufacturing industry.

Source: adapted from Aggestam et al. (2021) and Aggestam and Giurca (2022)

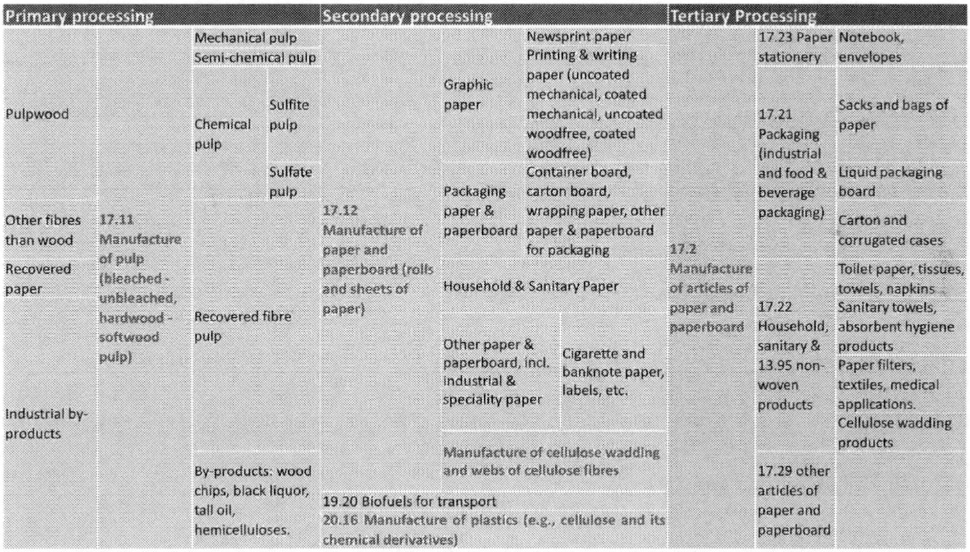

Figure 5.5 The pulp, paper, and cellulose value chain.

Source: adapted from Aggestam et al. (2021) and Aggestam and Giurca (2022)

chains provide a comprehensive coverage of the forest-based industries, offering a systemic overview of these sectors.

It is also important to note that the representation of products under each product group is limited. Some products can have more than one raw material component, for instance, wood-based panels that can be produced using a combination of sawnwood, recovered wood, and by-products. This can result in multiple variations of the specified value chains; for example, pulp and oriented strand boards can be manufactured using pulp logs. The value chains

Table 5.1 Cross-cutting factors.

Factors	*Description*
Integration across value and supply chains	Coordination and alignment of activities across different value and supply chain stages.
Knowledge, education, and awareness	Knowledge, education, and awareness about sustainable practices within the industry and among stakeholders.
Certification, quality standards and labelling	Using certifications, quality standards, and labels to promote and recognise sustainable practices.
Policy frameworks (regulatory and legislative frameworks)	Legal and regulatory frameworks governing the industry's sustainable practices.
Innovation and technological development	Advancing and adopting new technologies and innovations that support sustainable practices.

present the diversity and intricacy of products within a particular product group within each sector, highlighting the distinctiveness of the sectors even though they rely on the same raw material. Finally, the findings presented in the next section concerning the respective value chains (Figures 5.4 and 5.5) are based on a comprehensive review of both academic and non-academic literature, with a focus on messages regarding the circular (bio)economy found in outputs produced by relevant stakeholder organisations.

4. Results and analysis

One challenge encountered during the analysis was categorising cross-cutting issues into the relevant loops shown in Figure 5.3. For instance, 'access to information' is a general issue that applies to value chains, loops, and all 'Rs'. This highlights that some specific issues can be tackled systematically across all forest-based industries, such as 'promoting a circular lifestyle' or, within the relevant loops, for all producers and users of wood-based products, as shown in Table 5.1.

This is equally applicable under the respective loops where cross-cutting strategies can be adopted to promote the circular economy in the forest-based industry. Moreover, these strategies are interrelated and implementing one can have a positive (or negative) impact on others.

Under the user-to-user loop:

- Re-designing systems and products to make them more circular.
- Improving the effectiveness of the systems and processes used.
- Reducing the environmental impact of production by using more sustainable methods.
- Reducing competition over raw materials through more efficient use of resources.
- Encouraging a circular lifestyle by promoting the adoption of sustainable practices.

Under the user-to-business loop:

- Extending producer responsibility by making them more accountable for the end-of-life management of their products.

- Designing for circularity by creating products that can be easily repurposed, reused or recycled.
- Making repair and refurbishment economically viable by creating a market for second-hand products.
- Building trust in secondary materials and products by providing reliable information about the quality of these materials.

Under the business-to-business loop:

- Increasing the use of post-consumer waste streams by using more recycled materials in production.
- Expanding the available product mix by offering more circular products.
- Improving the infrastructure for collecting and recycling wood-based products by creating a more efficient system for waste management.
- Creating markets for waste streams by promoting the use of secondary materials in the production process.

4.1. The art of wood

Wood, as a raw material, is considered a naturally circular resource due to its renewable nature. This means that it can be replenished over time, making it a sustainable source of material. When wood is harvested and used, its nutrients are returned to the biosphere through decomposition. Although it may take many decades, this natural cycle highlights the environmental benefits of using wood as a material (Holmatov et al., 2019). Wood has several advantages and disadvantages compared to other materials, such as metals. One key benefit of wood is that it is bio-based and non-toxic, making it a safer and more environmentally friendly option. However, unlike other materials, wood cannot be transformed or renewed through closed production cycles. This is because cellulose fibres can only be reused a limited number of times. Because of these limitations, wood usage is often viewed through a cascading use and transformation rather than circular perspective. This means that wood is used in a series of stages, each leading to the material's transformation rather than being constantly reused and renewed in a closed loop. While this may limit the circularity of wood, it is still a valuable and sustainable material that plays an essential role in many industries.

In woodworking (Figure 5.4), primary processing refers to the initial stages of transforming raw wood into finished products. This includes sawing, planing, and chipping the wood into smaller pieces. Primary processing aims to prepare the raw wood for further processing and to create a uniform size and shape that can be used to produce finished goods. Secondary processing refers to the next stage of transforming the wood into more specific products. This includes activities such as drying, glueing, and sanding. Secondary processing aims to improve wood products' quality and uniformity and prepare them for the final stages of production. Finally, tertiary processing refers to the final stages of transforming wood into finished goods, such as furniture, flooring, and cabinets. This includes activities such as assembly, finishing, and packaging. Tertiary processing aims to produce high-quality, finished products that meet customer specifications. Together, these three stages form the complete value chain of the woodworking industry, from the raw material to the finished product. Each stage requires specialised skills and knowledge, and the industry's success depends on the efficiency and quality of each processing stage.

4.1.1. Sawnwood

The sawnwood sector (NACE 16.1) plays a crucial role in the solid wood value chain, a resource-intensive industry requiring maximum resource efficiency to be economically viable. The sector involves the production of sawnwood, a basic raw material used in various applications. This includes producing wood-based panels, solid wood products, construction materials such as beams, windows, and doors, and bioenergy production through wood chips and sawdust. In addition to the sawnwood production, the sawnwood sector also generates wood waste during production. This waste material can be used as a raw material to produce particle boards or pulp, which adds to the industry's efficiency by reducing waste and maximising the utilisation of resources. The quantity of wood co-products directed towards specific side streams, such as particle board production or bioenergy, depends on the type of wood being sawed, as indicated in Figure 5.4.

The sawnwood sector can become more circular by implementing practices promoting closed-loop use of resources and minimising waste. To achieve this, the sector can reduce waste by improving production processes and using waste as raw material for other products, as outlined above. The sawnwood sector can also increase the reuse and recycling of materials, such as utilising scrap wood to produce particle boards or recycling sawdust and wood chips as fuel. Alternatively, it can optimise resource use by selecting wood species well suited for specific products and using more efficient production methods that use less raw material. Finally, the sector can also work with other industries, such as the furniture and construction industries, to increase the circularity of wood use. Some additional considerations for how the sawnwood sector can become more circular across the respective loops are set out below:

Under the user-to-user loop:

- *Enhance product design to maximise the use of sawnwood while efficiently utilising waste products, such as sawdust*: Products should be designed to enable the use of sawnwood for as long as possible, reducing waste and maximising resource utilisation. In addition, waste products, such as sawdust, can be reused in other applications, reducing the environmental impact of the manufacturing process.
- *Implement sustainable forestry practices and labelling to promote responsibly sourced wood*: Sustainable forestry practices should ensure that the wood used in production is sourced from responsibly managed forests. In addition, labelling can be introduced to inform consumers about the wood's origin and the production process's sustainability.
- *Address consumer behaviour and perception towards the use of sawnwood, encourage the adoption of eco-friendly business models, and promote wood from sustainably managed forests*: Consumers should be made aware of the importance of using wood from sustainably managed forests and the benefits of eco-friendly business models, such as renting modular and recycled wood-based products. This will help shift the perception of sawnwood and encourage more sustainable practices.

Under the user-to-business loop:

- *Ensure the effective cascading of sawnwood and reduce negative impacts, such as water and noise pollution*: Sawnwood should be cascaded down, mainly because of material deterioration, to minimise waste and reduce negative impacts, such as water and noise pollution. Incineration should be considered a leakage out of the system and reduced as much as possible.
- *Foster collaboration with related industries to maximise material utilisation, for example, utilising sawdust for energy production and returning ash to the forest.* Interactions with related

sectors and actors should be facilitated to maximise the use of materials and minimise waste. For example, sawdust can produce energy in the mill, and the ash can be recycled back into the forest.
- *Encourage the recovery and reuse of post-consumer wood waste, depending on its quality, for recycling or as a raw material for biofuel*: Post-consumer wood waste, such as discarded wood, should be recovered and utilised for recycling or as a raw material for biofuel, depending on its quality. This will help reduce the waste generated from the production process and improve resource utilisation.

Under the business-to-business loop:

- *Design processes to minimise waste and harmful practices and balance material and energy utilisation of waste products, incinerating only those that cannot be downcycled*: Processes should be designed to minimise waste and harmful practices, such as reducing dust generated by woodworking machines. The balance between material and energy utilisation of waste products should be optimised, with only those that cannot be downcycled being incinerated.
- *Establish connections between producers and consumers to identify potential up- or downstream loops and valorise waste products*: Producers and consumers should be connected to identify potential loops and side streams up or downstream. This will help valorise waste products and minimise waste.
- *Upgrade capacities and infrastructure on both the consumer and producer side to improve the recycling and reuse of sawnwood*: Capacities and infrastructure should be upgraded on both the consumer and producer side to improve the recycling and reuse of sawnwood. This will help reduce waste, minimise the environmental impact of the production process, and improve resource utilisation.

4.1.2. Construction

The wood construction sector (NACE 16.23) plays a vital role in the global economy and significantly impacts the environment. In order to reduce the waste generated by this sector and promote a circular bioeconomy, systemic improvements are necessary for the sorting, separating, and recovery of post-consumer wood. For example, one of the critical challenges facing the wood construction sector is the efficient recycling and demolition of construction waste. The end-of-life recycling of construction materials is critical in ensuring that valuable resources are not lost, and that waste can be efficiently cycled back into production. Increased collaboration and integration across the entire value chain is required. This involves working closely with all stakeholders in the sector, including deconstruction operators, municipalities, architects, designers, builders, and end-users. By fostering greater collaboration and cooperation between these groups, the wood construction sector can improve its sustainability and contribute to developing a circular bioeconomy. However, logistical and infrastructural challenges must also be addressed to promote the efficient recovery and recycling of wood waste. For example, metal objects must be manually removed from contaminated wood before recycling, and metal detection processes must be in place to prevent damage to milling tools during further processing.

Aside from the topics addressed above, the wood construction sector can become more circular by encouraging designs that facilitate the disassembly and separation of materials at end-of-life, making it easier to recover and recycle them. The sector could also promote using recycled and recovered materials in new construction projects, reducing the demand for virgin materials and promoting a closed-loop system. Some additional considerations for how the sector could become more circular across the respective loops are set out below:

Under the user-to-user loop:

- *Design and detailing of mass timber buildings for greater durability*: Ensure that mass timber buildings are built to last and withstand the test of time; it is crucial to consider the design and detailing of the building components. This includes considering the materials' strength, stability, rigidity, joints, connections, and fasteners that hold the building together.
- *Changing the mindset of people in the construction industry and end-users*: To achieve greater sustainability and durability in the construction industry, it is essential to change how people think about building design, construction, and use. This means shifting the focus from short-term thinking and fast-paced building cycles to a more holistic and long-term approach that considers the entire life cycle of a building.
- *Making new housing models available for consumers*: To make sustainable and durable housing more accessible, exploring and promoting alternative housing models is important. One example is housing cooperatives, where a group collectively come together to build and own a building. This approach can reduce costs, increase affordability, and promote community involvement in the building process.

Under the user-to-business loop:

- *Design for combined manufacture and assembly and disassembly*: Facilitate the efficient and sustainable use of building materials; it is important to consider the entire life cycle of the building from the design stage. This means designing buildings with modular elements made of massive wood that can be manufactured, assembled off-site and later disassembled for reuse or recycling.
- *Sourcing raw materials from sustainably managed forests*: To ensure the sustainability of mass timber buildings, it is crucial to source raw materials from sustainably managed forests. This means selecting wood that has been harvested in a responsible and environmentally friendly manner and considering the social and economic impact of the forest on local communities.
- *Changing construction sector business models to promote 'designing for disassembly'*: Encourage efficient recycling and reuse of building materials; changing how the construction industry operates is important. This means shifting the focus from a traditional build-and-dispose model to a more sustainable and circular approach, emphasising the importance of 'designing for disassembly'. This means considering how buildings can be dismantled and recycled at the end of their life cycle rather than being demolished and thrown away.
- *Enhancing sorting, separation, and recovery*: To maximise the potential for recycling and reusing building materials, it is vital to develop new approaches to sorting, separating, and recovering materials from buildings. This means investing in developing new technologies and systems that can efficiently sort, separate, and recover materials like wood, metal, and plastic.

Under the business-to-business loop:

- *Decreasing emissions and reducing material waste*: To reduce the environmental impact of construction, it is essential to minimise emissions, such as carbon and water, and eliminate material waste as much as possible. This can be achieved by improving material efficiency, implementing material audits, and optimising logistics to reduce waste and minimise the use of harmful materials.

- *Integrating smart design, construction techniques, and building service technologies*: To create buildings that are sustainable, efficient, and durable, it is important to consider smart design, construction techniques, and building service technologies. This means considering the routing of building service technologies outside modular elements, reassembling wooden buildings, and making more efficient use of side products.
- *Accommodating cross-cutting and networked systems*: Ensure mass timber buildings' long-term sustainability and durability.

4.1.3. Bioenergy

Bioenergy is an important aspect of the woodworking value chain, and its inclusion as secondary processing is significant (NACE 16.29), as outlined in Figure 5.4. Biofuels can originate from various points in the supply chain, both upstream and downstream, with wood-based energy primarily coming from wood processing residues. One example of using wood residues for energy generation is seen in sawmills that have integrated biomass power plants. These power plants use the waste generated by the milling process to supply energy for the sawmill's operations. From the sawmill's perspective, this approach allows them to add value to what was once considered waste and reduce production costs. However, while using waste residues for energy generation may provide economic benefits, there may be more optimal approaches from a circular economy perspective. On the other hand, using the waste residues locally, without the need for additional transport or processing, can lead to a reduction in the environmental impact of the sawmill's operations and align with sustainability goals. This highlights a trade-off between circularity and sustainability, as optimising material reuse may sometimes lead to environmental costs, such as increased emissions. While circularity is a desirable goal, bioenergy production demonstrates that it is important to consider the potential environmental impacts and ensure that the optimisation of material reuse does not compromise sustainability.

The bioenergy sector can become more circular by adopting a closed-loop approach to energy generation, which involves using waste and residues from other sectors as feedstocks for energy production. Some ways to achieve this include improving the efficiency of energy generation processes, encouraging the integration of the bioenergy sector with other sectors, such as agriculture, forestry, and waste management, and implementing sustainable practices in energy production. Some additional considerations for how the sector could become more circular across the respective loops are set out below:

Under the user-to-user loop:

- *Improving labelling of bioenergy products*: Bioenergy labelling must be more comprehensive and transparent for producers and consumers. This can be achieved by implementing clear standards and regulations for labelling and monitoring procedures. This will help ensure that consumers are fully informed about bioenergy products' origin, quality, and environmental impact.
- *Addressing non-economic objectives*: The bioenergy sector should focus on meeting non-economic objectives such as creating jobs, promoting environmental sustainability, and preserving recreational spaces. This can be achieved by improving the utilisation of residues from forestry and finding resource-efficient combinations of biomass sources, conversion technologies, and energy end-uses.

Under the user-to-business loop:

- *Addressing competition for biomass feedstocks*: The competition for biomass feedstocks is a significant challenge for the bioenergy sector. It is crucial to prioritise long-life material uses over short-term energy uses to ensure that bioenergy does not compete with other important uses of wood.
- *Improving forest conditions*: Bioenergy should be used to improve forest conditions and increase the sustainability and resilience of the forest landscape. This can be achieved by using bioenergy to increase the value of managed stands and promote SFM practices.
- *Improving ash management*: The bioenergy industry should ensure that ash is appropriately managed and utilised for recycling nutrients in the biosphere. This will help reduce the environmental impact of bioenergy production and promote sustainable resource management.

Under the business-to-business loop:

- *Reducing environmental impact*: Efforts should be made to reduce the environmental impact of biomass production. This can be achieved by sustainably growing biomass for energy on land that is underutilised or unused for other purposes.
- *Expanding the product mix*: The bioenergy sector should aim to expand its product mix and engage with downstream value chains to ensure the sustainable extraction of biomass. The unpredictable quality of raw materials and the high cost of transportation are significant challenges that need to be addressed.
- *Adopting cascading principles*: The bioenergy sector should adopt cascading principles, which prioritise the use of biomass for its most valuable function first. Bioenergy production should be restricted to end-of-life uses, primarily as an alternative to disposal, and efforts should be made to improve biomass partitioning. This will help maximise the value of each portion of biomass and promote more sustainable resource management practices. However, this also presumes that energy production and cascading use are integrated into the circular economy model.

4.2. The paper trail

The transition towards circularity by the pulp and paper industry (NACE 17.11, 17.12 and 17.2) is challenging due to its complexity (Figure 5.5). Despite the high recycling rate of the sector, there is still room for improvement in reducing fibre loss and the use of virgin resources in paper production. To achieve this goal, the value of recycled fibres must be preserved through partnerships with related industries, such as producers of inks, dyes, and glue. One way to achieve this is by co-designing additives that can be easily separated from paper during recycling. Additionally, using non-toxic by-products from other industrial processes to produce paper can significantly contribute to the sector's sustainability. However, improving coordination across the value and supply chains is necessary to make these changes. This requires collaboration and investment in recycling facilities that can handle new materials. With these efforts in place, the pulp and paper industry can take a significant step towards circularity, reducing its environmental impact while maintaining its market competitiveness.

The paper and pulp industry has traditionally been linear, relying on the 'take-make-use-dispose' model in which raw materials are extracted, processed into products, used, and disposed of. For example, the industry can become more circular by ensuring that raw materials are sustainably sourced, such as using responsibly managed forests. Another way to make the paper and pulp industry more circular is to adopt closed-loop production methods, where waste from one process is used as feedstock for another. This is, however, happening to a large extent already as the industry has high recycling rates; it has significantly reduced the amount of energy required to produce paper and reduced the use of water by implementing water treatment systems that recycle water within their operations. Additional considerations for how the sector could become more circular across the respective loops are set out below:

Under the user-to-user loop:

- *Enhance end-user recycling behaviours*: To increase recycling participation, it is crucial to understand the factors that influence end-user behaviour. For example, a lack of accurate information, consumer misconceptions, and a lack of incentives can all act as barriers to recycling. Researching end-user behaviours and attitudes towards recycling may be helpful to address these barriers. This information can then be used to develop targeted educational programs and incentives to encourage more people to recycle.
- *Strengthen recyclable demand*: Ensuring a strong demand for recyclables is essential for a successful recycling program. Improving the accuracy of information about recycling and addressing consumer concerns, such as mistrust or risk aversion, could help to build demand.
- *Certification*: Certification programs like the Forest Stewardship Council (FSC) are important in reducing deforestation and environmental impacts and protecting high conservation value forests. By promoting the use of certified products, consumers can support sustainable practices and help to build demand for recyclables.
- *Improve traceability*: Implementing traceability systems for existing and new materials, such as packaging, is critical to ensure that recyclables are correctly sorted, collected, and processed. This can help to ensure that the materials are recycled effectively and that the recycling process is transparent and accountable.

Under the user-to-business loop:

- *Durability*: Enhancing the durability of paper products, such as optimising their physical structure or improving the compatibility of inks, can also help extend their lifespan and make them easier to recycle.
- *Streamline paper recycling*: Reducing fibre losses and minimising the use of virgin resources in paper production is an important step in making paper recycling more efficient and effective.
- *Standardising fibre-based materials*: Ensure that new fibre-based materials can be effectively recycled; it is essential to standardise these materials across different value chains. This can help ensure that recycling plants are equipped to handle these materials and that they are correctly sorted and processed.

Under the business-to-business loop:

- *Cooperation*: Encouraging cross-industry cooperation, such as between the ink and glue industries, to co-design products that are easier to recycle can help to reduce waste and conserve resources. Additives that are easily separable from paper, for example, can make it easier to recycle paper products and increase their value as recycled fibre.
- *Reduce waste generation*: Reducing waste generation is crucial in promoting sustainable practices. Implementing strategies that increase recycling and improve resource efficiency, such as reducing water and energy consumption in the papermaking process, can help minimise waste and conserve resources.

- *Material consumption and efficiency*: Coupling increased recycling with improved materials and energy recovery can help reduce waste discharge to the environment and promote a circular economy.

4.3. From trees to textiles and plastics

In recent years, there has been a significant increase in the use of cellulose-based fibres in various industries, including textiles, packaging, and construction. The advancements in cellulose fibre technology have expanded the potential applications of wood-based materials, adding value to the paper and pulp sector and meeting the growing demand for environmentally friendly, recyclable, and sustainable fibres. Cellulose-based fibres offer several advantages over synthetic fibres, including biodegradability. Biodegradability is an important factor in assessing the environmental impact of fibres as it ensures that the fibres will decompose in the natural environment, reducing the amount of waste in landfills. Using cellulose-based fibres can also help reduce non-renewable resources, as they are derived from renewable sources, such as trees and plants. However, producing cellulose-based fibres, particularly viscose, presents serious environmental concerns. The production process involves numerous chemicals and generates waste, which can pose a risk to the environment. Additionally, transporting the raw materials and chemicals used in production can result in significant GHG emissions.

Bio-based plastics provide an alternative to traditional fossil-fuel-based plastics, being non-toxic, renewable, and biodegradable in most cases. These properties make them attractive for companies looking to reduce their environmental impact. However, it is important to note that bio-based plastics can be grouped into three categories: (1) bio-based and non-biodegradable, (2) bio-based and biodegradable, and (3) fossil-based and biodegradable. The first category of bio-based and non-biodegradable plastics is not a sustainable solution, as they will persist in the environment for a long time and contribute to the accumulation of plastic waste. The second category, bio-based and biodegradable plastics, can solve the problem of plastic waste, but only if they are appropriately managed and disposed of in a way that allows them to biodegrade. The third category, fossil-based and biodegradable plastics, while biodegradable, still contribute to the depletion of finite fossil fuels and GHG emissions.

Cellulose-based fibres and bio-based plastics can become more circular by reducing harmful chemicals and minimising waste. Additionally, using closed-loop systems and renewable energy sources can reduce the carbon footprint of the production process. However, properly disposing of cellulose-based fibres and bio-based plastics is critical to ensure they do not contribute to environmental degradation. Additional considerations for how the cellulose-based fibres and bio-based plastics could become more circular across the respective loops are set out below:

Recent innovations in the use of cellulose-based fibres (NACE 20.16) have expanded the potential use of wood-based materials, adding value to the forest-based sector and addressing the growing demand for recyclable, responsible, and ecologically sustainable fibres. For example, cellulose-based fibres may deliver better environmental benefits than synthetic fibres in terms of biodegradability. However, producing cellulose-based fibres, particularly viscose, requires many chemicals in a process that raises serious environmental concerns. The same arguments apply to bio-based plastics, which provide an alternative to fossil-based plastics and are, in most cases, non-toxic, renewable and biodegradable. Nevertheless, bio-based plastics do not imply that they are better for the environment.

Under the user-to-user loop:

- *Increase awareness*: Improve knowledge among end-users about the benefits of using wood-based textiles and plastics, including their recyclability and eco-friendliness. This can be achieved by improving access to information about these products.
- *Improve traceability*: Ensure that the textiles and plastics are traceable throughout production. This will increase transparency and help promote sustainable production and consumption patterns.
- *Standardisation and certification*: Ensure that the textiles and plastics are of high quality and environmentally friendly by implementing standardisation and eco-design certification. This will include certifying that the products are biodegradable and compostable.

Under the user-to-business loop:

- *Eco-design*: Encourage businesses to focus on the eco-design of multi-materials and maximise product recyclability. This includes considering the entire life cycle of the textiles and plastics, including their disposal and environmental impact.
- *Recycling schemes*: Set up recycling schemes and improve recycling technologies to address the complex waste streams generated by textiles and plastics. The goal is to make the recycling process environmentally friendly, non-hazardous and efficient.
- Pollution *prevention*: Address the pollution caused by textile production, particularly the treatment and dyeing process. Producing cellulose fibres also requires toxic and explosive chemicals, which raises serious environmental concerns.

Under the business-to-business loop:

- *Collaboration across supply chains*: Foster collaboration among businesses in the supply chain to promote sustainability. This could include integrating the production of wood-based textiles with solid wood products or developing networks for waste management across the value chain.
- *Competition for raw materials*: Address the competition for raw materials in producing cellulose fibres. The impact of wood-based textile/plastics production on other supply chains should be considered, and opportunities for synergies should be explored.
- *Expand resource base*: Expand the resource base for producing cellulose fibres to reduce dependence on a limited number of raw materials.
- *Improve resource efficiency*: Reduce the amount of raw materials needed for textile and plastic production to reduce costs, materials, and energy. This will improve resource efficiency and reduce the environmental impact of production.

5. Implementing circular practices in the forest-based sector

Pursuing circularity in the forest-based sector is a complex and multi-faceted challenge, with varying limitations, challenges, and opportunities across different value chains and sectors. This chapter analysed two value chains (Figure 5.4 and 5.5) to explore the potential for circularity in the forest-based sector, encompassing five sectors: sawnwood, bioenergy, construction, pulp, paper, and cellulose manufacturing. The analysis highlights the diversity and intricacy of the forest-based ecosystem, with a range of circular approaches that can be adopted depending on the specific needs of each value chain and sector. For instance, the use of resources, energy, and waste management practices in woodworking and pulp and paper production varies significantly, demonstrating the need for tailored solutions. The chapter also emphasises that no universal approach can be applied to the forest-based sector. Instead, different value chain dependencies must be considered when developing strategies for retaining value within systems. By carefully considering the unique characteristics of each sector and value chain, the forest-based industry can work towards a more sustainable and circular future.

The forest-based industry has a significant challenge in transitioning towards a circular economy. Despite efforts to apply cascading use principles, there is a high degree of 'lock-in' or path dependency, meaning that established practices and investments often hinder the shift towards circularity. This is evident in the common underutilisation of sawmill residues, which are often turned into low-value products or used for energy production instead of being fully utilised. Unfortunately, investments in the industry continue to focus mainly on improving efficiency rather than promoting circularity. As a result, the value chains within the industry remain linear (Jarre et al., 2020) instead of moving towards a more circular model (D'Amato et al., 2020, Henrysson and Nuur, 2021).

The forest-based industry is also facing other challenges. For example, even though wood is a renewable resource, there is a common misconception that this automatically makes the sector circular (Hetemäki et al., 2020, Toppinen et al., 2020, Winkel, 2017). This misunderstanding not only overlooks the need for further innovation to overcome critical barriers to circularity, such as the degree of recyclability, but also restricts the spread of awareness, acceptance, and adoption of circular models (D'Amato et al., 2020). This highlights the need for a change in perspectives and mindsets towards wood, which is crucial for the industry to become truly circular on both the supply and demand sides. Without this change, the challenges faced by the industry will not improve. In addition to the misconceptions surrounding circularity, the irregular supply of post-consumer wood and its varying quality make it a low-value product. Furthermore, when considering the costs associated with transportation and the environmental sustainability of reintroducing wood into value chains, it becomes clear that wood is often not competitive in commercial operations. To overcome these challenges, a more comprehensive understanding of what circularity means for the wood-based industry is needed. This includes a greater emphasis on innovation, awareness, and adoption of circular models, as well as a shift in attitudes and mindsets towards wood. The wood-based industry can achieve a more sustainable and circular future by addressing these issues.

Another significant challenge is the potential for negative externalities, such as emissions generated by transportation in the wood supply chain. The transportation sector is one of the most significant contributors to GHG emissions (Kühmaier et al., 2022), making it a crucial area to focus on to minimise the negative impact of circular practices. A range of technical and organisational solutions are required to address these challenges. Eco-design and eco-innovation can help reduce the industry's environmental footprint while extending producer responsibility. Investing in collection infrastructure and sorting technologies can improve the efficiency of circular practices. Additionally, the geographical proximity of production facilities to waste stream users can reduce emissions from transportation and increase the sustainability of the supply chain. In conclusion, implementing circular practices in the wood-based industry requires a comprehensive approach that considers various challenges and opportunities. The industry can move towards a more circular future while ensuring economic viability by addressing the potential for negative externalities and working towards sustainable and environmentally friendly practices.

Pursuing innovation in the forest-based industry is crucial for the transition to a circular bioeconomy; however, it is essential to understand that not all innovations are necessarily sustainable or eco-friendly (Horbach et al., 2013). To be considered eco-innovations, these developments must lower the environmental hazards, pollution, and emissions produced throughout their life cycle (Horbach et al., 2013). This can be achieved by creating novel products, services, procedures, management systems, organisational structures, or business models specifically designed with sustainability in mind. Moreover, it is crucial to be aware of the potential unintended consequences of these innovations. For example, deforestation in other countries, or the indirect land use change that may result from such activities, can conflict with sustainable development goals (Ronzon and Sanjuán, 2020). These negative impacts are often called spillovers and must be carefully considered and addressed. It is, therefore, vital to ensure that the transition to a circular bioeconomy is innovative, sustainable, and environmentally responsible. This requires a comprehensive approach considering spillovers and challenges across sectors and supply chains and a commitment to responsible and sustainable practices. Further research and analysis should be conducted to identify these potential impacts and find solutions to mitigate them.

5.1. Navigating limitations and identifying opportunities

From a research perspective, it is important to note that applying value retention loops (Figure 5.3) was challenging. Specifically, it was sometimes difficult to categorise issues within the assigned loops. In future studies, a more systematic approach could be beneficial, for example, by integrating cross-cutting issues more comprehensively (Potting et al., 2017, UNEP, 2018). One solution is to simplify the loops by having only two covering users and consumers and one covering producers and enterprises. Additionally, including a cross-cutting dimension, as was done in this paper, could help avoid overlaps and provide a more holistic view of the challenges and opportunities in the industry. By taking these steps, future research and practice in the field could be improved and better inform the transition to a circular bioeconomy.

Another aspect worth considering is the general nature of the results in this chapter. Although the approach provides valuable insights into differences across various value chains, future research could benefit from a more in-depth examination of specific case studies. Future work should consider utilising more complex value chain models based on real-world examples to enhance the comparability of results and allow for a more thorough analysis. Additionally, by limiting the number of sectors studied, researchers can delve deeper into the complexities of each case and provide more comprehensive comparisons and analyses. By taking these steps, future research in the field can provide more nuanced insights into the challenges and opportunities of implementing a circular bioeconomy in the forest-based sector.

The results presented in this chapter offer comprehensive insights into the potential for circularity within the various forest-based sectors. The findings highlight that each sector faces unique challenges and opportunities regarding circular efficiency and significant variations in the solutions required to achieve circularity (Kirchherr et al., 2017, Nobre and Tavares, 2021). This underscores the importance of tailoring governance solutions at the sector and enterprise level. For example, upstream actors must work more closely with downstream actors to better coordinate decisions and activities within the value chain. In addition, incentives must be put in place to encourage circular practices, and awareness and knowledge about circular principles must be improved across the sector to ensure that circular solutions are understood and applied effectively. Overall, the results highlight the need for a systemic change within the forest-based sector to move towards a circular bioeconomy. This requires coordination and cooperation across the supply chain and a commitment to circular principles and practices.

Additionally, end-of-life valorisation, which aims to minimise the waste generated by forest-based industries, will play a crucial role in their journey towards circularity (den Hollander et al., 2017, MahmoumGonbadi et al., 2021, Diaz et al., 2022). This requires producers to implement effective resource management strategies that reduce resource consumption and increase the value of waste and by-products. One promising solution to achieve this is through the development of industrial symbioses. These symbiotic systems allow the different actors in the supply chain to collaborate and optimise the use of resources, leading to a more efficient and circular process. Biorefineries, for instance, have been identified as a promising option to implement large-scale industrial symbioses in the forest-based sector, thus making a significant contribution to the development of a circular bioeconomy (Arâmburu et al., 2022, Cervantes et al., 2020).

In conclusion, it is imperative to acknowledge the limitations of the circular bioeconomy concept when applied to the forest-based sector. While circular efficiency can improve material utilisation and reduce waste, it is limited by the natural regenerative capacity of forests. This means that SFM is crucial to ensure the long-term availability of wood as a raw material (Holmatov et al., 2019). Additionally, it is crucial to address the need for more consensus and clarity around the definition of circularity in the context of the forest-based sector. The sector needs to have a shared understanding of what it means to be circular and the implications for SFM, raw material sustainability, and the technical cycle of manufacturing and use of wood. To achieve a circular bioeconomy in the forest-based sector, addressing these challenges and filling the knowledge gaps is crucial. This will require a clear and shared conceptual understanding of circularity and its requirements for the sector and a concerted effort to implement SFM practices and promote the use of end-of-life valorisation options such as biorefineries.

6. The complexities of a circular transformation

This chapter has delved into the complexities of implementing circularity in forest-based industries, wood being a crucial raw material in the bioeconomy. Implementing circularity presents a unique challenge due to the nature of the material itself, which deteriorates over time and cannot be recycled in the same way as other non-bio-based materials like glass and metal. While the analysis presented in this chapter has been limited in scope, focusing only on two value chains, it demonstrates that implementing circularity is a complex process that requires tailored approaches across value chains. To achieve circularity, a combination of macro and micro solutions is necessary. This includes cross-cutting solutions that need to be addressed at a policy level, as well as enterprise-level solutions tailored to each sector's specific needs. Implementing circularity also requires adopting new technologies, innovative business models, increased connectivity and collaboration across sectors, and new management tools. In addition to these technical solutions, increasing awareness and understanding of circularity is essential, both on the supply and demand side. This requires increased training and education on circularity and the various implementation approaches.

The findings of this chapter highlight the complex and multi-faceted nature of implementing circularity in the forest-based sector. Despite its potential benefits, circular bioeconomy practices require a nuanced approach, considering the specific characteristics of wood as a raw material and its social and environmental implications. One of the critical challenges the sector faces is clarifying the relationship between circularity and sustainability (Kirchherr et al., 2017). The notion that circularity automatically equates to sustainability is misleading, as a purely circular approach to wood-based materials may lead to increased emissions or other negative externalities (Haupt and Hellweg, 2019, Schroeder et al., 2019, Mair and Stern, 2017). Moreover, the forest-based sector will need to account for the limitations of wood as a resource and understand that sustainability criteria must be integrated into the design of circular systems. There is no one-size-fits-all solution to achieving circularity in the forest-based sector, as the approach will need to be tailored to each value chain's unique characteristics and requirements.

Finally, the field of circular bioeconomy requires more practical tools and methods for analysing circularity. The limitations of the approach taken in this chapter, characterised by using three value retention loops (Potting et al., 2017), highlight the need for further development of analytical approaches. While it is crucial to make distinctions between different value chains, such as user-to-user, user-to-business, and business-to-business, it is crucial to work closely with the industry to develop frameworks that can measure circularity effectively. Policymakers need the necessary information to optimise measures, while enterprises need practical tools to assess their practices and find the best solutions. The results underscore the need for a holistic and integrated approach to implementing circularity in the forest-based sector. Therefore, it is imperative to bring together the scientific community, policymakers, and industry to collaborate in developing effective methods for analysing circularity in the wood-based sector.

Want to find out more?

For those seeking more general knowledge on the circular bioeconomy, please see the references provided below or some of the many references throughout the text.

- Circularity Gap Report (CGRI, 2023).
- Squaring the Circle: Policies from Europe's Circular Economy Transition (WB, 2022).
- Exploring new visions for a sustainable bioeconomy (Giuntoli et al., 2023).
- Towards a sustainable European forest-based bioeconomy- assessment and the way forward (Winkel, 2017).
- Circularity concepts in forest-based industries (Aggestam et al., 2021).

The set of questions listed below may also help guide further inquiry into specific areas of interest:

1. How can the forest-based sector contribute to creating a sustainable system within the circular bioeconomy?
2. What is the relationship between circularity and sustainability in the forest-based sector?
3. What are the key strategies of the 9Rs model, and how do they apply to resource use in the forest-based sector?
4. What are the key challenges in implementing circular bioeconomy practices in the forest-based sector?
5. How can circularity in the forest-based sector be achieved by tailoring it to meet the unique characteristics and requirements of different value chains?

7. Assignment V: Designing circular bioeconomy solutions

Description

Based on Chapter 5, this assignment is meant to help the student understand and apply the principles of a circular bioeconomy within the forest-based sector, identify challenges and opportunities, and propose innovative solutions to develop their research skills.

Purpose

Students will be required to select a forest-based sector, such as sawnwood, bioenergy or construction, and analyse its value chain and current practices in the context of the circular bioeconomy. They should identify specific challenges and opportunities to implement circularity and consider innovative solutions, considering technological, socio-economic, and environmental barriers or enabling factors.

Outline of task

- *Sector selection and research*: Choose a forest-based sector and research its role in the circular bioeconomy, considering current practices, challenges, and opportunities for circularity.
- *Identify challenges and opportunities*: Outline a generic value chain to identify this sector's main challenges and opportunities in transitioning towards a circular model.
- *Develop solutions*: Propose solutions to address the identified challenges and opportunities, describing how these solutions can be implemented, their expected outcomes regarding circularity, and any potential barriers to adoption or uptake.
- *Presentation*: Prepare a presentation or report outlining your findings and proposals, including a critical analysis of how your solutions contribute to the circular bioeconomy.

Suggested additional reading

- Aggestam, F., Kacprzak, A. & Wollnik, R. (2021) *Circularity Concepts in Forest-Based Industries*. Geneva: United Nations Economic Commission for Europe.
- Widmark, C., Persichina, M. & Martinez-Cruz, A. L. (2022) The bioeconomy, circularity, and sustainability – How the concepts are conceptualised in the forestry sector. CERE Working Paper, 2022:03.
- Hassegawa, M., Van Brusselen, J., Cramm, M., & Verkerk, P. J. (2022) Wood-Based Products in the Circular Bioeconomy: Status and Opportunities towards Environmental Sustainability. *Land*, 11(12): 2131.

Deliverable

- Prepare a concise report (1000–1500 words) or a 15-minute presentation (15–20 slides) outlining the sectoral analysis (including a generic value chain), identified challenges and opportunities, and proposed solutions for the development of a circular bioeconomy.

8. Notes

1 The 2023 Circularity Gap Report highlights that the global economy extracted and used more over the last six years than in the twentieth century.

2 The forest-based sector refers to any industry whose primary raw material is wood or wood-based products, such as round or sawnwood, softwood, pulpwood, plywood, veneers, wood chips, and pulp. This sector encompasses various activities, including forestry, logging, sawmilling, wood processing, and paper production.

3 See https://ellenmacarthurfoundation.org (accessed 2 February 2023).

4 Alliance for Beverage Cartons and the Environment (ACE); European Biomass Association (AEBIOM); The European Confederation of woodworking industries (CEI-Bois); Confederation of European Paper Industries (CEPI); International Confederation of Paper and Board Converters in Europe (CITPA); European Furniture Industries' Confederation (EFIC); European Organisation of the Sawmill Industry (EOS); European Panel Federation (EPF); European Federation of Wooden Pallet and Packaging Manufacturers (FEFPEB); International confederation for printing and allied Industries (INTERGRAF); and European Furniture Manufacturer's Federation (UEA).

9. References

AGGESTAM, F. & GIURCA, A. 2022. Implementing circular-bioeconomy principles across two value chains of the wood-based sector: A conceptual approach. *Land* [Online], 11.

AGGESTAM, F., KACPRZAK, A. & WOLLNIK, R. 2021. Circularity concepts in forest-based industries. *In:* ECE (ed.). Geneva: United Nations Economic Commission for Europe.

ARÂMBURU, A., LUNKES, N., DE CADEMARTORI, P. H. G., GATTO, D. A., MISSIO, A. L. & DELUCIS, R. A. 2022. Forestry wastes: Technical concepts, economic circularity, and sustainability approaches. *In:* JACOB-LOPES, E., QUEIROZ ZEPKA, L. & COSTA DEPRÁ, M. (eds.) *Handbook of Waste Biorefinery: Circular Economy of Renewable Energy.* Cham: Springer International Publishing.

BESSERER, A., TROILO, S., GIRODS, P., ROGAUME, Y. & BROSSE, N. 2021. Cascading recycling of wood waste: A review. *Polymers*, 13, 1752.

BOULDING, E. K. 1966. The economics of the coming spaceship Earth. *In:* JARRETT, H. (ed.) *Environmental Quality in a Growing Economy.* Baltimore: Johns Hopkins University Press.

BRAUNGART, M. & MCDONOUGH, W. 2002. *Cradle to Cradle: Remaking the Way We Make Things.* North Point Press.

BRUNDTLAND COMMISSION 1987. *Our Common Future.* World Commission on Environment and Development.

CALISTO FRIANT, M., VERMEULEN, W. J. V. & SALOMONE, R. 2020. A typology of circular economy discourses: Navigating the diverse visions of a contested paradigm. *Resources, Conservation and Recycling*, 161, 104917.

CARUS, M. & DAMMER, L. 2018. The circular bioeconomy – Concepts, opportunities, and limitations. *Industrial Biotechnology*, 14.

CATTON, W. R. 1980. *Overshoot: The Ecological Basis of Revolutionary Change*, University of Illinois Press.

CERVANTES, G., TORRES, L. G. & ORTEGA, M. 2020. Valorization of agricultural wastes and biorefineries: A way of heading to circular economy. *In:* SALOMONE, R., CECCHIN, A., DEUTZ, P., RAGGI, A. & CUTAIA, L. (eds.) *Industrial Symbiosis for the Circular Economy: Operational Experiences, Best Practices and Obstacles to a Collaborative Business Approach.* Cham: Springer International Publishing.

CGRI 2022. Circularity Gap Report 2022. Davos: Circularity Gap Reporting Initiative.

CGRI 2023. Circularity Gap Report 2023. Davos: Circularity Gap Reporting Initiative.

D'AMATO, D., DROSTE, N., ALLEN, B., KETTUNEN, M., LÄHTINEN, K., KORHONEN, J., LESKINEN, P., MATTHIES, B. D. & TOPPINEN, A. 2017. Green, circular, bio economy: A comparative analysis of sustainability avenues. *Journal of Cleaner Production*, 168, 716–734.

D'AMATO, D., VEIJONAHO, S. & TOPPINEN, A. 2020. Towards sustainability? Forest-based circular bioeconomy business models in Finnish SMEs. *Forest Policy and Economics*, 110, 101848.

DEN HOLLANDER, M. C., BAKKER, C. A. & HULTINK, E. J. 2017. Product design in a circular economy: Development of a typology of key concepts and terms. *Journal of Industrial Ecology*, 21, 517–525.

DIAZ, A., REYES, T. & BAUMGARTNER, R. J. 2022. Implementing circular economy strategies during product development. *Resources, Conservation and Recycling*, 184, 106344.

DIRECTIVE 1893/2006. Establishing the statistical classification of economic activities NACE Revision 2. *L 393/1.* Brussels: Official Journal of the European Union.

EC 2015. Closing the loop – An EU action plan for the Circular Economy. *In:* EUROPEAN COMMISSION (ed.) *COM(2015) 614.* Brussels.

EC 2016. CASCADES: Study on the optimised cascading use of wood. *In:* VIS, M., MANTAU, U. & ALLEN, B. (eds.) *No 394/PP/ENT/RCH/14/7689. Final report.* Brussels: European Commission.

EC 2020. A new Circular Economy Action Plan: For a cleaner and more competitive Europe. *COM(2020) 98 final.* Brussels: European Commission.

ECKHART, R. 2021. Recyclability of Cartonboard and Cartons – Translation of study in 'Wochenblatt für Papierfabrikation 11/2021'. Graz: Technische Universität Graz.

EMF 2012. Towards the Circular Economy: Economic and business rationale for an accelerated transition. www.ellenmacarthurfoundation.org: Ellen MacArthur Foundation.

EMF 2015. Towards a Circular Economy: Business rationale for an accelerated transition. www.ellenmacarthurfoundation.org: Ellen MacArthur Foundation.

EMF 2020. Definitions List. Cowes, United Kingdom: Ellen MacArthur Foundation.

EUROSTAT 2008. *NACE Rev. 2 – Statistical classification of economic activities in the European Community.* Luxembourg: Office for Official Publications of the European Communities.

EUROSTAT 2017. NACE Rev. 2 – Structure, Explanatory Notes and Caselaw. Brussels: Eurostat, Unit B5 Central Data and Metadata Services.

EUROSTAT 2019. PRODCOM List 2019. Brussels: Eurostat.

FAO 2007. Guidelines for Rapid Appraisals of Agrifood Chain Performance in Developing Countries. *In:* DA SILVA, A. C. & DE SOUZA FILHO, M. H. (eds.) *Agricultural Management, Marketing and Finance Occasional Paper 20.* Rome, Italy: Food and Agriculture Organization of the United Nations.

FAO 2013. Value chain analysis for policy making: Methodological guidelines and country cases for a quantitative approach. *In:* BELLÙ, L. G. (ed.) *EASYPol Series 129.* Rome: Food and Agriculture Organization of the United Nations.

FEARNE, A., GARCIA MARTINEZ, M. & DENT, B. 2012. Dimensions of sustainable value chains: implications for value chain analysis. *Supply Chain Management: An International Journal*, 17, 575–581.

FULLERTON, J. 2015. Regenerative capitalism. Greenwich: Capital Institute.

GEISSDOERFER, M., SAVAGET, P., BOCKEN, N. M. P. & HULTINK, E. J. 2017. The Circular Economy – A new sustainability paradigm? *Journal of Cleaner Production*, 143, 757–768.

GEORGESCU-ROEGEN, N. 1975. Bio-economics aspects of entropy. *In:* KUBAT, L. & ZEMAN, J. (eds.) *Entropy and Information in Science and Philosophy.* Amsterdam.

GHISELLINI, P., CIALANI, C. & ULGIATI, S. 2016. A review on circular economy: the expected transition to a balanced interplay of environmental and economic systems. *Journal of Cleaner Production*, 114, 11–32.

GIAMPIETRO, M. 2019. On the circular bioeconomy and decoupling: Implications for sustainable growth. *Ecological Economics*, 162, 143–156.

GIUNTOLI, J., BARREDO, J. I., AVITABILE, V., CAMIA, A., CAZZANIGA, N. E., GRASSI, G., JASINEVIČIUS, G., JONSSON, R., MARELLI, L., ROBERT, N., AGOSTINI, A. & MUBAREKA, S. 2022. The quest for sustainable forest bioenergy: win–win solutions for climate and biodiversity. *Renewable and Sustainable Energy Reviews*, 159, 112180.

GIUNTOLI, J., OLIVER, T., KALLIS, G., RAMCILOVIK-SUOMINEN, S. & MONBIOT, G. 2023. Exploring new visions for a sustainable bioeconomy. *In:* GIUNTOLI, J. & MUBAREKA, S. (eds.). Luxembourg: Publications Office of the European Union.

HASSEGAWA, M., KARLBERG, A., HERTZBERG, M. & VERKERK, H. 2022. Innovative forest products in the circular bioeconomy. 2, 19.

HAUPT, M. & HELLWEG, S. 2019. Measuring the environmental sustainability of a circular economy. *Environmental and Sustainability Indicators*, 1–2, 100005.

HENRYSSON, M. & NUUR, C. 2021. The role of institutions in creating circular economy pathways for regional development. *The Journal of Environment & Development*, 30, 149–171.

HETEMÄKI, L. 2014. Future of the European forest-based sector – Structural changes towards bioeconomy. *What Science Can Tell Us No. 6.* EFI.

HETEMÄKI, L., HANEWINKEL, M., MUYS, B., OLLIKAINEN, M., PALAHÍ, M. & TRASOBARES, A. 2017. Leading the way to a European circular bioeconomy strategy. *From Science to Policy 5.* Joensuu: European Forest Institute.

HETEMÄKI, L., PALAHÍ, M. & NASI, R. 2020. Seeing the wood in the forests. *Knowledge to Action 01.* Joensuu: European Forest Institute.

HOLMATOV, B., HOEKSTRA, A. Y. & KROL, M. S. 2019. Land, water and carbon footprints of circular bioenergy production systems. *Renewable and Sustainable Energy Reviews*, 111, 224–235.

HOLVOET, B. & MUYS, B. 2004. Sustainable forest management worldwide: a comparative assessment of standards. *The International Forestry Review*, 6, 99–122.

HORBACH, J., OLTRA, V. & BELIN, J. 2013. Determinants and specificities of eco-innovations compared to other innovations – An econometric analysis for the French and German industry based on the community innovation survey. *Industry and Innovation*, 20, 523–543.

HURMEKOSKI, E., JONSSON, R., KORHONEN, J., JÄNIS, J., MÄKINEN, M., LESKINEN, P. & HETEMÄKI, L. 2018. Diversification of the forest industries: role of new wood-based products. *Canadian Journal of Forest Research*, 48, 1417–1432.

HURMEKOSKI, E., LOVRIĆ, M., LOVRIĆ, N., HETEMÄKI, L. & WINKEL, G. 2019. Frontiers of the forest-based bioeconomy – A European Delphi study. *Forest Policy and Economics*, 102, 86–99.

JANKOVSKÝ, M., GARCÍA-JÁCOME, S. P., DVOŘÁK, J., NYARKO, I. & HÁJEK, M. 2021. Innovations in forest bioeconomy: A bibliometric analysis. *Forests* [Online], 12.

JARRE, M., PETIT-BOIX, A., PRIEFER, C., MEYER, R. & LEIPOLD, S. 2020. Transforming the bio-based sector towards a circular economy – What can we learn from wood cascading? *Forest Policy and Economics*, 110, 101872.

KAMPELMANN, S. 2020. Wood works: how local value chains based on urban forests contribute to place-based circular economy. *Urban Geography*, 41, 911–914.

KAPUR, A. & GRAEDEL, T. E. 2004. Industrial ecology. *In:* CLEVELAND, C. J. (ed.) *Encyclopedia of Energy.* New York: Elsevier.

KHAW-NGERN, K., PEUCHTHONGLANG, P., KLOMKUL, L. & KHAW-NGERN, C. 2021. The 9Rs strategies for the circular economy 3.0. *Psychology and Education*, 58, 1440–1446.

KIRCHHERR, J., REIKE, D. & HEKKERT, M. 2017. Conceptualizing the circular economy: An analysis of 114 definitions. *Resources, Conservation and Recycling*, 127, 221–232.

KORHONEN, J., HONKASALO, A. & SEPPÄLÄ, J. 2018. Circular economy: The concept and its limitations. *Ecological Economics*, 143, 37–46.

KÜHMAIER, M., KRAL, I. & KANZIAN, C. 2022. Greenhouse gas emissions of the forest supply chain in Austria in the year 2018. *Sustainability*, 14.

LATOUCHE, S. 2009. *Farewell to growth*, Polity.

LAZDINIS, M., ANGELSTAM, P. & PÜLZL, H. 2019. Towards sustainable forest management in the European Union through polycentric forest governance and an integrated landscape approach. *Landscape Ecology*, 34, 1737–1749.

LIEDER, M. & RASHID, A. 2016. Towards circular economy implementation: a comprehensive review in context of manufacturing industry. *Journal of Cleaner Production*, 115, 36–51.

LIER, M., KÖHL, M., KORHONEN, K. T., LINSER, S. & PRINS, K. 2021. Forest relevant targets in EU policy instruments – can progress be measured by the pan-European criteria and indicators for sustainable forest management? *Forest Policy and Economics*, 128, 102481.

LINSER, S., WOLFSLEHNER, B., BRIDGE, S. R. J., GRITTEN, D., JOHNSON, S., PAYN, T., PRINS, K., RAŠI, R. & ROBERTSON, G. 2018. 25 years of criteria and indicators for sustainable forest management: How intergovernmental C&I processes have made a difference. *Forests*, 9, 578.

MAHMOUMGONBADI, A., GENOVESE, A. & SGALAMBRO, A. 2021. Closed-loop supply chain design for the transition towards a circular economy: A systematic literature review of methods, applications and current gaps. *Journal of Cleaner Production*, 323, 129101.

MAIR, C. & STERN, T. 2017. Cascading utilization of wood: A matter of circular economy? *Current Forestry Reports*, 3, 281–295.

MAZUR-WIERZBICKA, E. 2021. Circular economy: advancement of European Union countries. *Environmental Sciences Europe*, 33, 111.

MCCORMICK, K. & KAUTTO, N. 2013. The bioeconomy in Europe: An overview. *Sustainability* [Online], 5.

MCPFE 2005. *Sustainable Forest Management and Ecosystem Approach.* Ministerial Conference on the Protection of Forests in Europe Laison Unit Warsaw, Outcome of the MCPFE and EfE/PEBLDS ad hoc Working Group on Development of the Pan-European Understanding of the Linkages of Ecosystem Approach and Sustainable Forest Management. Session held 19–21 April, 2005, in Krakow, Poland.

MEADOWS, D. H., MEADOWS, D. L., RANDERS, J. & BEHRENS III, W. W. 1972. *Limits to Growth.* Universe Books.

NOBRE, G. C. & TAVARES, E. 2021. The quest for a circular economy final definition: A scientific perspective. *Journal of Cleaner Production*, 314, 127973.

PEARCE, D. W. & TURNER, R. K. 1990. *Economics of Natural Resources and the Environment.* Baltimore, MD: Johns Hopkins University Press.

POTTING, J., HEKKERT, M., WORRELL, E. & HANEMAAIJER, A. 2017. Circular economy: Measuring innovation in the product chain. www.pbl.nl/sites/default/files/cms/publicaties/pbl-2016-circular-economy-measuring-innovation-in-product-chains-2544.pdf (accessed 20 November 2020).

REIKE, D., VERMEULEN, W. J. V. & WITJES, S. 2018. The circular economy: New or refurbished as CE 3.0? Exploring controversies in the conceptualization of the circular economy through a focus on history and resource value retention options. *Resources, Conservation and Recycling*, 135, 246–264.

RIFKIN, J. 2011. *The Third Industrial Revolution: How Lateral Power is Transforming Energy, the Economy, and the World*. Basingstoke: Macmillan.

RITZÉN, S. & SANDSTRÖM, G. Ö. 2017. Barriers to the circular economy – Integration of perspectives and domains. *Procedia CIRP*, 64, 7–12.

RIVERA LEÓN, L., BOUGAS, K., AGGESTAM, F., PÜLZL, H., ZOBOLI, E., RAVET, J., GRINIECE, E., VERMEER, J., MAROULIS, N., ETTWEIN, F., VAN BRUSSELEN J. & GREEN, T. 2016. An assessment of the cumulative cost impact of specified EU legislation and policies on the EU forest-based industries. Brussels: DG GROW.

RONZON, T. & SANJUÁN, A. I. 2020. Friends or foes? A compatibility assessment of bioeconomy-related Sustainable Development Goals for European policy coherence. *Journal of Cleaner Production*, 254, 119832.

SCHROEDER, P., ANGGRAENI, K. & WEBER, U. 2019. The relevance of circular economy practices to the Sustainable Development Goals. *Journal of Industrial Ecology*, 23, 77–95.

SFC 2015. Sustainable Forest Management: Criteria and Indicators *Final Report.* Brussels: Standing Forestry Committee Ad Hoc Working Group on Sustainable Forest Management.

SHABANI, N., AKHTARI, S. & SOWLATI, T. 2013. Value chain optimization of forest biomass for bioenergy production: A review. *Renewable and Sustainable Energy Reviews*, 23, 299–311.

SHERWOOD, J. 2020. The significance of biomass in a circular economy. *Bioresource Technology*, 300, 122755.

SOTIROV, M., HOOGSTRA-KLEIN, M., TRUBINS, R., SCHÜLL, E., RIEMER, A., ORAZIO, C., CORDERO, R., SALNÄSS, O. & ERIKSSON, O. 2015. Shaping the future of sustainable forest management. *EFI News*, 23, 10–11.

SPATHELF, P. 2009. *Sustainable Forest Management in a Changing World: a European Perspective.* Dordrecht, Springer.

STAHEL, W. R. & CLIFT, R. 2016. Stocks and flows in the performance economy. *In:* CLIFT, R. & DRUCKMAN, A. (eds.) *Taking Stock of Industrial Ecology.* Cham: Springer International Publishing.

STEGMANN, P., LONDO, M. & JUNGINGER, M. 2020. The circular bioeconomy: Its elements and role in European bioeconomy clusters. *Resources, Conservation & Recycling: X*, 6, 100029.

TOPPINEN, A., D'AMATO, D. & STERN, T. 2020. Forest-based circular bioeconomy: matching sustainability challenges and novel business opportunities? *Forest Policy and Economics*, 110, 102041.

TRAINER, T. & ALEXANDER, S. 2019. The simpler way: envisioning a sustainable society in an age of limits. *Real-World Economics Review*, 87, 247–260.

UBA 2017. BIOMASS CASCADES. Increasing resource efficiency by cascading use of biomass – from theory to practice. Summary. *TEXTE 53/2017.* Dessau-Roßlau: Federal Ministry for the Environment, Nature Conservation, Building and Nuclear Safety.

UN 1992. Rio Declaration on Environment and Development. *A/CONF.151/26.* Rio de Janeiro: United Nations.

UN 2015. Transforming our world: the 2030 Agenda for Sustainable Development *A/RES/70/1.* New York: United Nations.

UNEP 2018. Re-defining value: The manufacturing revolution-remanufacturing, refurbishment, repair and direct reuse in the circular economy. *In:* NASR, N., RUSSELL, J., BRINGEZU, S., HELLWEG, S., HILTON, B., KREISS, C. & VON GRIES, N. (eds.) *International Resource Panel Reports.* Nairobi, Kenya: United Nations Environment Programme.

VIS M., MANTAU, U. & ALLEN, B. 2016. Study on the optimised cascading use of wood. Final report. *No 394/PP/ENT/RCH/14/7689.* Brussels: European Commission.

VIVIEN, F. D., NIEDDU, M., BEFORT, N., DEBREF, R. & GIAMPIETRO, M. 2019. The hijacking of the bioeconomy. *Ecological Economics*, 159, 189–197.

WB 2022. *Squaring the Circle: Policies from Europe's Circular Economy Transition.* Washington: World Bank.

WIEDMANN, T. O., SCHANDL, H., LENZEN, M., MORAN, D., SUH, S., WEST, J. & KANEMOTO, K. 2015. The material footprint of nations. *Proceedings of the National Academy of Sciences*, 112, 6271–6276.

WINKEL, G. 2017. Towards a sustainable European forest-based bioeconomy- assessment and the way forward. *What Science Can Tell Us 8.* Joensuu: European Forest Institute.

WWF 2022. Living Planet Report 2022: Building a nature positive society. *In:* ALMOND, R. E. A., GROOTEN, M., JUFFE BIGNOLI, D. & PETERSEN, T. (eds.). Gland: World Wildlife Fund.

6 Forest ownership and the provision of forest ecosystem services

1. Introduction

Forests act as keystones not only for environmental conservation but also for the economy as well as societal and public well-being. Within the complex web of factors affecting forests and their management, forest ownership emerges as a crucial consideration, influencing forest resource conservation, management, and use. This chapter digs into some of the complexities surrounding forest ownership and its impact on providing forest ecosystem services (FES), attempting to shed light on the interactions between people and forests.

Understanding forest ownership goes beyond who holds the deed to a tract of forest land. It covers a spectrum of rights and responsibilities shaped by different legal frameworks, policy instruments, societal norms, and environmental narratives. This spectrum is apparent across Europe, home to roughly 16 million forest owners, both private and public (Mendes et al., 2006, ECE/FAO, 2010, EFI, 2013, Sarvašová et al., 2015, ECE, 2020). Within the European Union (EU), the forest land is divided into approximately 60% private ownership and 40% public, the latter encompassing forests commonly managed by the state (ECE/FAO, 2010, ECE, 2020, FOREST EUROPE, 2015, 2020). Private forest owners come in different forms, from traditional to urban to industrial owners. Some ownership types blur traditional boundaries, like community-owned forests, reflecting historical communal use or social movements (Hogl et al., 2005, Blanco et al., 2015, Khanal et al., 2017, Ficko et al., 2017). Property sizes also vary significantly, with nearly 90% of private forest lands covering less than 10 hectares (ha) and a significant portion even smaller, especially in Eastern and South-Eastern Europe. Large forest areas are commonly held by public entities, religious organisations, and industrial enterprises (EFI, 2013, Weiss et al., 2019, Ficko, 2019).

This ownership diversity reflects Europe's historical, legal, and social diversity, leading to significant regional and national variations (Quiroga et al., 2019). For instance, around 70% of forests in Northern Europe are privately owned, while in South-Eastern Europe, public ownership can be as high as 90%. The rights associated with these forests also vary significantly, with Western Europe generally allowing more freedom in forest management compared to more stringent regulations in Eastern Europe (Nichiforel et al., 2020). Whether through the lens of a private citizen in Sweden, exercising their rights within the bounds of the 'allemansrätten' (everyman's right), or a community in Bulgaria, collectively managing their ancestral woodlands, the nuances surrounding forest use and ownership are diverse.

With a focus on Europe, this chapter aims to explore what forest ownership means and outline what impact ownership may have on the provision of FES. The subsequent sections will introduce various forms of ownership, consider the influence of socio-economic and legislative frameworks, and evaluate their impact on FES provision.

DOI: 10.4324/9781003198987-6

This chapter draws upon findings from the 'Forest Land Ownership Change in Europe: Significance for Management and Policy (FACESMAP)' project conducted as part of a COST Action.[1] This initiative investigated the dynamics of forest ownership changes across Europe, a topic of significant importance when considering forest policy and management. This research was partly conducted in partnership with the United Nations Economic Commission for Europe (UNECE) and the Food and Agriculture Organization of the United Nations (FAO) in the context of forest ownership within the pan-European region (ECE, 2020).

1.1. What is a forest owner?

The concept of a 'forest owner' might seem straightforward; however, its interpretation varies significantly depending on the country to which it is applied. For instance, as per the FAO Forest Resources Assessment (FAO, 2015, 2020), forest ownership 'generally refers to the legal right to freely and exclusively use, control, transfer, or otherwise benefit from a forest. Ownership can be acquired through transfers such as sales, donations, and inheritance' (FAO, 2018, p. 16). This definition suggests that forest ownership means complete and exclusive legal authority over the forest resource, encompassing the rights to fully exploit, manage, and transfer those rights to others. In reality, such rights are seldom exercised by forest owners, mostly since forest ownership rights are subject to varying degrees of constraints imposed by national legislative frameworks and social norms linked to the specific forest land in question. This suggests that forest ownership should be viewed as a system of relationships between the legally recognised holder of the resource and the associated rights and responsibilities relating to that forest resource (Bouriaud and Schmithüsen, 2005, Nichiforel et al., 2020).

Moreover, this perspective underscores the importance of outlining the distinct components of forest ownership, differentiating between ownership, tenure, property rights, and forest management. It emphasises the significance of understanding how the terms 'public' and 'private' ownership are utilised in different contextual settings. These are all crucial factors to consider when comparing what it means to be a forest owner across different countries and regions.

Forest ownership effectively encompasses a system of interconnected elements, covering institutional frameworks, property rights, the nature of ownership (e.g., land use policies), the type of owner (e.g., public or private), and the regulations governing forest management (Lähdesmäki and Matilainen, 2014, Khanal et al., 2017, Takala et al., 2017, Sotirov et al., 2019). For each element, the role of regulatory bodies is pivotal, as they have the authority to impart a more pronounced public or private character to forest ownership (e.g., user rights) through relevant legislative mechanisms. Within this perspective, forest ownership involves different forms of tenure that depend on the national configuration of property rights. These rights, regardless of whether formal or informal, pertain to the recognised holder of the resource (FAO, 2011, Bouriaud and Schmithüsen, 2005, Siry et al., 2015). However, it is furthermore vital to recognise that forest ownership, property rights, and tenure arrangements are intricately interwoven concepts (Schlager and Ostrom, 1992, Ostrom and Hess, 2007):

- Formal institutional structures and frameworks governing forest ownership include policies, laws, technical standards, and operational directives. These elements exert influence across various levels of ownership. They encompass not solely policies directly related to forests but broader, cross-sectoral policies that shape the allocation of rights concerning different goods and services derived from forest ecosystems. As such, the regulatory framework for forest tenure plays a decisive role in defining the rights of individuals or entities to access and utilise forest resources (e.g., who has access to what resources, for how long, and under what specific conditions).

- Land use policies typically establish the framework through which ownership rights are conferred upon public or private entities, encompassing elements such as land ownership, trees, and other components within the forest ecosystem. National legislation and contractual agreements commonly delineate property rights concerning various aspects of land use, which can then be allocated to different stakeholders. For instance, national laws may regulate the selection of contractors for specific forest management activities, such as reforestation or timber harvesting, or mandate that the development of forest management plans be the prerogative of public authorities. The main point here is that the right to decide about forest resource use is allocated differently to public authorities, legal owners, or forest users, depending on the national context.
- User rights of individuals or enterprises to access and utilise forestland and its associated resources can be categorised into different types of land rights. These rights may be granted to specific individuals or groups, such as landowners, hunters, or neighbours, or they may be open to the broader public. Examples include rights related to biodiversity conservation, recreational use, or access to non-wood forest products (NWFPs) such as berries and mushrooms. Additionally, forest owners may have the authority to confer user rights, either in their entirety or for specific services, through formal contractual arrangements such as leases, licenses, or permits, as well as through informal agreements.
- Property rights are frequently fragmented, with some rights held by the landowner and others distributed to public authorities or other stakeholders. An effective way to conceptualise this division is through the 'bundle of rights' framework developed by Schlager and Ostrom (1992). This framework categorises property rights into five distinct categories, as follows. (1) **Access rights**: These pertain to the rights to enter forested areas. (2) **Withdrawal rights**: These encompass the rights to harvest or extract resources such as timber, firewood, and NWFPs. (3) **Management rights**: These involve the rights to plan and conduct internal forest activities, including transformative actions within the forest. (4) **Exclusion rights**: These relate to the rights to prevent others from accessing and harvesting wood or NWFPs in a designated area. (5) **Alienation rights**: These extend to the rights to sell forestland and forest products, as well as to lease or transfer management and exclusion rights to others. These categories help to understand how different rights associated with forest resources can be allocated and managed.
- Tenure is a broad concept that covers several property rights categories, as outlined above. In essence, tenure arrangements depend on the degree of control exercised by the legitimate holder of a resource, such as a forest. They are furthermore commonly categorised into five groups, as follows. (1) **Owners**: These are the rightful holders of a resource and possess all property rights associated with it. (2) **Proprietors**: Proprietors cannot exercise alienation rights, meaning they cannot sell the land, although they retain other property rights. (3) **Claimants**: Claimants do not hold exclusion or alienation rights but have the authority to make decisions regarding management and withdrawal rules. (4) **Authorised users**: These individuals or entities can solely access resource benefits as granted by holders of exclusion and alienation rights. (5) **Authorised entrants**: Authorised entrants are granted the right to enter forestland for enjoying non-subtractive benefits (e.g., talking a walk). These groups delineate the varying levels of rights and responsibilities.

It is also important to delve into the distinction between private and public ownership, as it carries significant implications, particularly as definitions of forest ownership can vary widely between countries (Box 6.1). Concerning the entity with ownership rights over a forest, five legal ownership categories typically emerge (McKean, 2000, Ostrom, 2000, Cole and Grossman, 2002):

- **State property**: These are publicly owned forests managed by the government.
- **Municipal or communal property**: Forests owned at the local or regional level, overseen by communes, towns, municipalities, or other administrative bodies.
- **Common property**: Privately owned forests managed by co-owners who establish a governance structure to allocate rights and responsibilities.
- **Private property**: Forests under the ownership of an individual or a legal entity that possesses control over their usage.[2]
- **Open access property**: Forests that lack formal ownership are open to unrestricted access by anyone.

Furthermore, challenges arise when attempting to define specific ownership categories and deciding regardless of whether they should be categorised as public or private (Bromley, 1991, Hanna et al., 1996, Vatn, 2005, Bouriaud and Schmithüsen, 2005). A common source of confusion resides in distinguishing between municipal or communal forests, representing public ownership, and community or common properties, which fall under private ownership. This confusion is exemplified by the classification of municipal forests (Aggestam et al., 2020a). For instance, in Bulgaria, the Czech Republic, and Latvia, they are categorised as private, while in Estonia, Poland, and Romania, they are considered public (ECE, 2020). These variations underscore the importance of understanding them and their impacts in various contexts. It is also relevant when considering data on forest ownership, as the respective categories may vary across countries and regions. This implies that relevant terms characterising forest ownership are not consistently used across countries and institutions. This has, in turn, implications in terms of using official national statistics on forest ownership.

Box 6.1 Country examples concerning forest ownership categories

Slovenia

The Slovenian Forest Act sets out forest owners' rights, considering specific factors for the country. For instance, given the high degree of fragmentation in private forest land, the Forest Act imposes measures to prevent further division, such as prohibiting splitting forest plots smaller than 5 ha. The national categorisation of forest owners encompasses the following groups. (1) **Private forest owners**: Individuals or collectives who hold ownership of a forest, with their property documented in the Land Register of the Republic of Slovenia. (2) **Family farms**: These are owned and managed by one or multiple individuals who collectively oversee the land following relevant land use criteria. (3) **Non-family forest holdings**: Comprising all households that either exclusively own forest land or do not meet the criteria for land use. (4) **Municipal forests**: These forests are owned by municipalities, and their management approaches can vary, ranging from direct involvement to simply holding a representative role in forest management decisions. (5) **State forests**: Forests owned by the state and officially recorded in the Land Register of the Republic of Slovenia. Notably, the rights for exploiting state forests are often delegated to various forest enterprises. In this context, the Forest Act is instrumental in defining and safeguarding the rights of these different forest owner categories, particularly in the context of land fragmentation and ownership diversity.

Source: Krč et al. (2015)

Turkey

In Turkey, most of the forest land, approximately 99.5%, is under state ownership, while the remaining 0.5% is held by other forest owners, such as state-owned forests leased to local communities. Under the Constitution of the Republic of Turkey, it is explicitly stated that ownership of state forests, whether officially registered or undergoing cadastral processes, cannot be transferred to any other entity or individual. Consequently, no prevailing forest property regime in Turkey allows for private ownership of forests.

Source: Gubbuk et al. (2015)

United Kingdom

The Land Registration Act maintains the official record of land ownership in the United Kingdom. This legislation mandates that individuals purchasing land must register their ownership. However, for land not being actively offered for sale, registration is voluntary and is optional for land acquired through inheritance. Consequently, the United Kingdom needs more reliable statistics on forest ownership due to this registration system.

On land under freehold tenure, the ownership of trees rests with the freeholder. In contrast, the Waste Act of 12674 vests tree rights on tenanted land with the landlord. This act prohibits tenants from engaging in activities such as felling or selling trees without obtaining explicit written permission from the landowner. This permission can either be integrated into the terms of the lease agreement or may need to be sought on a case-by-case basis. Securing the landlord's agreement is a prerequisite for obtaining a felling license throughout the United Kingdom. Additionally, planting new woodlands also necessitates the landlord's express permission.

This arrangement, where the landlord retains tree ownership and management rights, exacerbates the challenge of identifying forest landowners. This challenge is particularly pronounced in cases where an estimated 40% of farmland in England and Wales is under tenancy arrangements. In such instances, information about the land's occupier may not necessarily reveal the true owner of the land and, consequently, the right to cut trees.

Source: Wong et al. (2015).

Finally, the level of autonomy granted to forest owners in making decisions about and executing different forest management practices[3] is an important factor distinguishing regulatory frameworks across Europe (Nichiforel et al., 2018). Regardless of the ownership category, the regulation and oversight of forest management affect the property rights of forest owners. State intervention takes various forms, encompassing legal constraints or directives related to forest management, the assignment of public responsibilities in forest management planning, and the zoning of forest land into distinct protection classes, each entailing specific management restrictions for protected or specialised-purpose forests. Notably, there exists a spectrum of regulations governing forest management across the European region (ECE, 2020). These encompass limitations on altering forests for alternative land uses, obligations to regenerate forests following clearcutting, and significant variations in requirements for formally approved Forest Management Plans, as well as the organisational structures tasked with overseeing and implementing these plans. Legal frameworks also diverge in terms of the freedom accorded to forest owners in selecting service providers for various forest management tasks, such as designing management plans or conducting harvesting operations, in addition to the legal constraints imposed on forest management (Živojinovic et al., 2015, ECE, 2020).

2. The relationship between forest ecosystem services and ownership

Forests, with their diverse range of natural resources, offer an array of outputs and values, often conceptualised through the lens of ecosystem goods and services. However, the specific nature and categorisation of these services can often be anything but clear-cut. For instance, the Millennium Ecosystem Assessment sets out four main categories of ecosystem services: provisioning, regulating, cultural, and supporting, acknowledging both the overt and covert benefits society derives from ecosystems (Alcamo et al., 2005). Similarly, the Economics of Ecosystems and Biodiversity (TEEB) categorises these services into provisioning, regulating, habitat, and cultural (TEEB, 2008, Ring et al., 2010, TEEB, 2010). Adding to these are classifications from the European Commission's (ECs) Mapping and Assessment of Ecosystems and their Services (MAES) initiative and the Common International Classification of Ecosystem Services (CICES), both tailored for accounting frameworks (EC, 2013, 2014).

Box 6.2 Defining forest ecosystem services

In this context, '*ecosystem services*' is equivalent to 'ecosystem goods and services'. Forest Ecosystem Services (FES) refers explicitly to those benefits arising from forest ecosystems. However, it is crucial to acknowledge that definitions of FES also diverge across different classification systems. A recent analysis considered five different FES classification schemes, including the Millennium Ecosystem Assessment, MAES, CICES, FORVALUE, and TEEB (Forest Europe, 2014). Commonalities between the schemes include wood products, which are typically categorised to accommodate accounting for marketable and non-marketable products and are often classified as regulating services. Cultural services generally encompass benefits linked to recreation, culture, and education provided by forest ecosystems.

For the sake of uniformity in discourse, this chapter will use the categorisation utilised in the Millennium Ecosystem Assessment (Alcamo et al., 2005) to distinguish between different types of FES:

- **Provisioning services**: These are the tangible products obtained from ecosystems, including but not limited to food, water, fuel, fibre, biochemicals, and genetic materials.
- **Regulating services**: These services are the advantages of the natural control of ecosystem processes. They encompass climate and disease control, water and carbon sequestration, water cleansing, erosion control, flood prevention, and pollination, among others.

- **Cultural services**: These intangible benefits include spiritual and religious enrichment, recreational experiences, eco-tourism, aesthetic enjoyment, inspiration, educational enhancement, cultural heritage, and a sense of belonging.
- **Supporting services**: These services are essential to produce all other ecosystem services and include soil formation, nutrient cycling, primary production, and provision of habitat.

2.1. Wood and non-wood forest products

Products derived from forests, both wood and non-wood, usually fall under provisioning services. The categorisation for wood and wood-based products typically extends from the initial harvesting of trees to the production of primary and secondary products. For example, timber is primarily used in construction, the paper industry uses secondary products like wood pulp, and the energy sector uses biomass for energy production (like pellets). Innovations are also leading to the use of wood in non-traditional sectors such as textiles. The Food and Agriculture Organization (FAO) defines non-wood forest products (NWFP) as 'products of biological origin, besides wood, sourced from forests, other wooded land, and trees outside forests' (FAO, 1999). Other commonly used terms include non-timber forest products (NTFP), encompassing plant and animal products, excluding wood, sourced from forest ecosystems or forest species.

2.2. The interplay between forest ecosystem services and ownership

The presence of forests naturally facilitates the generation of FES, regardless of the management status of the forest land. However, the extent of FES provision is deeply connected with forest ownership. Forest owners, whether public or private, are instrumental in deciding the management and usage of forest land under varying political, legal, and socio-economic conditions. Consequently, forest owners' perspectives, behaviours, priorities, and objectives are pivotal in shaping forest use and the provision of FES. For example, contemporary or urban forest owners, who might not rely on their forests for income, may not focus on timber production as traditional forest owners do (Blanco et al., 2015, Ficko et al., 2017, Ní Dhubháin et al., 2007). Thus, beyond the physical and biological features that underpin the provision of FES, forest use is dependent on the forest owner's relationship with their land, making 'ownership' a critical factor in FES provision (Vedel et al., 2015, Irvine et al., 2016, Matilainen et al., 2015, 2019, Aggestam et al., 2020b).

In light of social and demographic shifts, including post-restitution changes in Central and Eastern European countries, the structure of forest ownership assumes even greater significance. This is particularly relevant when considering NWFPs, as the provision of non-market products and services can be intricately linked to different forest ownership types (Vedel et al., 2015). Even more, these observations do not consider the legislative differences across Europe, which also influence the provision of such FES (Nichiforel et al., 2018).

Forest ownership structures are, therefore, crucial in determining the actual delivery of FES. Understanding the forest owner and their relationship with their land (e.g., attitudes towards forest management) can provide insights into the effects of different forest ownership on forest use. For example, the shift in forest ownership due to the restitution processes in certain European countries has altered the national ownership patterns and the management and usage of forests (Weiss et al., 2019). This implies that enhancing our comprehension of the nexus between FES and forest ownership, both public and private, can help in understanding the effects of these changes.

Finally, the government's role in delineating forest management guidelines is also relevant to FES provision. Legal structures, from regulations detailing property rights and tenure to institutional rules defining the authorities managing public and private forests, are vital in outlining forest ownership and use. Legal structures provide the operational framework for FES and are integral to grasping the interrelation between forest ownership structures and FES provision (Aggestam, 2015). Considering these considerations, the rest of this chapter will probe the nexus between FES provision and forest ownership.

2.3. Approach

This section extends upon Chapter 4.3 in 'Who owns our forests? Forest ownership in the ECE region' (ECE, 2020), specifically focusing on forest ownership's role in delivering wood and other forest ecosystem services in Europe (Aggestam et al., 2020b).

2.3.1. Methods and data

The principal dataset for exploring forest ownership and the delivery of forest ecosystem services comes from the FACESMAP/UNECE/FAO Enquiry,[4] an inquiry purposefully constructed to delve into the evolving characteristics and trends of forest ownership, the influence of governance and societal frameworks on forest owners, users, and forest management approaches (Aggestam et al., 2020b, ECE, 2020, Živojinović et al., 2015). This survey provided both numerical and descriptive information from 32 countries in the pan-European region. Additional information was sourced from the FACESMAP National Reports (Živojinović et al., 2015), which encompass 28 countries, with 22 from the EU and six outside the EU, and is further supported by background documents related to FACESMAP (Živojinović et al., 2015, Weiss et al., 2019).

The survey's creation was a joint effort involving the UNECE/FAO Forestry and Timber Section and COST Action FP1201 – FACESMAP, alongside multiple other entities. Collaborators included the Confederation of European Private Forest Owners (CEPF), the European Forest Institute (EFI), the European State Forest Association (EUSTAFOR), the Federation of European Communal Forest Owners (FECOF), the U.S. Forest Service, and the Unión de Selvicultores del Sur de Europa (USSE).

Given the lack of numerical data regarding the relationship between FES, NWFP, and forest ownership configurations from the surveys, the primary focus is on forests accessible for wood provision, forest ownership, and forest production (e.g., net growth and harvests) concerning different ownership categories.

2.4. Forests available for wood supply

Reviewing the data from countries that participated in the FACESMAP/UNECE/FAO Enquiry, it can be noted that the total forest land within the ECE region was reported to be approximately 1,577 million ha. Out of this, the forests earmarked for timber harvest were about 1,107 million ha as of 2015. This suggests that roughly 70% of the total pan-European forest land is allocated for timber production. However, these figures exhibit significant variation between countries (Figure 6.1). In situations where forest ownership details are available, it can be noted that around 765 million ha (representing 81.5%) are governed by public authorities, 209 million ha (or 18.3%) are in private hands. The ownership of 1.61 million ha (about 0.2%) remains unspecified (Figure 6.2).

Additionally, the variation of forest land earmarked for timber harvesting shows significant variations based on European ownership categories. For instance, countries like Cyprus and Turkey, where forests are entirely state-owned, represent 69.4% of the total forest land

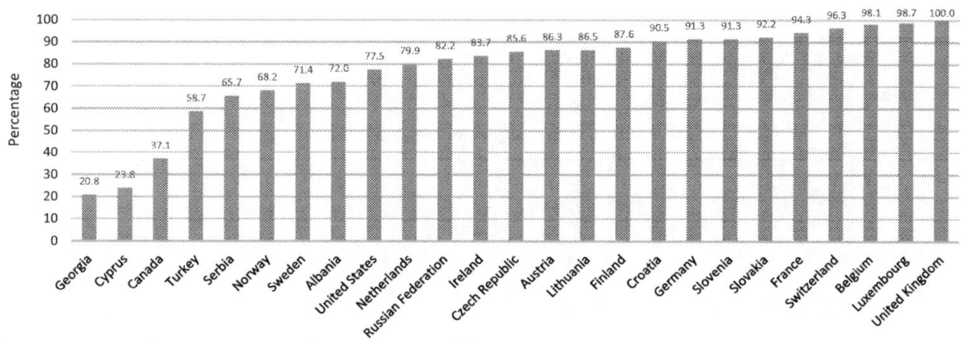

Figure 6.1 Forest land available for wood supply by country.

Source: adapted from ECE (2020)

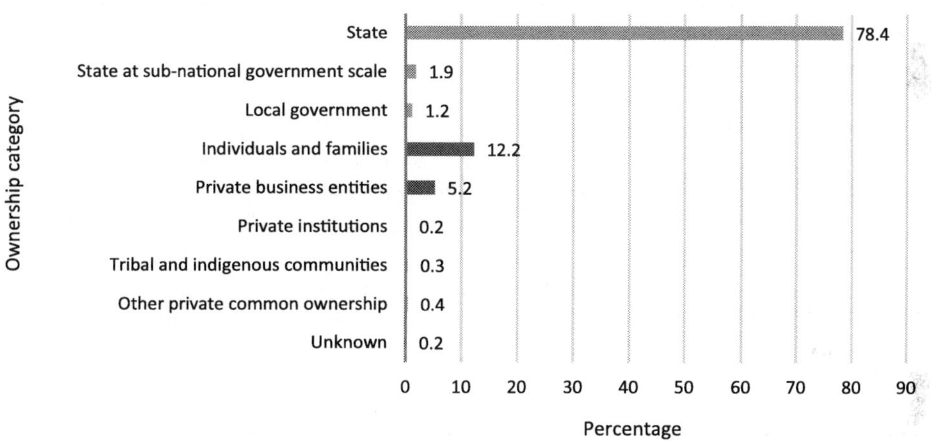

Figure 6.2 Forest land available for wood supply by ownership category.

Source: adapted from ECE (2020)

documented. This percentage significantly shifts the comprehensive data on harvesting rates associated with different forest ownership categories, as the difference between European countries is pretty pronounced (Figure 6.3).

During the period extending from 1990 to 2015, there was an evident shift in the forest land being designated for timber production, with public forest owners seeing a reduction, while there was an increase in private forest owners (e.g., linked to restitution processes[5] throughout Eastern Europe (Bouriaud and Schmithüsen, 2005, Sarvašová et al., 2015). While broadly applicable to the European region, this trend does not hold universally for every country. In certain countries, there was an increase in forest land available for timber provision from public and private forest owners, such as in the Czech Republic, France, Germany, Norway, Slovakia, and Sweden (Živojinović et al., 2015). These variations stress the significant national differences in timber production amid diverse forest ownership frameworks (Figure 6.3). It would also

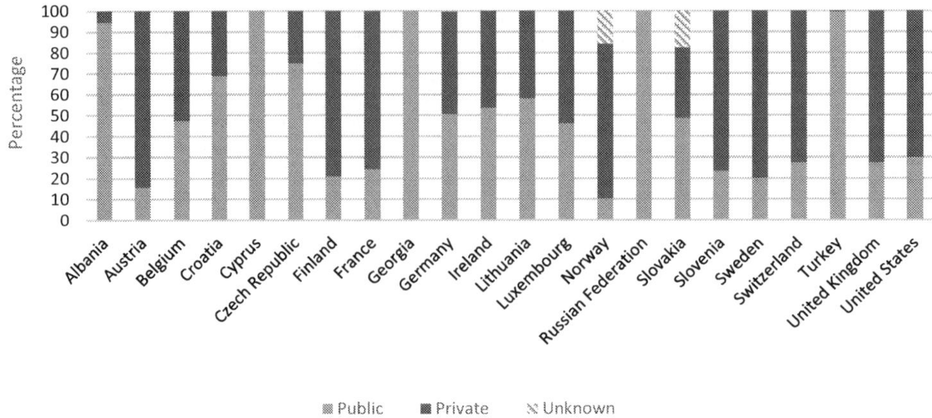

Figure 6.3 Forest land available for wood supply according to ownership category and country.

Source: adapted from ECE (2020)

suggest the need to formulate strategies tailored to stimulate the production of timber or NWFP throughout the region. In other words, there is no 'one-size-fits-all' strategy. For example, one commonly highlighted factor relates to price elasticity, which is frequently intertwined with forest ownership categories, with outputs from public forests often exhibiting less price elasticity (Koch et al., 2013, Shahi et al., 2023).

These results demonstrate the prominent variance in demand and supply dynamics between different forest owner categories and countries. It also underscores the important role forest ownership can play when evaluating the potential for offering different forms of FES, the uptake of different forest management practices, and overall prospects surrounding timber supply. For instance, the prevalence of provisioning services may be more pronounced in private forests, while regulatory, cultural, and support services might find greater representation in public forests. However, when most forest land falls under public ownership, it is logical to suggest that public forests contribute significantly to provisioning services. Nevertheless, this implies that the equilibrium between private and public forests within a country potentially influences the type of FES offered beyond the ownership structures. However, it should also be noted that these assumptions cannot be verified with the data used in this chapter.

2.5. Growing stock, annual fellings, and wood removal

The variations in the quantity of growing stock and the annual rate of fellings are markedly pronounced between publicly and privately owned forests, as depicted in Figures 6.4 and 6.5. Observations demonstrate that a substantial portion of the growing stock is on privately owned forest land in countries such as the USA, France, Sweden, Finland, Austria, and Norway (Živojinović et al., 2015). In contrast, countries like Canada, the Russian Federation, Turkey, the Czech Republic, Bulgaria, and Slovakia are characterised by a dominance of growing stock in publicly owned forests (Whiteman et al., 2015). Please note that the figure omits the Russian Federation, the USA, and Canada – holders of a combined 88.7% of the total growing stock – to ensure the legibility of Figure 6.4.

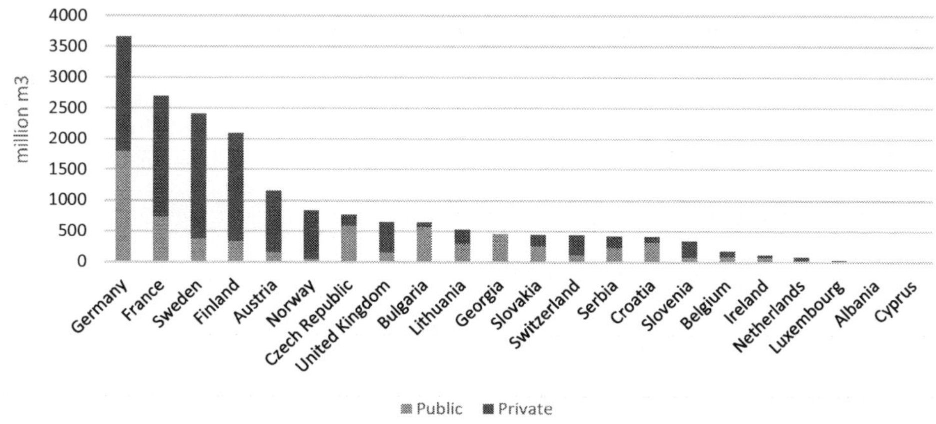

Figure 6.4 Growing stock in public or private forests in 2015.

Source: adapted from ECE (2020)

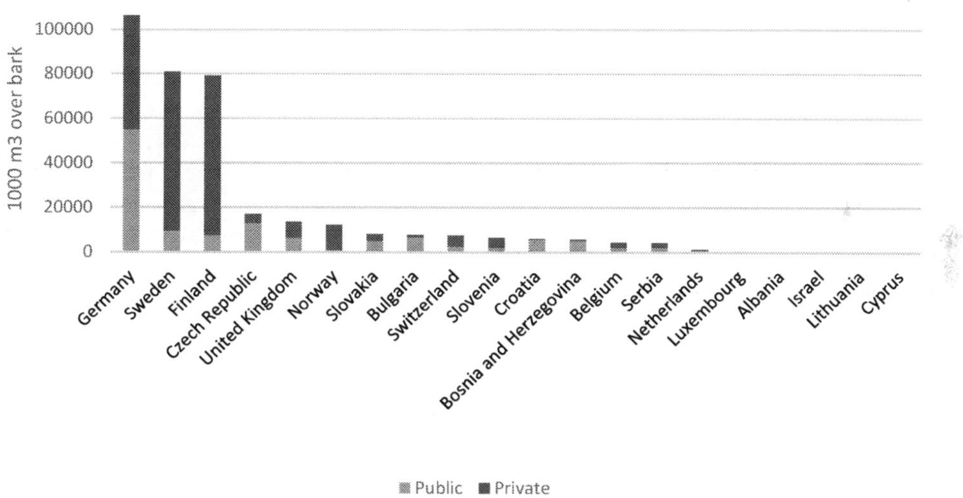

Figure 6.5 Annual fellings in public and private forests in 2015.

Source: adapted from ECE (2020)

When assessing annual fellings through the lens of forest ownership, a trend emerges in alignment with the growing stock distribution (Figure 6.5). A majority of the annual fellings from forests and other wooded areas – 63% to be precise – are attributed to private lands, excluding statistics from the Russian Federation. This trend is illustrated in the data, with private forest owners in Sweden, Finland, and Germany contributing 71.2 million m3, 71.8 million m3, and 51.3 million m3, respectively, towards a cumulative total of 375 million m3 in 2015 (ECE, 2020). In this instance, too, the figure excludes data from the Russian Federation and the USA for clarity purposes. Examining the wood supply, similar trends are evident between the different categories of forest ownership. Notably, 52.7% of all wood removals can be traced back to private forest owners in Sweden, Finland, Germany, and France. In a broader sense, private owners are responsible for an estimated 65.4% of the activity. This dynamic shift, however, once data from the Russian Federation and Canada is incorporated, leading to a scenario where 64.2% of all wood removals are sourced from publicly owned forests (ECE, 2020, Živojinović et al., 2015).

Expanding this analysis, it is relevant to consider some of the underlying factors and implications. For example, ownership structures can directly influence forest management strategies, conservation efforts, and local and global timber markets (ECE, 2020, ECE/FAO, 2010). To illustrate, privately owned forests in the USA, especially in the southern regions, are often managed with a strong emphasis on timber production (Wear and Greis, 2013). In contrast, public forests, like those under the US Forest Service or Canada's Crown forests, commonly have multifunctional objectives, including conservation, recreation, and indigenous people's rights, alongside timber production (Luckert et al., 2011). These variations in forest management have broader environmental and economic implications. For instance, the sustainable management of private forests in Finland has been crucial in supporting the country's timber industry and ensuring ecological balance (Korhonen et al., 2013). Similarly, management policies must carefully balance economic outputs with national and global responsibilities in countries with significant public forest ownership, such as Canada or the Russian Federation (Ulybina and Fennell, 2013).

2.6. Forest use and ownership across regions

The use of forests, expressed as felling as a proportion of net annual increment over time, is a key metric (Figure 6.6). This measure reflects shifts in forest management strategies and has implications for the sustained provision of FES. For instance, the net annual increment of forests can be influenced by an array of factors, including climate, the nature of the forest, biodiversity, the age distribution of trees, and the expansion of forest land.

Data from the collaborative inquiry by FACESMAP, UNECE, and FAO, as documented in their respective reports, shows that a broad spectrum of countries have felling rates spanning from 50 to 100%, encompassing both privately and publicly owned forests, with Albania being an exception (Figure 6.6). These figures underscore the variation in felling rates between countries, which is a direct consequence of diverse climatic conditions and the tree species composition, which are further reflected in the varying rates of net growth and fellings under different ownership regimes.[6]

The FACESMAP/UNECE/FAO Enquiry highlights substantial variations in the rate at which the net annual increment is utilised, a trend that contrasts when comparing private and public forest lands. For instance, public forests in countries like Austria and Slovenia exhibit the most pronounced felling rates. In contrast, for private forests, countries including Sweden and Belgium report higher felling rates (ECE, 2020, Živojinović et al., 2015). On average, the indicator for private forest owners is significantly higher at 68.9%, as opposed to 29.9% for forests under public ownership. However, excluding data from the Russian Federation sees the felling rates in publicly owned forests leap to 61.2%. The data also emphasises the variability among countries regarding how felling rates have evolved over the years (Figure 6.6). For the period spanning 1990 to 2015, the Nordic nations (including Norway, Finland, and Sweden), alongside Luxembourg, demonstrate negligible variations (ECE, 2020, Živojinović et al., 2015). However, a different scenario plays out in Central and Western Europe, where countries like Austria and the United Kingdom witnessed noticeable shifts in felling rates across different ownership types. In contrast, Eastern European countries presented no consistent trends, except for unique situations like Albania, which underwent significant shifts within the same period.

The period spanning 1990 to 2015 in Europe saw an overall increase in felling rates, notably among private forest owners. However, this trend has varied trajectories across countries (Figure 6.6). There are no uniform regional trajectories concerning felling rates related to forest ownership types. This would suggest that the variations in forest use are rather interlinked with conditions unique to each country, such as the intricacies of restitution processes in Eastern Europe or the overall status of national forest-based industries. For instance, delving into the FACESMAP/UNECE/FAO Enquiry's findings, the Nordic countries (comprising Norway, Finland, and Sweden) alongside Luxembourg did not undergo any substantial shifts within the last 25 years. In contrast, in South-Eastern Europe, while there is not a unifying trend, countries like Albania have experienced dramatic shifts and even trend reversals within the same timeline (Figure 6.7). Another example, in Central Western Europe, the United Kingdom emerged as a front-runner in terms of change, witnessing an increase of 29.4% in the net increment within public forests and a 14.6% increase in private forests between 1990 and 2010. This phase was succeeded by another period of 12.3% growth in the subsequent five years (NFI, 2014, ECE, 2020). On the one hand, these examples underscore the argument that regional trends are elusive and, on the other hand, suggest that these variations are linked to national circumstances. For example, in the case of the United Kingdom, the increase in forest cover could be attributed to reforestation initiatives in the mid-90s, which helped to increase the national forest cover from 5% to 12% (NFI, 2014).

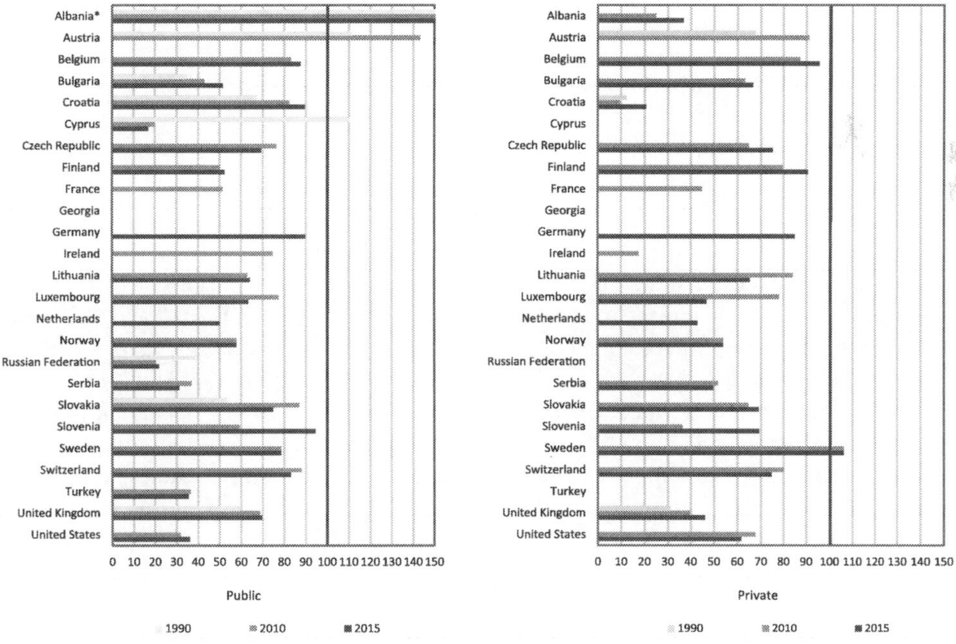

Figure 6.6 Private and public fellings as a percentage of net annual increment.

* Albania reported public fellings rates as 194.6% in 2015, 717.8% in 2010, and 233.5% in 1990.

Source: adapted from ECE (2020)

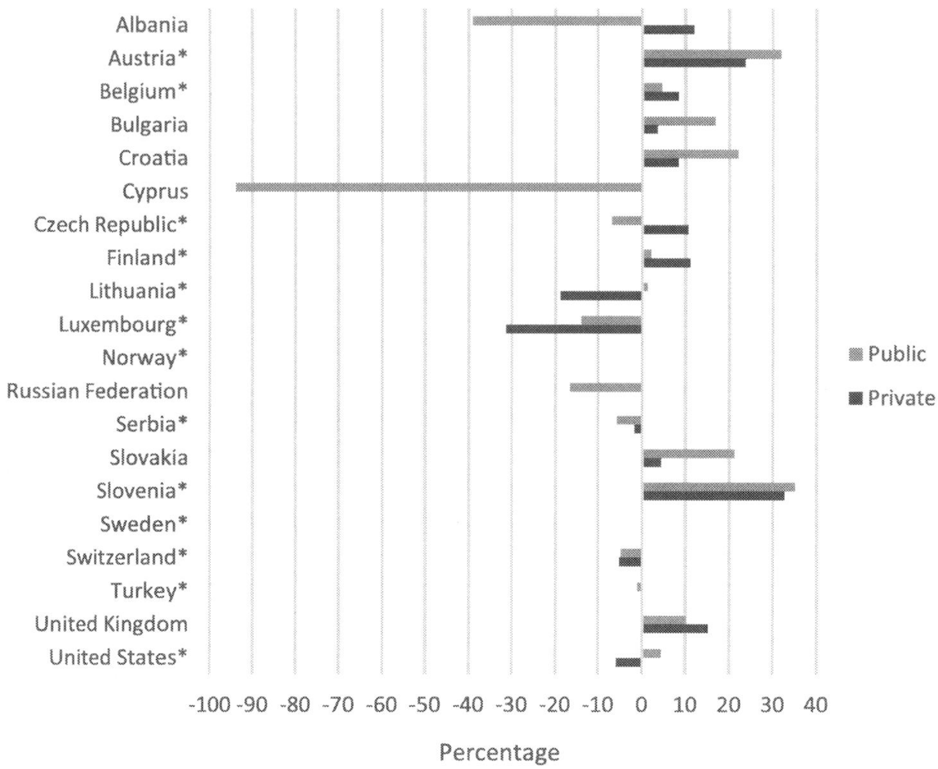

Figure 6.7 Changes in the net increment felling rates over the 1990 to 2015 period.

* Data missing for 1990.

Source: adapted from ECE (2020)

In assessing the forestry trends from 1990 to 2015, it is evident that private forest owners have primarily driven the increase in felling rates. For example, between 2010 and 2015, the total annual felling volumes increased by 7.5%, with private forests registering an 8.4% increase. Conversely, public forests observed a 4.3% decrease in the same duration. However, the story behind public forests is less linear. While many have ramped up their harvesting rates over the years, several others, including Albania, Cyprus, Czech Republic, Luxembourg, Russia, Serbia, Switzerland, and Turkey, appear to go against this trend. These discrepancies in felling rates between countries and privately and publicly owned forests raise pertinent questions regarding the sustainability and conservation strategies employed by different ownership categories. Reduced felling rates in publicly owned forests may point towards a growing commitment to sustainable practices, or they may be driven by conservation goals. However, while this may be one explanation, it is also important to acknowledge the data limitations that make these conclusions speculative, especially concerning private forests.[7] The lack of adequate data underscores the need for more comprehensive and consistent data collection methods on forest ownership. Nevertheless, the results demonstrate the need to consider each area's ecological, historical, and socio-economic particularities (Box 6.3).

Box 6.3 Regional differences in utilisation rates

South-eastern Europe

The period from 1990 to 2015 marked a significant shift in private felling rates, with certain countries exhibiting particularly noteworthy trends:

- Bulgaria stands out as a unique case; here, felling rates in public forests surpassed those under private ownership, going against the general trend observed elsewhere. This deviation could be attributed to specific national policies or economic pressures, underscoring the influence of local factors in forest management practices.
- Albania presents a striking scenario: felling rates in private forests surged by 11.9%, while those in public forests declined by 38.9% over the 25 years. This could be traced back to changes in regulatory frameworks, economic incentives, or shifts in domestic demand for forest products.
- Croatia displayed an uptick in felling rates across both the private and public sectors, with increases of 23% and 8.2%, respectively. This concurrent rise might indicate a growing industrial demand for timber or a strategic response to forest health challenges such as pest infestations or disease outbreaks.
- Slovenia witnessed an upsurge in both sectors, with public forest felling rates growing by 35.1% and private ones by 32.6%, possibly reflecting an expanding forestry sector or a shift in forest management strategies.
- Serbia reduced felling intensity in both domains, decreasing by 5.7% in public forests and 1.8% in private holdings. This reduction could indicate successful conservation efforts, changes in market demand, or the impact of global economic conditions on local forestry practices.

These results highlight the importance of considering local contexts when analysing forestry data.

Central and eastern Europe

Central and eastern Europe exhibit a trend like those observed elsewhere, with a singular deviation: felling rates in forests under public ownership generally declined.

- Both the Czech Republic and Slovakia witnessed an increase in felling rates within privately managed forests, contrasted with a downturn in those overseen by public entities. Specifically, Slovakia experienced a sharp 21.2% decline in public forest fellings, juxtaposed with a 4.3% increase in the private sector. The Czech Republic's public forests similarly saw a reduction of 6.9%, while private felling rates increased by 10.6%.
- Austria presents a distinct scenario where both sectors observed significant increases – 31.8% in public forests and 23.6% in private holdings. This uniform growth could be attributed to various factors, including but not limited to national policy shifts encouraging timber production, economic incentives for more aggressive forestry practices, or ecological strategies, such as increased felling to combat infestations or diseases.

Considering the various socio-economic and ecological drivers at play, it is relevant to consider these patterns in the broader context of forest management. For instance, while increased felling might point to industrial growth, it could also be a response to forest health challenges or shifts in global market demands. Without more data, these possibilities remain speculative.

Northern and western Europe

In northern Europe, the forest felling rates have largely maintained their status quo over the period in question.

- In countries like Sweden and Norway, there has been an observable consistency in the felling rates, indicating a stable approach to forest utilisation.
- Finland exhibits a modest increase in its felling rates, with a 2% increase in public forests and a more noticeable 11% increase in private forests. This could reflect Finland's increasing demand for timber, possibly driven by its growing bio-economy sector or infrastructural developments.
- In Lithuania, felling rates among private forest owners have significantly decreased by 18.8%, whereas those in public forests have marginally increased by 1.3%. This reduction in private fellings could be due to changes in market dynamics and timber prices or new regulatory frameworks.

In Western Europe, felling rates appear more homogenised, with negligible differences discerned between the felling rates of private and public forest owners. This uniformity might come from consistent forestry management policies across different ownership types or these countries' mature and stable market conditions.

Divergences at the regional level have significant weight and are often attributed to contextual factors, including climatic, cultural, and policy variations, to name a few (Box 6.3). For example, the restitution initiatives in Central and Eastern Europe have profoundly influenced national strategies and priorities in forest management, creating disparities between private and public forest owners. Additionally, the rich historical and cultural practices related to forestry shape the ownership paradigms, determining whether forests are held by governmental bodies (whether local or national), individual or familial owners, corporate entities, indigenous or tribal groups, or other forms of communal private ownership throughout Europe. Moreover, it is critical to recognise the considerable heterogeneity in the forest land designated for wood supply under private and public ownership, which differs notably across countries. This suggests that, while discrepancies between privately and publicly managed forests might be pronounced nationally, they contribute minimally to the regional aggregate of annual felling. Take Slovakia, for instance, where the national annual felling constitutes a mere 1.5% of the regional sum, in stark contrast to the Russian Federation, which represents a significant 31.7% (with the exclusion of Canada and the USA). This underlines how critical it is also to consider scale.

2.7. Forest ecosystem services

The role of forest ownership in the provision of FES is multifaceted and influenced by a variety of factors, including the owner's **objectives and motivations** (e.g., individuals or corporate entities often own forests with profitability as a primary motive), **regulatory and policy contexts** (e.g., incentives and support programs, such as grants for reforestation, tax breaks for conservation efforts, or payments for ecosystem services), **cultural and historical factors** (e.g., traditional and indigenous communities might own or manage forests following centuries-old practices), and **economic considerations and market forces** (e.g., carbon trading schemes can incentivise forest conservation and the enhancement of carbon stocks) (Torralba et al., 2020, Maier et al., 2021, Sorge et al., 2022). Understanding this dynamic is crucial for developing policies and practices supporting diverse ecosystem services. For instance, while timber production is a common focus, there is a growing trend towards diversifying income through NWFPs, recreational services (like eco-tourism) and carbon sequestration programs. NWFPs constitute about 15% of the global value of all forest products (FAO, 2010, 2015). Considering this, the spectrum of NWFPs important to European forest owners is broad, encompassing items like fungi, berries, herbs with medicinal properties, and wildlife resources. Regions with substantial data on these products include Europe, North America, and Northern Africa. Even if somewhat out-of-date, records show that in 2005, Europe was responsible for nearly half of the worldwide production of NWFPs (FAO, 2010, 2015).

The FACESMAP/UNECE/FAO Enquiry provides information on several NWFPs reported on by European countries. For example, mushrooms are deemed significant in countries such as Bulgaria, the Czech Republic, Finland, France, Greece, Italy, Lithuania, Slovenia, Spain, and Switzerland. Specifically, in the Czech Republic, mushrooms constituted up to two-thirds of the total weight of edible items harvested in the country in 2012 and about 60% of the commercial value of NWFPs in 2010 (FAO, 2016, Forest Europe, 2020). Furthermore, activities like hunting and fishing are prevalent in countries like Bulgaria, Estonia, France, Greece, Latvia, Serbia, Spain, and the United Kingdom. In France, for example, venison accounts for nearly 80% of the recorded value (FAO, 2014, 2016, Forest Europe, 2020). Other edible items classified as NWFPs include fruits, honey, and maple products, which comprise nearly three-quarters of Canadian NWFPs in value (FAO, 2014).

It can also be noted that the ability to monetise FES, such as mushrooms, differs significantly across countries due to factors including legal frameworks, cultural practices, and historical land use rights. The comparison between Italy and Sweden regarding mushroom picking as a source of income illustrates these differences. In Italy, mushroom picking can be lucrative, primarily as local authorities regulate mushroom picking. For example, pickers may need to acquire a license (e.g., there are often rules about the quantity and types of mushrooms that can be harvested. In contrast, Sweden's 'allemansrätten' or 'everyman's right' presents a different scenario. The 'everyman's right' is a unique right of public access to nature that allows everyone to roam freely in forests, provided they respect the natural environment and private property. This right includes picking mushrooms, berries, and certain other natural products. This also implies that forest owners cannot charge pickers for these services.

In Portugal, NWFPs comprise nearly half of the forest output, with cork production being a primary activity, contributing 26% to the overall gross value (Mendes et al., 2004, Mendes, 2012). Cork is also significant in countries like Spain and France. Additionally, resin is produced in Portugal, as well as in countries like Slovenia and Greece. Countries in the Alpine and northern European regions include Christmas trees in their list of NWFPs. Various nuts are harvested in countries like Bulgaria and Switzerland and the Caucasus and Central Asian areas, with Spanish chestnuts seeming particularly significant. Another common category of NWFPs includes medicinal and aromatic plants, with Turkey's significant contributions valued at 95 million USD in 2010. Examples include anise, sage, thyme, and rosemary. These examples illustrate the relevance of a broad spectrum of NWFPs across countries.

Data linking forest ownership structures with FES provision is limited. However, insights into this relationship can be found in the FACESMAP Country Reports (Živojinović et al., 2015) and the Global Forest Resources Assessment (FAO, 2010, 2015, 2020). One trend is the purchase of forest land by NGOs, charities, and associations, with aims such as preserving biodiversity or other regulatory ecosystem services (Živojinović et al., 2015). For example, in Slovakia, the WOLF Forest Protection Movement established a reserve encompassing 1,037 ha (Živojinović et al., 2015), while in the UK, charitable entities own 3.5% of the forest land (Živojinović et al., 2015). These instances demonstrate the strategic acquisition of forest lands by organisations intending to foster specific FES beyond timber production. Other initiatives include the preservation of cultural heritage sites situated within historic estates or attempts to enhance local economies by offering employment to underprivileged demographic segments (FAO, 2010, 2015, 2020), thus indicating that forest management can also pursue distinct social objectives. This further emphasises that the relationship between forest ownership and FES is complex and context-specific.

Moreover, forests play a crucial role in combating climate change, suggesting a symbiotic relationship between the delivery of FES and services related to climate change mitigation. For instance, expanding forest coverage in the European region is a positive indicator for enhanced carbon sequestration. Countries like Serbia, Switzerland, and Luxembourg, which have reported a decline in felling rates over the past quarter-century (Figure 6.6), have thus indirectly contributed to increased carbon sequestration. There is also a concerted push in Europe for climate-smart forestry and Land Use, Land-Use Change, and Forestry (LULUCF) initiatives as part of comprehensive strategies to balance out emissions (Ellison et al., 2014, Paquel et al., 2017, Sotirov et al., 2021). Another critical forest-related service is water management, which is often a high priority for local governments. Forests serve as instruments enabling municipalities to supply crucial water-related services to the populace, such as maintaining water purity, ensuring water availability, and controlling floods. Hence, forests, regardless of their ownership status, underpin many ecological functions and provide vital services that extend beyond the supply of timber and NWFPs.

While not applicable across all countries and forest owners, it can be noted that publicly owned forests may, at times, account for a broader set of forest management objectives than found in privately owned forests. This would imply that alternative FES are sometimes prioritised, such as biodiversity conservation, public health or recreation, over timber production. Public forests may also promote other services, such as mushroom or fruit production. This is, however, not to say that private forest owners do not focus on providing FES, as privately owned forests are often used for collecting and producing NWFPs. One typical example is the use of private forest land for hunting. For instance, in Belgium, many new forest owners have purchased forest lands, among other things, to be able to hunt. This illustrates that the management and use of FES are ultimately subject to national forest ownership and user rights systems. Access rights are, for example, a fundamental aspect of the NWFP sector, reflecting historical institutional developments, land use patterns and the availability of forest resources. Another example is countries with more small-scale private forest owners, which often implies less intensive and more diverse forest management driven by different objectives and management schemes. The extent to which a country's forest ownership structure is dominated by large, medium or small-scale forest owners consequently plays an important role.

Although it is not a universal rule, public forests sometimes encompass a broader spectrum of management goals than their private counterparts. This might suggest a greater emphasis on alternative FES over timber production, such as biodiversity conservation, public health, and recreational activities. Public forests might also foster other services, like the production of mushrooms or fruits. However, this does not imply that private forest owners neglect FES provision, as private forests are frequently utilised to collect and produce NWFPs. A typical example is the use of private forest land for hunting activities. For instance, in Belgium, many new forest owners have acquired land, partly for hunting purposes. This demonstrates that the administration and use of FES are ultimately determined by the national frameworks governing forest ownership and user rights. Access rights, for instance, are a crucial element of the NWFP sector, mirroring historical institutional developments, land use norms, and the accessibility of forest resources. Another example is countries with a prevalence of small-scale private forest owners, often leading to less intensive, more diverse forest management driven by various objectives and strategies. Consequently, the prevalence of large, medium, or small-scale forest owners within a country's forest ownership structure plays a significant role.

3. Conclusions

Wood products hold a central position in the realm of forest production, a fact underscored by the allocation of approximately 70% of Europe's forested land for such activities (EEA, 2016, FAO, 2020, Forest Europe, 2020). In an observation that speaks to the sustainability efforts in the region, the ratio of fellings to growth – or increment – has not exceeded 80% for most European countries from 1990 to 2015. However, the equilibrium between increment and fellings is a finely tuned balance influenced by ecological, political, and socio-economic factors. For example, climate change exerts an influence in that warmer temperatures and altered rainfall patterns affect tree growth rates and health, subsequently affecting sustainable yield calculations and harvesting strategies. Additionally, market demands for wood products can trigger fluctuations in felling rates. At the same time, advancements in forestry technologies may increase harvesting efficiency without increasing the number of trees felled. Forest ownership structures present another layer, as they inherently bring varied objectives, management strategies, and long-term visions. Public forests might be managed with a stronger emphasis on conservation, recreation, and biodiversity, often leading to more conservative felling strategies than privately owned forests, which might sometimes be under more pressure to deliver short-term economic returns. However, as the results have shown, this is not a universal rule; many private forest owners advocate for sustainable management practices. Furthermore, policies and regulations at both the national and EU level play a critical role. For example, environmental directives, conservation programmes, carbon sequestration targets, and even subsidies for forest landowners who adopt eco-friendly practices affect felling rates.

Felling rates emerge as an important indicator of the interplay between societal demands and environmental management. It is a straightforward metric that sits at the nexus of forest management. For instance, data from the FACESMAP/UNECE/FAO Enquiry indicates that forest use intensity has diverged considerably between countries. The felling rate is a telling indicator of each country's circumstances and forest owners. These rates might be attributed to factors ranging from restitution processes and evolving ownership frameworks to introducing new legislative and policy instruments affecting FES delivery or the shifting balance between private and public forest owners, among others. As such, the results reflect a spectrum of ecological and socio-economic determinants affecting the forest-based sector on the national and international level. Nonetheless, the findings do not allow for regional generalisations, pointing instead to many country-specific factors shaping forest management. Moreover, one limitation of felling rates is their inability to offer insights into other FES, like alternative biomass forms or NWFP.

The findings underscore a need for more exhaustive data or information concerning forest ownership frameworks and FES offerings. This deficit is further highlighted by the need for more detailed exploration into forest ownership and FES beyond timber production in this segment. The challenge extends beyond data scarcity to the nonexistence of unified categorisation systems for forest ownership types and FES across nations. This has inevitably constrained the scope of analysis regarding the interplay between ownership configurations and FES provision in this section. Essentially, there is a stark paucity of analogous data at hand. This poses implications for the comprehensive compilation of data on private and public ownership frameworks. Consequently, there is an ongoing demand for alignment in the methodologies employed for categorising and assembling data on forest ownership frameworks and FES throughout the European region.

The insights gathered furthermore reveal a gap in the availability of comprehensive data and information about forest ownership structures and FES. This absence of data is, in part, accentuated by the apparent omission of an in-depth discourse on the dynamics of forest ownership and the diversity of FES, extending beyond the conventional focus on timber production within this discourse. However, this is complicated by more than just the absence of data, as no standardised, cohesive system for classifying the varied forms of forest ownership exists. In other words, the need for harmonisation of classification paradigms across different countries and regions hampers comparability and research. This fragmentation in data – both in terms of quality and categorisation– underscores the need for a harmonised approach when considering the taxonomies and data-gathering methodologies employed for European forest ownership research. Moreover, without reliable, standardised data, formulating effective, evidence-based policies and management strategies becomes a challenge, potentially leading to suboptimal forest use or missed opportunities for conservation at the EU level.

As noted above, the field of research concerned with forest ownership and FES is affected by several methodological challenges. These challenges extend beyond the classification of forest ownership types to defining, categorising, and quantifying ecosystem services. This task becomes more complex when trying to align with international standards or protocols. With this in mind, attributing specific ecosystem services to specific forest ownership types is far from straightforward. For instance, services such as biodiversity conservation, carbon sequestration, or cultural and recreational values might manifest differently. They may not be easily quantifiable or directly attributable to the ownership type per se but rather to a mix of management practices, ecological variables, and policy environments. This complexity is further deepened when considering services that are even more difficult to measure – such as the intrinsic value of biodiversity, the cultural significance of forests, or the potential of forests in mental health improvement. Therefore, an important area for future research involves digging into the impacts and potential benefits of various forest ownership frameworks on providing FES. This research should particularly focus on those FES for which data is scarce or non-existent. Such research might involve developing new metrics, applying emerging technologies for data collection and analysis, and the integration of different fields of knowledge.

Want to find out more?

The body of work from the FACESMAP project, partially listed below, stands as one of the more comprehensive efforts to study forest ownership trends in Europe in recent years. Interested readers are referred to the content below for more information:

- Who owns our forests? (ECE, 2020).
- Forest ownership changes in Europe (Weiss et al., 2019).

- European private forest owner typologies (Ficko et al., 2017).
- How private are Europe's private forests? (Nichiforel et al., 2018).

In addition, the set of questions listed below may help guide further inquiry into specific areas of interest:

1. How can the type of forest ownership affect the management, conservation, and use of forests across countries?
2. What role do different legal frameworks play in shaping the rights and responsibilities of European forest owners?
3. How can forest ownership influence the application of different forest management practices and the provision of FES?
4. How might the values and perspectives of forest owners evolve, and what implications could this have for future forest management and policymaking?
5. What factors, such as the size of the forest land owned, the owner's relationship to the forest, and their economic reliance on it, contribute to different forest owner typologies?

4. Assignment VI: Analysing the impact of forest ownership on forest ecosystem services (FES) provision

Description

Building on Chapter 6, this assignment is focused on the implications of different forest ownership structures concerning the provision of forest ecosystem services (FES). It is also meant to introduce students to the challenge of drafting a policy brief that offers strategic recommendations, targeting a specific target audience (e.g., policymakers or forest managers).

Purpose

This exercise tasks students with analysing the relationship between forest ownership types and FES provision in Europe. Drawing on Chapter 6 and other materials, students are expected to identify key challenges and opportunities associated with different ownership models, for example, by using a case study approach. The aim is for the student to develop a policy brief that outlines policy recommendations for a specific target group (to be selected by the student).

Outline of task

- *Review*: Carry out a literature review on how forest ownership influences FES provision, highlighting differences, such as between private, public, and communal ownership.
- *Analysis and synthesis*: Identify key challenges and opportunities associated with different ownership models affecting FES provision. Consider factors such as legislative frameworks, management practices, and societal norms.
- *Policy recommendations*: Develop recommendations for enhancing FES provision. Propose policy interventions, incentives, or regulatory changes to improve FES provision across forest ownership structures.
- *Policy brief*: Develop a policy brief, including an executive summary, background, analysis, policy recommendations, and a call to action for policymakers or specific stakeholders.

Suggested additional reading

- UNECE (2020) Forest Ownership in the ECE Region. United Nations Economic Commission for Europe.
- Weiss, G., Lawrence, A., Hujala, T., Lidestav, G., Nichiforel, L., Nybakk, E., Quiroga, S., Sarvašová, Z., Suarez, C. & Živojinović, I. (2019) Forest ownership changes in Europe: State of knowledge and conceptual foundations. *Forest Policy and Economics*, 99, 9–20.
- Mölder, A., Tiebel, M. & Plieninger, T. (2021) On the interplay of ownership patterns, biodiversity, and conservation in past and present temperate forest landscapes of Europe and North America. *Curr. Forestry Rep.*, 7, 195–213.

Deliverable

- Prepare a policy brief (1000–1500 words) that outlines the relationship between forest ownership patterns and FES provision. It should include an executive summary, key findings from the analysis, and policy recommendations for specific policy actions that may optimise FES provision through forest ownership policies. The brief should be designed to inform policy decisions, using evidence-based insights to propose strategic interventions.

5. Notes

1 COST Actions are interdisciplinary research networks typically comprised of researchers from academia, SMEs, public institutions, and other relevant organisations (see www.cost.eu/).
2 Various categories of private forest owners can also be identified, including traditional, new, and urban ones (Hogl, et al. 2005, Blanco, et al. 2015, Ficko, et al. 2017). These classifications of forest ownership types are closely tied to the knowledge, objectives, or management approaches individual owners adopt.
3 The forest management concept can extend beyond practical forestry operations, covering forest stewardship's planning, decision-making, and other aspects (ECE, 2020).
4 See https://facesmap.boku.ac.at/.
5 The restitution processes refer to the post-communist era initiatives that aimed to redress the nationalisation or confiscation of privately owned land and assets during the communist period. It is important to note that the exact outcomes and influence of these processes can be quite intricate and country-specific (Živojinović et al. 2016, Sarvašová et al. 2016 and Weiss et al. 2019).
6 It is important to interpret these percentages within the context of the net increment rate, as they represent the actual proportion of forest stock felled.
7 A significant caveat in this analysis is the limited data availability spanning the entire 1990 to 2015 timeframe, particularly concerning private forest holdings.

6. References

AGGESTAM, F. 2015. Framing the ecosystem concept through a longitudinal study of developments in science and policy. *Conservation Biology*, 29, 1052–1064.

AGGESTAM, F., KROTT, M. & HAUCK, M. 2020a. Municipal forests. *In:* UNECE (ed.) *Who Owns Our Forests? Forest Ownership in the ECE Region.* Geneva: United Nations Economic Commission for Europe.

AGGESTAM, F. & PÜLZL, H. 2018. Coordinating the uncoordinated: the EU Forest Strategy. *Forests*, 9, 125, DOI: https://doi.org/10.3390/f9030125.

AGGESTAM, F., QUIROGA, S., SUÁREZ, C., MARTÍNEZ, P. & SARVASOVA, Z. 2020b. Forest ownership and the provision of wood and other forest ecosystem services. *In:* UNECE (ed.), *Who Owns Our Forests? Forest Ownership in the ECE Region.* Geneva: United Nations Economic Commission for Europe.

ALCAMO, J., BENNETT, E. M. & MILLENNIUM ECOSYSTEM ASSESSMENT (PROGRAM) 2005. *Millennium Ecosystem Assessment: Ecosystems and Human Well-being: A Framework for Assessment.* Washington, DC: Island Press.

BLANCO, V., BROWN, C. & ROUNSEVELL, M. 2015. Characterising forest owners through their objectives, attributes and management strategies. *European Journal of Forest Research*, 134, 1027–1041.
BOURIAUD, L. & SCHMITHÜSEN, F. 2005. Allocation of property rights on forests through ownership reform and forest policies in central and eastern European countries. *Schweizerische Zeitschrift fur Forstwesen*, 156, 297–305.
BROMLEY, D. W. 1991. *Environment and Economy: Property Rights and Public Policy.* Basil Blackwell.
COLE, D. H. & GROSSMAN, P. Z. 2002. The meaning of property rights: Law versus economics? *Land Economics*, 78, 317–330.
EC 2013. Mapping and assessment of ecosystems and their services: An analytical framework for ecosystem assessments under Action 5 of the EU Biodiversity Strategy to 2020. *Technical Report 2013–067.* Brussels. doi: 10.2779/12398
EC 2014. Mapping and assessment of ecosystems and their services. *Technical Report 2014–080.*
ECE 2020. Who owns our forests? Forest ownership in the ECE region. Geneva, Switzerland: United Nations Economic Commission for Europe.
ECE/FAO 2010. Private forest ownership in Europe. *In:* SCHMITHÜSEN, F. & HIRSCH, F. (eds.). Geneva: United Nations Economic Commission for Europe.
EEA 2016. European forest ecosystems – State and trends. Denmark: Copenhagen: European Environment Agency.
EFI 2013. Mapping the distribution of forest ownership in Europe *In:* PULLA, P., SCHUCK, A., VERKERK, P. J., LASSERRE, B., MARCHETTI, M. & GREEN, T. (eds.) *EFI Technical Report.* European Forest Institute.
ELLISON, D., LUNDBLAD, M. & PETERSSON, H. 2014. Reforming the EU approach to LULUCF and the climate policy framework. *Envronmental Science and Policy*, 40, 1–15.
FAO 1999. Non-wood Forest Products and Income Generation: Towards a harmonised definition of non-wood forest products. *Unasylva*, 3.
FAO 2010. Global Forest Resources Assessment 2010. *FAO FORESTRY PAPER 163.* Rome: Food and Agriculture Organization of the United Nations.
FAO 2011. Reforming forest tenure: Issues, principles and process. *FAO Forestry Paper 165.* Rome: Food and Agriculture Organization of the United Nations.
FAO 2014. State of the world's forests 2014. *Forests and Agriculture: Enhancing the Socio-economic Benefits from Forests.* Rome.
FAO 2015. Global forest resources assessment 2015. Rome: Food and Agriculture Organization of the United Nations
FAO 2016. State of the world's forests 2016. *Forests and Agriculture: Land-Use Challenges and Opportunities.* Rome.
FAO 2018. FRA 2020 Terms and definitions. *Forest Resources Assessment Working Paper 188.* Rome: Food and Agriculture Organization of the United Nations.
FAO 2020. Global forest resources assessment 2020: Main report. Rome, Italy: Food and Agriculture Organization of the United Nations.
FICKO, A. 2019. Private forest owners' social economic profiles weakly influence forest management conceptualizations. *Forests*, 10, 956.
FICKO, A., LIDESTAV, G., NÍ DHUBHÁIN, Á., KARPPINEN, H., ZIVOJINOVIC, I. & WESTIN, K. 2017. European private forest owner typologies: A review of methods and use. *Forest Policy and Economics.*
FOREST EUROPE 2014. Expert group and workshop on valuation of forest ecosystem services. *Group of Expert (2012–2014) & Belgrade Workshop (Republic of Serbia), 24–25 September 2014.* Madrid: Ministerial Conference on the Protection of Forests in Europe.
FOREST EUROPE 2015. State of Europe's forests 2015. Madrid, Spain: Ministerial Conference on the Protection of Forests in Europe.
FOREST EUROPE 2020. State of Europe's forests 2020. Bratislava: Ministerial Conference on the Protection of Forests in Europe.
GUBBUK, H., CETINAY, S., BASARAN, S., GULER, S., ETLI, T. & BALKIC, R. 2015. Forest land ownership change in Turkey. *COST Action FP1201 FACESMAP Country Report.* Vienna: European Forest Institute Central-East and South-East European Regional Office.
HANNA, S. S., FOLKE, C. & MALER, K. G. 1996. *Rights to Nature: Ecological, Economic, Cultural and Political Principles of Institutions for the Environment.* Washington, DC: Island Press.
HOGL, K., PREGERNIG, M. & WEISS, G. 2005. What is new about new forest owners? A typology of private forest ownership in Austria. *Small-scale Forest Economics, Management and Policy*, 4, 325–342.

IRVINE, K. N., O'BRIEN, L., RAVENSCROFT, N., COOPER, N., EVERARD, M., FAZEY, I., REED, M. S. & KENTER, J. O. 2016. Ecosystem services and the idea of shared values. *Ecosystem Services*, 21, 184–193.

KHANAL, P. N., GREBNER, D. L., MUNN, I. A., GRADO, S. C., GRALA, R. K. & HENDERSON, J. E. 2017. Typology of nonindustrial private forest landowners and forestry behavior: Implications for forest carbon sequestration in the southern US. *Small-scale Forestry*, 16, 419–434.

KOCH, S. P., SCHWARZBAUER, P. & STERN, T. 2013. Monthly wood supply behavior of associated forest owners in Austria – Insights from the analysis of a micro-econometric panel. *Journal of Forest Economics*, 19, 331–346.

KORHONEN, K., HUJALA, T. & KURTTILA, M. 2013. Diffusion of voluntary protection among family forest owners: Decision process and success factors. *Forest Policy and Economics*, 26, 82–90.

KRČ, J., PEZDEVŠEK MALOVRH, Š., FICKO, A., ŠINKO, M., PREMRL, T., BOGATAJ, N. & UDOVČ, A. 2015. Forest land ownership change in Slovenia. *COST Action FP1201 FACESMAP Country Report.* Vienna: European Forest Institute Central-East and South-East European Regional Office.

LÄHDESMÄKI, M. & MATILAINEN, A. 2014. Born to be a forest owner? An empirical study of the aspects of psychological ownership in the context of inherited forests in Finland. *Scandinavian Journal of Forest Research*, 29, 101–110.

LUCKERT, M. K., HALEY, D. & HOBERG, G. 2011. *Policies for Sustainably Managing Canada's Forests Tenure, Stumpage Fees, and Forest Practices*. University of British Columbia Press.

MAIER, C., HEBERMEHL, W., GROSSMANN, C. M., LOFT, L., MANN, C. & HERNÁNDEZ-MORCILLO, M. 2021. Innovations for securing forest ecosystem service provision in Europe – A systematic literature review. *Ecosystem Services*, 52, 101374.

MATILAINEN, A., KOCH, M., ZIVOJINOVIC, I., LÄHDESMÄKI, M., LIDESTAV, G., KARPPINEN, H., DIDOLOT, F., JARSKY, V., PÕLLUMÄE, P., COLSON, V., HRICOVA, Z., GLAVONJIC, P. & SCRIBAN, R. E. 2019. Perceptions of ownership among new forest owners – A qualitative study in European context. *Forest Policy and Economics*, 99, 43–51.

MATILAINEN, A., ŽIVOJINOVIĆ, I., NEDELJKOVIĆ, J. & SCRIBAN, R. 2015. Traditional or not? Analysis of forest owner interviews from 10 European countries. *In:* ŽIVOJINOVIĆ, I., LIDESTAV, G., FELICIANO, D., HUJALA, T., LAWRENCE, A. & WEISS, G. (eds.) *Concepts, methods and findings in forest ownership research in Europe. Mid-term Proceedings of the COST Action FP1201 Forest Land Ownership Changes in Europe: Significance for Management and Policy.* Vienna: University of Natural Resources and Life Sciences, Vienna (BOKU).

MCKEAN, A. M. 2000. Common property: What is it, what is it good for, and what makes it work? *Forests, Trees and People Programme. Phase II Working Papers: W96I-24.* Indiana: Indiana University.

MENDES, A. 2012. Payments for forest environmental services in Portugal. The case of the Permanent Forest Fund. *Power Point Presentation.* Porto: Faculty of Economics and Management. Portuguse Catholic University.

MENDES, A., STØRDAL, S., ADAMCZYK, W., BANCU, D., BOURIAUD, L., FELICIANO, D., GALLAGHER, R., KAJANUS, M., MÉSZÁROS, K., SCHRAML, U. & VENZI, L. 2006. Forest owners' organisations across Europe: similarities and differences. *In:* NISKANEN, A. (ed.) *Issues affecting enterprise development in the forest sector in Europe.* Joensuu: University of Joensuu, Faculty of Forestry.

MENDES, A. M. S. C., FELICIANO, D., TAVARES, M. & DIAS, R. 2004. The Portuguese forests: Country level report. *EFFE – Evaluating Financing of Forestry in Europe.* Portugese Catholic University: Porto Regional Center

NFI 2014. 50-year forecast of softwood timber availability: Overview of the 50-year forecast of softwood timber availability. *NFI Statistical Analysis Report* Edinburgh: National Forest Inventory, Forestry Commission.

NÍ DHUBHÁIN, A., COBANOVA, R., KARPPINEN, H., MIZARAITE, D., RITTER, E., SLEE, B. & WALL, S. 2007. The values and objectives of private forest owners and their influence on forestry behaviour: The implications for entrepreneurship. *Small-Scale Forestry*, 6, 347–357.

NICHIFOREL, L., DEUFFIC, P., THORSEN, B.-J., WEISS, G., HUJALA, T., KEARY, K., LAWRENCE, A., AVDIBEGOVIĆ, M., DOBŠINSKÁ, Z., FELICIANO, D., GÓRRIZ-MIFSUD, E., HOOGSTRA-KLEIN, M., HRIB, M., JARSKÝ, V., JODŁOWSKI, K., LUKMINE, D., PEZDEVŠEK MALOVRH, S., NEDELJKOVIĆ, J., NONIĆ, D., KRAJTER OSTOIĆ, S., PUKALL, K., RONDEUX, J., SAMARA, T., SARVAŠOVÁ, Z., SCRIBAN, R. E., ŠILINGIENĖ, R., SINKO, M., STOJANOVSKA, M., STOJANOVSKI, V., STOYANOV, T., TEDER, M., VENNESLAND, B., WILHELMSSON, E., WILKES-ALLEMANN, J., ŽIVOJINOVIĆ, I. & BOURIAUD, L. 2020. Two decades of forest-related legislation changes in European countries analysed from a property rights perspective. *Forest Policy and Economics*, 115, 102146.

NICHIFOREL, L., KEARY, K., DEUFFIC, P., WEISS, G., THORSEN, B. J., WINKEL, G., AVDIBEGOVIĆ, M., DOBŠINSKÁ, Z., FELICIANO, D., GATTO, P., GORRIZ MIFSUD, E., HOOGSTRA-KLEIN, M., HRIB, M., HUJALA, T., JAGER, L., JARSKÝ, V., JODŁOWSKI, K., LAWRENCE, A., LUKMINE, D., PEZDEVŠEK MALOVRH, S., NEDELJKOVIĆ, J., NONIĆ, D., KRAJTER OSTOIĆ, S., PUKALL, K., RONDEUX, J., SAMARA, T., SARVAŠOVÁ, Z., SCRIBAN, E. A., ŠILINGIENĖ, R., SINKO, M., STOJANOVSKA, M., STOJANOVSKI, V., STOYANOV, N., TEDER, M., VENNESLAND, B., VILKRISTE, L., WILHELMSSON, E., WILKES-ALLEMANN, J. & BOURIAUD, L. 2018. How private are Europe's private forests? A comparative property rights analysis. *Land Use Policy*, 76, 535–552.

OSTROM, E. 2000. Private and common property rights. *Encyclopedia of Law and Economics. Civil Law and Economics.* Cheltenham: Edward Elgar.

OSTROM, E. & HESS, C. 2007. Private and common property rights. *Workshop in Political Theory and Policy Analysis: W07–25* Indiana: Indiana University

PAQUEL, K., BOWYER, C., ALLEN, B., NESBIT, M., MARTINEAU, H., LESSCHEN, J. P. & ARETS, E. 2017. Analysis of LULUCF actions in EU Member States as reported under Art. 10 of the LULUCF Decision. Brussels: European Commission.

QUIROGA, S., SUAREZ, C., FICKO, A., FELICIANO, D., BOURIAUD, L., BRAHIC, E., DEUFFIC, P., DOBSINSKA, Z., JARSKY, V., LAWRENCE, A. & NYBAKK, E. 2019. What influences European private forest owners' affinity for subsidies? *Forest Policy and Economics*, 99, 136–144.

RING, I., HANSJÜRGENS, B., ELMQVIST, T., WITTMER, H. & SUKHDEV, P. 2010. Challenges in framing the economics of ecosystems and biodiversity: the TEEB initiative. *Current Opinion in Environmental Sustainability*, 2, 15–26.

SARVAŠOVÁ, Z., ZIVOJINOVIC, I., WEISS, G., DOBŠINSKÁ, Z., DRĂGOI, M., GÁL, J., JARSKÝ, V., MIZARAITE, D., PÕLLUMÄE, P., ŠÁLKA, J., SCHIBERNA, E., ŠIŠÁK, L., WOLFSLEHNER, B., ZALITE, Z. & ZALITIS, T. 2015. Forest owners associations in the central and eastern European region. *Small-scale Forestry*, 14, 217–232.

SCHLAGER, E. & OSTROM, E. 1992. Property-rights regimes and natural resources: A conceptual analysis. *Land Economics*, 68, 249–262.

SHAHI, K., SIRY, J., BETTINGER, P., YAGI, N. & SMITH, J. 2023. Estimating timber supply elasticity of private forest landowners in the US South. *Forest Policy and Economics*, 154, 103038.

SIRY, P. J., MCGINLEY, K., CUBBAGE. W.F. & BETTINGER, P. 2015. Forest tenure and sustainable forest management. *Open Journal of Forestry*, 5, 526–545.

SORGE, S., MANN, C., SCHLEYER, C., LOFT, L., SPACEK, M., HERNÁNDEZ-MORCILLO, M. & KLUVANKOVA, T. 2022. Understanding dynamics of forest ecosystem services governance: A socio-ecological-technical-analytical framework. *Ecosystem Services*, 55, 101427.

SOTIROV, M., SALLNÄS, O. & ERIKSSON, L. O. 2019. Forest owner behavioral models, policy changes, and forest management. An agent-based framework for studying the provision of forest ecosystem goods and services at the landscape level. *Forest Policy Econ.*, 103, 79–89. https://doi.org/10.1016/j.forpol.2017.10.015.

SOTIROV, M., WINKEL, G. & ECKERBERG, K. 2021. The coalitional politics of the European Union's environmental forest policy: Biodiversity conservation, timber legality, and climate protection. *Ambio*, 50, 2153–2167.

TAKALA, T., HUJALA, T., TANSKANEN, M. & TIKKANEN, J. 2017. Forest owners' discourses of forests: Ideological origins of ownership objectives. *Journal of Rural Studies*, 51, 1–14.

TEEB 2008. The Economics of Ecosystems and Biodiversity: Interim Report. Wesseling, Germany: Welzel+Hardt.

TEEB 2010. The Economics of Ecosystems and Biodiversity: Mainstreaming the Economics of Nature: A synthesis of the approach, conclusions and recommendations of TEEB. Malta: Progress Press.

TORRALBA, M., LOVRI, M., ROUX, J.-L., BUDNIOK, M.-A., MULIER, A.-S., WINKEL, G. & PLIENINGER, T. 2020. Examining the relevance of cultural ecosystem services in forest management in Europe. *Ecology and Society*, 25.

ULYBINA, O. & FENNELL, S. 2013. Forest certification in Russia: Challenges of institutional development. *Ecological Economics*, 95, 178–187.

VATN, A. 2005. *Institutions and the Environment.* Cheltenham: Edward Elgar.

VEDEL, S. E., JACOBSEN, J. B. & THORSEN, B. J. 2015. Forest owners' willingness to accept contracts for ecosystem service provision is sensitive to additionality. *Ecological Economics*, 113, 15–24.

WEAR, D. N. & GREIS, J. G. 2013. The Southern Forest Futures Project: technical report. US Department of Agriculture, Forest Service, Southern Research Station.

WEISS, G., LAWRENCE, A., HUJALA, T., LIDESTAV, G., NICHIFOREL, L., NYBAKK, E., QUIROGA, S., SARVAŠOVÁ, Z., SUAREZ, C. & ŽIVOJINOVIĆ, I. 2019. Forest ownership changes in Europe: State of knowledge and conceptual foundations. *Forest Policy Econ.*, 99, 9–20. https://doi.org/10.1016/j.forpol.2018.03.003.

WHITEMAN, A., WICKRAMASINGHE, A. & PIÑA, L. 2015. Global trends in forest ownership, public income and expenditure on forestry and forestry employment. *Forest Ecology and Management*, 352, 99–108.

WONG, J., LAWRENCE, A., URQUHART, J., FELICIANO, D. & SLEE, B. 2015. Forest land ownership change in United Kingdom. *COST Action FP1201 FACESMAP Country Report.* Vienna: European Forest Institute Central-East and South-East European Regional Office.

ŽIVOJINOVIĆ, I., DOBSINSKA, Z., SALKA, J., WEISS, G., JARSK, V., NEDELJKOVIC, J., PETROVIC, N. & SARVASOVA, Z. 2016. Actors and interests related to the restitution in the forestry sectors in transition. *In:* WEISS, G., DOBSINSKA, Z., FELICIANO, D., HUJALA, T., LAWRENCE, A., LIDESTAV, G., SARVASOVA, Z. & ZIVOJINOVIC, I. (eds.) *Book of Abstracts, Final conference of the COST Action FP1201 FACESMAP.* Vienna: University of Natural Resources and Life Sciences, Vienna.

ŽIVOJINOVIC, I., LIDESTAV, G., FELICIANO, D., HUJALA, T., LAWRENCE, A. & WEISS, G. 2015. Concepts, methods and findings in forest ownership research in Europe. *Forest Land Ownership Changes in Europe: Significance for Management And Policy (FACESMAP) – COST Action FP1201.* Austria: VIenna: European Forest Institute Central-East and South-East European Regional Office, Vienna.

ŽIVOJINOVIĆ, I., WEISS, G., LIDESTAV, G., FELICIANO, D., HUJALA, T., DOBŠINSKÁ, Z., LAWRENCE, A., NYBAKK, E., QUIROGA, S. & SCHRAML, U. 2015. Forest Land Ownership Change in Europe. COST Action FP1201 FACESMAP Country Reports, Joint Volume. *EFICEEC-EFISEE Research Report.* Vienna: University of Natural Resources and Life Sciences, Vienna.

7 Forest fires, resilience, and climate change

1. Introduction

Wildfires represent a challenging natural (and anthropogenic) hazard in Europe, generating substantial damage to forests, agricultural lands, and urban areas. The economic impact of these fires is steep, resulting in up to 1% of the gross domestic product (GDP) being lost in several European countries. Annually, the European Union (EU) Member States face the consequences of wildfires, which leave an average of around 500,000 hectares (ha) reduced to ash (Figure 7.1). Even more, statistics reveal a troubling trend with over 50,000 wildfires across Europe each summer (EC, 2018). For instance, in 2022, wildfires in Europe reached unprecedented levels. During the heatwave that year, it was estimated that forest fires destroyed a record-breaking 786,000 ha of forestland (Canelas da Silva et al., 2023; Rodrigues et al., 2023). This set a record since EU-wide tracking began in 2006 and signalled an escalation in the intensity and impact of these types of events. The rise of these so-called 'mega-fires' presents a severe example of the increased impact and frequency of extreme weather events across Europe and globally. This trend underscores the pressing need for an in-depth analysis of EU strategies and policies regarding forest fires and the broader approach to managing extreme weather events. Understanding how the EU addresses forest fires provides an avenue for enhancing preparedness, mitigating risks, and designing more resilient policies in the face of climate change.

Forest fires are the result of a complex web of factors, including, but not limited to, strategic land and urban development, traditional and cultural practices, and the variable influence of local and regional climate and weather conditions (Galiana et al., 2013, Tedim et al., 2014, JRC, 2022). This complexity extends to forest fire policy, where significant variations can be found across the EU. These variations are fundamentally rooted in the diverse risk exposure landscape, varied forest management strategies, national political agendas and forest governance frameworks unique to each EU Member State. For instance, different government agencies or ministries are commonly tasked with overseeing fire risk management, each with its specific mandate, role and responsibilities (EC, 2018, 2021). For example, in Spain, forest fire management is typically the responsibility of the regional governments, where each region has its forest fire agency. In France, within the Ministry of the Interior, the General Directorate for Civil Security and Crisis Management coordinates responses to forest fires, often in collaboration with local fire brigades. Whereas in Sweden, the Swedish Civil Contingencies Agency manages forest fires together with the Swedish Forest Agency. These examples showcase that forest fire governance is not harmonised across Europe and that there is much variation.

DOI: 10.4324/9781003198987-7

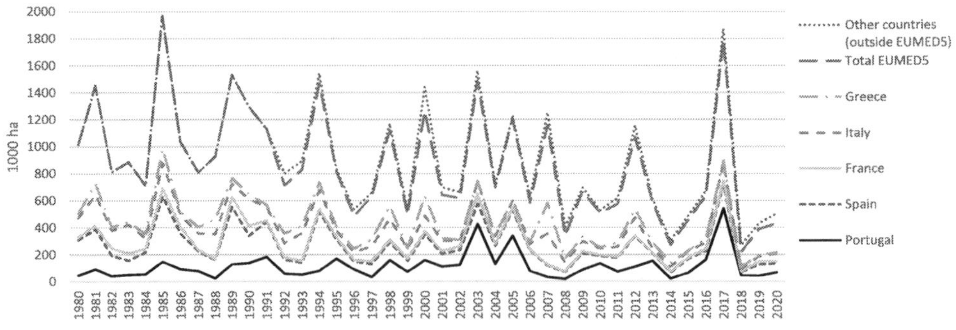

Figure 7.1 Area (in ha) burnt by forest fires in five European countries.

Source: European Environment Agency

The varied forest fire policy environment is further linked to the EU's principle of subsidiarity, which essentially means that the authority to develop forest-related policies resides with the individual EU Member State (Aggestam and Pülzl, 2018, 2020, Aggestam and Giurca, 2021). This allows forest fire policies to be tailored to national and regional conditions and legislative specificities. While this makes sense, it also means that the EU's role regarding forest fire policy is mainly supportive, aimed at improving and supporting the efforts of its Member States. This supportive role is exemplified through various EU initiatives, such as financial contributions to forest fire prevention projects within the framework of national Rural Development Programmes (RDPs), the provision of funds for recovery through the EU Solidarity Fund, and research support via EU Research Framework Programmes. Another significant example of EU-wide support is the establishment of the European Forest Fire Information System (EFFIS), a monitoring tool for forest fire information and analysis (McInerney et al., 2013). EFFIS has been integrated into the EU Copernicus program's emergency management services (Box 7.1).

In recent years, research into forest fires has increased substantially, and it is a broad and dynamic field that extends into many interdisciplinary areas. This growing body of research and critical analysis covers a wide range of topics, including wildfire prevention techniques, fire behaviour modelling, advanced fire mapping and monitoring technologies, post-fire rehabilitation and reforestation, and the development of control and integrated management systems (Silva et al., 2010, Dupuy et al., 2020, López-Vicente et al., 2021, Müller et al., 2020). Academics have also reviewed policy instruments and governance structures that form the core of EU forest fire policy (Mourao and Martinho, 2019). This includes in-depth reviews regarding the interplay between EU institutions, EU Member States, and local governments in the development and implementation of forest fire policies. For instance, Aguilar and Montiel (2011) have analysed forest fire governance mechanisms focusing on the Mediterranean region, where countries like Spain, France, Portugal, and Italy face increasing risks associated with wildfires.

Moreover, studies have investigated the evolution of wildfire legislation and policies within the European framework. These studies have observed a trend towards the harmonisation of EU policies and increased operational efficiency (Montiel-Molina, 2013, Ascoli et al., 2023, Neidermeier et al., 2023). EFFIS has also been the subject of comprehensive reviews by San-Miguel-Ayanz et al. (2013b) and McInerney et al. (2013), highlighting its essential role in the EU's environmental policy framework. EFFIS represents a significant development in collecting, analysing, and disseminating forest fire data to relevant agencies, policymakers, and the public (Box 7.1).

In parallel to this work, an extensive body of research focuses on the EU's Civil Protection Mechanism, for example, studied by Ekengren et al. (2006) and Parker et al. (2019). These studies demonstrate the civil protection mechanism's critical role in coordinating a unified EU response to forest fires, particularly since fires do not respect national boundaries.

Box 7.1 Understanding forest fire monitoring and the European Forest Fire Information System

Timely and accurate information is critical when addressing forest fires. Various systems and initiatives have been established to monitor, assess, and respond to the threat of forest fires. Among these, the European Forest Fire Information System (EFFIS) is an example of an integrated EU-based platform designed to address wildfires. It was established in response to the increasing frequency and intensity of forest fires in Europe, and it sets out to improve the EU's preparedness and capacity to respond to forest fires. Its primary function is to monitor and provide early warning of fires, using satellites and ground-based sensors to monitor forests in real-time. It also sets out to forecast fire danger, considering weather conditions, vegetation, and historical fire data to predict the likelihood and intensity of fire occurrences in different regions. One aspect of this work is disseminating critical information to relevant stakeholders, such as fire occurrences (see figure below), fire danger levels, and fire risk maps, to national and regional authorities and the public.

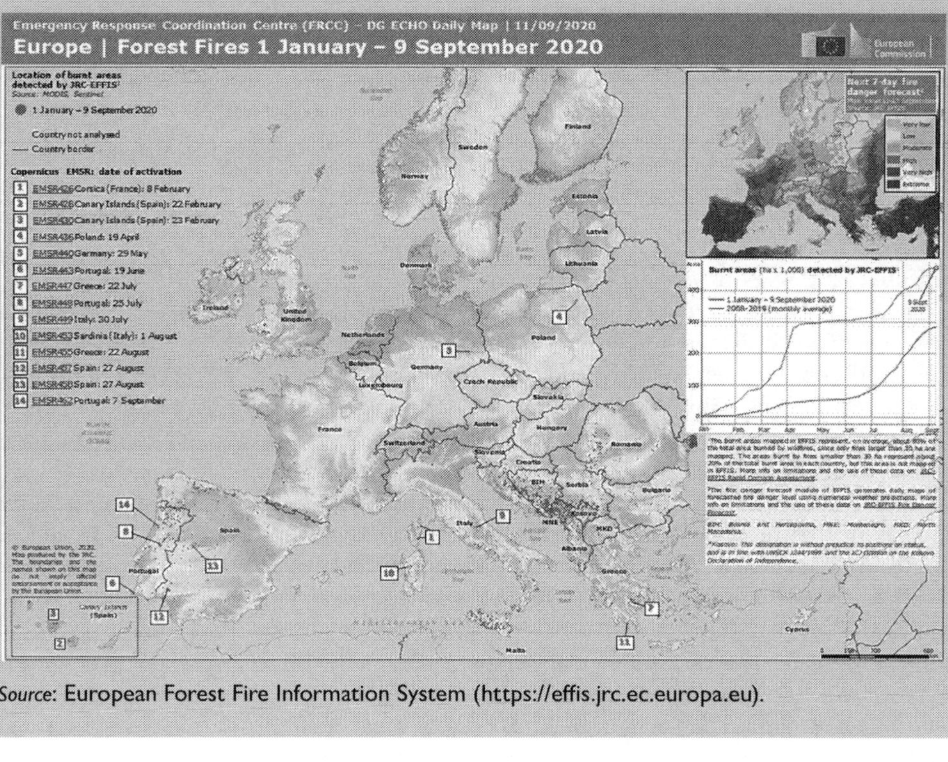

Source: European Forest Fire Information System (https://effis.jrc.ec.europa.eu).

Amidst the growing concern over climate change, integrating climate adaptation strategies into EU forest fire policies has also become an important study area. Researchers like Dupuy et al. (2020 and), Miezïte et al. (2022) have examined how policies are changing to address the increasing wildfire risks brought on by the changing climate. All this knowledge has been instrumental in informing policy decisions and operational strategies across the EU but also in helping to ensure that the response to forest fires is more proactive, efficient, and coordinated.

As can be noted, quite a few studies have investigated EU forest fire policy, its management, and complexities surrounding its monitoring systems more directly (San-Miguel-Ayanz et al., 2013b, Barbosa et al., 2008, Montiel-Molina, 2013, McInerney et al., 2013, San-Miguel-Ayanz, 2012). However, these studies have principally concentrated on the Mediterranean region, where the risk of forest fires is exceptionally high (Corona et al., 2015, Damianidis et al., 2021, San-Miguel-Ayanz et al., 2013a, Dupuy et al., 2020). Moreover, despite the extensive research on forest fires, there has, to date, yet to be a comprehensive analysis concerning how EU policy instruments are framing forest fires. The literature on framing, including empirical investigations into forest fires, has primarily been concerned with how the media represents extreme weather events. For instance, studies by Ekayani et al. (2016), Castelló and Montagut (2019), Anderson et al. (2018), and Nilsson and Enander (2020) have reviewed the media narratives surrounding forest fires, considering public opinion, responses, and perspectives on forest fires. Castelló and Montagut (2019), for instance, provide an insightful review of media coverage in Catalonia, finding five media frames: these encompass agricultural risk, climate change and weather conditions; imprudent and negligent attitudes; inappropriate fuel management and woodland conditions; and arson. Similarly, Seijo (2009) conducted an analytical review regarding how the media framed forest fires, focusing on the content and impact of an anti-forest fire campaign in Spain. While this research demonstrates the role of media framing in shaping public discourse and dialogue on forest fires, the EU policy instruments that are designed to address forest fires have yet to be explored. Such an analysis could complement insights into how the framing of forest fires may affect its management.

Research into framing in EU forest policy and forest fire management has led to substantive contributions across scientific disciplines, improving our understanding of how forests are perceived and addressed. Notably, Elomina and Pülzl (2021) have conducted an in-depth analysis of EU forest-related policies, identifying nine frames that reflect the diversity of perspectives across policy domains, including environment, bioenergy, and climate change. Their research demonstrates the varied nature of framing in EU forest-related policy. Adding to this area of work, Söderberg and Eckerberg (2013) considered the narratives surrounding the promotion of bioenergy. Their study outlined four frames that set out narratives associated with the ongoing debate on biofuel production. These frames cover a range of concerns, from energy security to rural development, demonstrating some of the complexities surrounding EU bioenergy policy. From a different perspective, Bosomworth (2015) took a new institutional perspective to review how frames influence the discourse around adaptation strategies in the forest-based sector, particularly regarding wildfires. This approach demonstrates how institutional frameworks and dominant narratives shape policy responses to forest fires.

Complementing these perspectives, Lidskog et al. (2019) conducted a study that aligned with framing analysis, which outlined public perceptions of wildfires in Sweden. Their findings emphasise the role of personal experiences, available resources, worldviews, and framing in shaping how the public understands and responds to natural disasters, emphasising the importance of socio-cultural factors regarding how forest fires are perceived and managed.

Similarly, Beland Lindahl et al. (2017) analysed how forests were framed in connection with Swedish policy frameworks, providing insights into the policy discourse at the national level. Interlinked with work on the bioeconomy, Kleinschmit et al. (2017) explored frames associated with the discourse around the bioeconomy, in this case, focusing on forests. This study contributes to a deeper understanding of how forest resources are conceptualised and used in relation to the bioeconomy sector, which depends on forests for raw materials.

In social science research, studies on forest fires have covered a diverse range of topics, such as risk management and public perception of forest fires (Calkin et al., 2014, Langer and McGee, Velez et al., 2017), forest owners' preparedness with regards to fire incidents (Valente et al., 2015), media coverage of forest fires, revealing how media narratives influence public opinion and policy (Berglez and Lidskog, 2019), and attitudes towards fire mitigation and adaptation measures, highlighting the importance of understanding community perspectives in developing effective fire management strategies (Lidskog and Sjödin, 2016). These studies reveal the complexity of forest fire management and the wide range of factors influencing its effectiveness. More importantly, they highlight the importance of interdisciplinary approaches in understanding and addressing forest fires in the context of EU policy.

This chapter introduces the reader to forest fire research by investigating the framing of forest fires within the context of EU policymaking. Firstly, the chapter will review how forest fires, as a natural hazard, are depicted in EU forest-related policy documents. Analysing how forest fires are portrayed can reveal underlying values, priorities, and biases in EU policy approaches. The aim will be to provide insights into how the framing of forest fires reflects EU perspectives on environmental and socio-economic issues. Secondly, the chapter aims to identify and characterise collective frames used to discuss forest fires across different EU policy domains over time. Forest fires are cross-sectoral, intersecting several policy domains such as the environment, agriculture, rural development, climate action, and civil protection. This chapter sheds light on the varied forest fire discourse by mapping collective frames. The chapter will critically engage with the existing literature on framing, aiming to differentiate the various types of frames used to contextualise and narrate forest fire issues. The intent will be to outline the discourse surrounding forest fires, moving beyond the focus of existing research on media framing and public discourse. The methodology will be a mixed-method, grounded approach, combining qualitative content analysis with a frame analysis. This methodological framework will enable an in-depth review of EU policy documents. The chapter seeks to complement and expand upon the existing framing literature on forest fires.

In fire management, terms such as 'hazards' and 'risks' play a crucial role. The search for a scientific foundation for evaluating forest fire hazards and liabilities dates back to 1916 (Hardy, 1983). Even today, multiple definitions exist for 'risks' and 'hazards', and the metrics used to quantify these concepts vary significantly. For instance, Bachmann and Allgower (2000) have pointed out that the term 'hazard' can encompass two distinct aspects: denote the conditions that precede a fire or refer to the fire itself. In the latter case, wildfires themselves are considered hazards. However, when combined with the word 'fire' (e.g., 'fire hazard'), it specifically refers to the conditions leading up to a fire. This highlights the critical importance of considering the terminology associated with forest fires. In the EU, and more specifically within the European Forest Fire Information System (EFFIS), the terminology revolves around 'fire danger', which is synonymous with hazards, and 'vulnerabilities' across three key categories: human, ecological, and economic values exposed in vulnerable areas (McInerney et al., 2013, San-Miguel-Ayanz, 2012, San-Miguel-Ayanz et al., 2013b). These terms form the basis for conducting forest fire risk assessments (Box 7.1). Risk assessments, in turn, integrate the evaluation of fire danger (or probability) and vulnerabilities (or potential outcomes). This approach results in a risk ranking that can be used to enhance fire management strategies.

1.1. EU policymaking on forest fires

The responsibility for managing and protecting forests principally falls to the EU Member State. However, the European Commission plays a supportive role in addressing forest fires through different Directorates-General (DGs). These DGs are involved in various activities, from firefighting to forest restoration initiatives. Key among them are the DGs for Environment, Research, Agriculture and Rural Development, Regional Policy, and European Civil Protection and Humanitarian Aid Operations. Each DG contributes towards the management and mitigation of forest fires.

In terms of policy, one of the earliest EU policy frameworks dedicated to combating forest fires was established through Regulation, 2158/92. This regulation played a role in supporting national efforts for forest fire prevention and restoration between 1992 and 2002,[1] marking a significant step in the EU's commitment to addressing forest fires. This effort was succeeded by the Forest Focus Regulation (Regulation, 2152/2003), which continued and expanded upon the work of its predecessor, supporting preventative measures against forest fires. The Forest Focus Regulation demonstrates the growing focus on proactive fire management strategies at the EU level. Following this, the LIFE programme (Regulation 614/2007), with its financial backing extended until 2027, took over the objectives covered by the Forest Focus Regulation. The LIFE programme has placed a strong emphasis on forest monitoring, the development of information systems, and the prevention of forest fires. This programme represents the EU's ongoing commitment to forest preservation and highlights the evolution of its approach to forest fire management.

In addition to the policies mentioned above, the Common Agricultural Policy (CAP) and the European Agricultural Fund for Rural Development (EAFRD) provide further support for forest fire management. The CAP is crucial in enhancing the EU's forest fire response framework, offering financial support for forest restoration and preventive actions against fires. More importantly, they mandate EU Member States to identify and manage high-risk fire zones within their forest management plans. This requirement underscores the EU's proactive approach, highlighting the importance of risk assessment and mitigation. Furthermore, the EU Solidarity Fund, established under Regulation 2012/2002, supports EU Member States after catastrophic events like forest fires. This fund demonstrates how the EU can assist countries during a disaster, such as providing financial aid to recover from the impacts of a forest fire. Moreover, operational linkages exist between this financial support and the EU Civil Protection Mechanism (Ekengren et al., 2006, Parker et al., 2019). This mechanism ensures the rapid mobilisation of emergency resources and personnel to respond to forest fires. Its significance is highlighted because wildfires accounted for 30% of all assistance requests received between 2007 and 2019. This statistic underscores the prevalence of forest fires as a significant challenge within the EU.

Concerning coordination and advisory services, the Commission's Emergency Response Coordination Centre (ERCC) is crucial in overseeing forest fire risk and emergency monitoring services. The ERCC integrates national efforts with European monitoring services, such as EFFIS (Box 7.1). The intent behind this initiative is to leverage both national capabilities and EU-wide resources. Further strengthening this framework is the Expert Group on Forest Fires, established by DG Environment in 1998. This group guides the European Commission and EU Member States on fire prevention and mitigation strategies. In addition to its advisory role, the Expert Group collaborates with EFFIS to deliver EU-level assessments before and after forest fires. These assessments include risk mapping and post-fire evaluations, essential for understanding fire impacts. For example, one tangible output from the Expert Group is the publication of the 2021 guidelines on land-based wildfire prevention. This document offers practical advice and best practices for preventing wildfires (EC, 2021).

1.2. Frames, framing, and forest fires

With disciplinary roots in psychology and sociology, frames are broadly understood as mechanisms by which individuals or institutions interpret their social and natural surroundings. They serve as a system or process that simplifies the management and classification of information. While there are various interpretations as to what constitutes a 'frame' and 'framing', a central theme common to frame theory is the concept of information filtering. This perspective, as introduced by researchers like Goffman (1974), Gamson and Modigliani (1989), Entman (1993), Pan and Kosicki (1993), and Reese et al. (2001), argues that frames help in the selective processing of information. It is widely acknowledged that frames influence how individuals and institutions perceive information and make decisions (Rein and Schön, 1996, Daviter, 2007). Framing, then, becomes an active process of applying a frame. For example, when a person tries to affect a discussion by highlighting a particular aspect of reality (Snow and Benford, 1988, Pan and Kosicki, 1993, Reese et al., 2001) or how people interpret experiences, facts, and events (Stern et al., 1999, Kaltenborn and Bjerke, 2002, Vugteveen et al., 2010).

This sense-making process, like simple heuristics, is partly innate (e.g., cognitive), partly socially constructed (e.g., through power relations), and partly learnt (e.g., via education) (Gigerenzer and Todd, 1999). As Bruno and Lay (2008) suggested, this combination forms the foundation for decision-making processes. Moreover, the multi-dimensional nature of framing explains why it has been extensively used to try and understand social conflicts (Benford and Snow, 2000, Dewulf et al., 2009). For instance, the impact of issue framing on environmental conflicts underscores the effects of framing in shaping perspectives and outcomes in environmental discourse (Shmueli et al., 2006, López-i-Gelats et al., 2009). In other words, in this era of alternative facts, framing can be a valuable tool to understand how individuals and institutions address complex social and environmental issues.

This chapter focuses on the concept of collective framing, drawing upon the work of Rein and Schön (Rein and Schön, 1996, Schön and Rein, 1994). In this context, frames are the lenses or narratives adopted by different stakeholder groups or institutions intended to influence or shape the interpretation of forest fires during the policymaking process. These frames capture the perspectives and agendas of different actors, ranging from government agencies and environmental interest groups to local communities and industry representatives. For example, when stakeholder groups stress specific aspects of forest fires, such as the causes, impacts, or potential solutions, they feed into a process aiming at developing a shared understanding or a collective frame. This collective frame is critical in shaping the policy discourse on forest fires (Rein and Schön, 1996). More importantly, creating a collective frame is a dynamic process emerging from the interplay and negotiation process between actors. This process commonly leads to a consolidated perspective in the final policy documents.

Individual frames are, in contrast, usually promoted by single people or smaller groups. These frames represent personal viewpoints and are often characterised by their directness and specificity. In other words, they are unmediated by the broader policy discourse. As a result, individual frames are usually more varied and individualistic than collective frames. For example, collective frames are most often shaped through negotiation and compromise, adjusting to the interests and perspectives of different stakeholders. In contrast, individual frames are more likely to reflect the perspectives and agendas of specific actors (Baumgartner and Mahoney, 2008, Coninx et al., 2015). Thus, while this chapter zooms in on collective framing, understanding both collective and individual frames is important to understand the full spectrum of perspectives that may shape policymaking.

Considering the policymaking process itself, it is reasonable to expect that adopted EU policy documents, which represent the finalisation of a negotiation process, will mainly represent collective frames. In other words, these documents are not just static texts; they are the end-products of a negotiation process reflecting the dominant framing of issues as shaped through the interactions between stakeholders (Schön and Rein, 1994, Donahue et al., 2011). However, it is important to recognise that these collective frames are not explicitly stated. Instead, they are constructs within the text, implying that reconstruction and interpretation are needed. With this in mind, this chapter aims to reconstruct and analyse these frames, using EU policy documents to characterise how forest fires are framed. One key aspect of this analysis will be assessing whether and how these frames have evolved. For instance, the transition in the EU's approach towards forest fire management from a principally reactive to a proactive and preventative position indicates a shift in collective framing over time (Montiel and Kraus, 2010, Tedim et al., 2014). This change is reflected in the evolution of policy language, changes in funding priorities, and the implementation of various initiatives. One example is EFFIS, which has played a critical role in monitoring and managing forest fire risks. The development and integration of EFFIS into EU policy suggests a movement towards a more comprehensive and informed approach regarding forest fire management (Ganteaume et al., 2013, San-Miguel-Ayanz, 2012, San-Miguel-Ayanz et al., 2013a). Thus, by conducting an in-depth analysis of these policy documents, this study hopes to uncover the dominant collective frames that have influenced and shaped EU policy on forest fires. This is important not only for explaining policy responses to forest fires but also as a way to identify potential areas for improvement.

2. Approach

This chapter builds on a recent publication – Through smoke to policy: Framing the EU forest fire policy landscape (Aggestam, 2024). The current text has been extended and revised to reflect additional elements that could not be covered in the original work.

2.1. *Methodological framework*

2.1.1. Building a database of EU policy documents

The initial phase focused on constructing a database of relevant EU policy documents. This was accomplished by using EUR-LEX,[2] the official repository for EU law and public documents. A preliminary search with the keywords 'forest' and 'fire' yielded 4,362 results. To refine this set of policy documents, two advanced search queries were carried out, distinguishing between 'soft' (non-legally binding) and 'hard' (legally binding) policy documents.

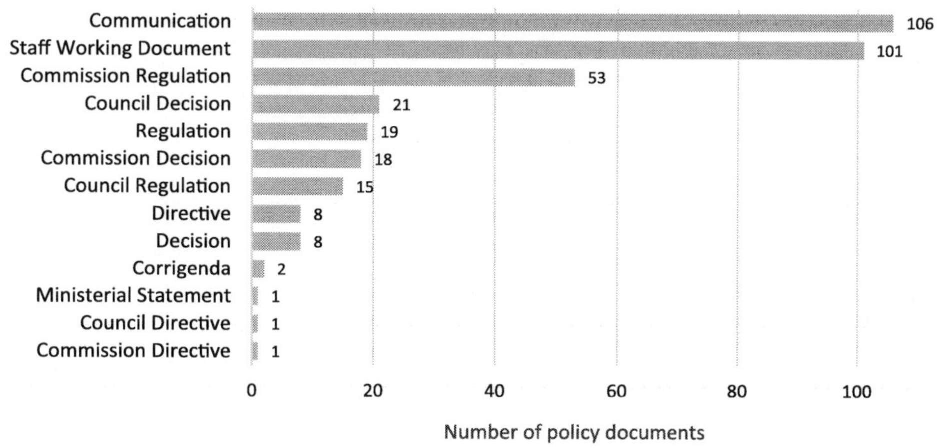

Figure 7.2 Types of EU documents included following the first screening.

Source: own figure

This refined search produced 354 policy documents, comprising 208 soft and 146 hard documents (Figure 7.2).[3] The timeframe for these documents spanned from 1982 to 2021 for soft policies and from 1966 to 2021 for hard policies. Following the advanced search, all relevant policy documents were compiled into an Atlas.ti 22 database for in-depth analysis.

More detailed information about the screening process, search queries, and the composition of the sample is provided in Aggestam (2024). This approach to creating a database ensured a suitable representative sample of EU policy documents for the analysis.

2.1.2. Carrying out a content analysis, assigning relevance, and coding content

The next phase of the study focused on evaluating the relevance of the 354 policy documents. This evaluation was conducted through a two-step process:

1. **Frequency count of 'fire'**: Each document underwent a frequency count for 'fire'. Based on the number of times 'fire' appeared, each document was preliminarily categorised with a low, medium, or high relevance score. These scores were intended to provide an initial assessment of potential relevance to the study.
2. **Content verification**: Each document was reviewed to assess the accuracy of the preliminary relevance score. This involved examining the context in which 'fire' was mentioned to assess its importance.

As a result of this process, 94 policy documents were assessed as not directly related to forest fires and were excluded from further analysis. This approach ensured that the dataset only included documents that substantively addressed the subject of forest fires.

After the initial relevance assessment, the study moved on to a more detailed content analysis of the 263 policy documents (Krippendorff, 2004, Elo et al., 2014, Erlingsson and Brysiewicz, 2017). This phase involved a review of all explicit references to forest fires within all policy documents. Utilising the text search functionality of Atlas.ti, a targeted search was conducted for paragraphs containing the keywords 'forest' and 'fire', along with their inflected forms. The advanced model for English language text analysis guided this search. The initial purpose was not to contextualise relevant paragraphs but to quantify and locate them within the documents. This search yielded 608 paragraphs in the soft policy instruments and 235 in the hard policy instruments, covering all variations of the terms 'forest' and 'fire'. This step was crucial to isolate text that addressed forest fires, setting the stage for further analysis.

Each paragraph referencing forest fires was then read and screened for inductive codification. This codification process was grounded, meaning codes were assigned iteratively based on the emerging findings. This approach aligns with the grounded approach outlined by Strauss and Corbin (1990). The primary purpose was to identify the thematic focus of each citation (Erlingsson and Brysiewicz, 2017, Friese, 2019, Krippendorff, 2004). This evolving coding scheme eventually comprised 64 codes, which enabled a comprehensive categorisation of all citations (Aggestam, 2024). Moreover, this approach facilitated the analysis of relationships within the primary text material. The study identified patterns and trends associated with forest fires by categorising the content based on themes and concepts.

2.1.3. Developing collective frames

After the codification process, a co-occurrence analysis was conducted to determine the frequency of code combinations (Friese, 2019). This analysis assessed the prevalence of codes within different categories of policy documents, which were divided into six groups based on the type (soft or hard) and relevance (low, medium, and high). The results were compiled into two tables, one covering frequency counts by document type and another showing co-occurrences across the 64 codes.

This analysis revealed trends in code allocation. For soft policy documents, 1,227 codes were assigned, with 830 (67.6%) corresponding to highly relevant documents. This significant concentration of codes in only 38 documents (23.2% of the sample) highlights the correlation between the assigned coding and the documents' relevance. Similarly, in the hard policy documents, 242 out of 429 codes (56.4%) were allocated to highly relevant categories. Additionally, 54 search results were labelled as irrelevant, covering 19 soft policy documents (3.1%) and 35 hard policy documents (8.1%). These included instances such as web links or source references containing the words 'forest' or 'fire'. Notably, the proportion of citations considered irrelevant was higher among hard policy documents, mainly due to the inclusion of reporting templates.

The frame-building process began with the thematic grouping of categories that frequently co-occurred, distinguishing between hard and soft policy documents. For instance, categories such as 'forest management', 'forest protection', and 'forestry measures' were commonly found together with 'agriculture' and 'rural development' in both documents. The next step involved ranking these thematic groups based on the frequency of their co-occurrence. An example of a prominent theme was 'climate change'. After establishing these rankings, the associated citations for each thematic group were re-examined. This re-evaluation aimed to outline how 'forest fires' were characterised within these themes, particularly regarding consistency in language use. Additionally, this stage assessed variances across different types of documents (soft versus hard) and between the different relevance groups (low, medium, high). The purpose was to understand how the framing of 'forest fires' differed or aligned across these categories, providing an overview of how forest fires were addressed and prioritised within each EU policy document.

The final stage of the study involved constructing the collective frames, considering the variations observed across different types of policy documents (Daviter, 2018). This step entailed a detailed review and analysis of the thematically grouped content identified through the grounded approach, which provided a foundation for the frames. Furthermore, the co-occurrence analysis played an important role in this process, enabling the determination of the relative dominance of each frame. This was done by quantifying the frequency with which each frame appeared across the hard and soft policy documents.

While the analysis focused on the most prevalent set of frames, it is important to recognise that all identified frames are significant. Each has contributed to shaping the discourse on forest fires, evolving from initial agenda-setting to policy formulation. This perspective aligns with arguments outlined by Björnehed and Erikson (2018), suggesting that the prominence of a frame is not solely determined by its frequency in the current discourse but also by its historical role in policy discussions. Therefore, even frames considered less dominant in the current context have played a role in developing relevant policy documents.

2.1.4. Methodological considerations

It is important to note the challenges inherent in characterising frames within hard policy documents. These documents are typically the result of negotiations and are crafted by legal professionals from different epistemic communities (Zito, 2001, Dunlop, 2013, Polman, 2018). As a result, they often lack emotionally charged language or argumentative framing, making it more difficult to detect underlying frames. This subtlety in legal texts raises the possibility of researchers' subjectivities influencing the interpretation of the document. This issue is a well-known concern in text-based frame analysis ((Erlingsson and Brysiewicz, 2017, Krippendorff, 2004), and while it is a common challenge, it requires careful consideration.

Furthermore, textual analysis naturally involves complexities, such as subjective interpretations of language and difficulties differentiating between framing devices and broader frames or themes. However, this study's grounded and methodical approach was designed to systematically identify and categorise specific narratives related to forest fire policy. This approach aimed to mitigate potential biases and enable a comprehensive understanding of the policy framing process.

Moreover, this study focused deliberately on EU policy documents that address forest fires. It is, however, important to note that by concentrating exclusively on finalised policy documents, the scope of the analysis is inherently limited. More specifically, this approach needs to cover the dynamic aspects of the policy negotiation process. Acknowledging this limitation, it is recognised that the paper cannot reconstruct the dialogue and discussions that led to the formation of the collective frames. Mainly, it does not capture the frames that were part of the negotiation process but did not make it into the final policy document. The analysis of such dynamics would require a much broader methodological framework. For instance, this would involve examining draft documents, meeting minutes, stakeholders' inputs, and potentially conducting interviews with key participants in the policymaking process.

While this study did not consider the deliberative process, this represents a valuable avenue for future research. Exploring these aspects could generate more detailed insights into the discourse surrounding EU forest fire policy. Such future research would be relevant to complement and enrich the findings of this review, providing a more complete picture of the policy framing process.

3. Collective forest fire frames

3.1. Framing in soft versus hard policy documents

The analysis reveals that soft or non-legally binding policy documents, such as communications, staff working documents, and reports, mainly address forest fires using a descriptive approach. For example, 12.7% of the coded content addresses forest fires factually. This includes outlining statistical impacts, like the area of land affected by forest fires (measured in ha lost or restored), funding allocated to fire prevention programs, risk management implications, and links between fires and other environmental challenges, such as climate change and natural hazards. In these references to forest fires, soft policies generally do not issue prescriptive directives to countries. Instead, they provide narratives highlighting particular perspectives, such as escalating socio-economic challenges brought on by climate change. This descriptive framing has likely played a role in shaping the overall discourse around forest fires by focusing on factual and impact-oriented narratives; it may also reflect the role that epistemic communities (or expert committees) play in EU policymaking.

In comparison, hard policy documents, characterised by their legally binding nature, such as directives, regulations, and decisions, tend to be more prescriptive. These documents frequently contain specific calls to action, including mandates associated with forest fire monitoring, prevention, and management, as well as research into innovative technologies, such as remote sensing and earth observations. For instance, while soft policies typically share findings from evaluations, research, or project outcomes, hard policies are more likely to prescribe concrete actions or interventions. This can include directives for national strategies to protect the public from forest fires. This distinction highlights the more action-oriented nature of hard policies compared to soft policies' informational and advisory role.

The data underscores another significant trend: reporting requirements related to forest fires, covering everything from implemented strategies to firefighting efforts, is the most prevalent subject in hard policies. This topic represents 11.4% of all the coded content in these documents. Such a finding aligns with the expected differences between hard and soft policy documents. As outlined, hard policies, like regulations, often employ direct and action-oriented language. In contrast, soft policies like communications usually feature more nuanced and persuasive language. This distinction underscores a notable variation in both language and framing between these two types of policy documents, reflecting their respective purposes and roles in EU policymaking.

> In the EU, policy documents on forest fires exhibit distinct framing approaches associated with either soft or hard policies. Soft policies, typically prepared by different expert-based Commission services, undergo fewer negotiations and lack legal enforcement, often featuring language that allows for more interpretive freedom. This includes value-laden narratives that might not have survived the negotiation process associated with hard policy documents. This difference indicates that the linguistic choices and framing in soft policies are likely designed to influence the discourse on forest fires in a manner that hard policies, constrained by their formal and legal implications, cannot. Further, this suggests the use of language in soft policies to shape perspectives and discussions around forest fire management.

3.2. Frames associated with forest fires

The collective frames, as identified through the content analysis, provide insights into how forest fire management aspects are conceptualised and prioritised within EU policy documents. Table 7.1 synthesises six collective frames, capturing the thematic topics addressed by each frame. This synthesis emerged from categorising various topics under specific themes, such as agriculture and rural development, which were then further analysed to inform the structure of the frame. Three observations should be noted in this context:

1. **Interconnectedness of frames**: The frames, while distinct for analysis, are naturally interconnected, reflecting diverse perspectives expressed as part of the forest fire discourse. For example, discussions around climate change often overlap with sustainable energy concerns; however, they are treated separately here.
2. **Reflections on policy negotiations**: The frames are derived from adopted policy documents, representing the final step of the policymaking and adoption process. It is important to note that these frames consequently represent the end results and do not cover the negotiation process underlying these documents.
3. **Framing distinct policy domains**: The grounded approach means that the frames do not necessarily align precisely with the EU policy domains. Some frames cover broader or cross-cutting issues, such as public access to information, which does not fit into conventional frames.

Table 7.1 Six collective frames covering forest fires in EU policy documents.

Climate Adaptation and Resilience

Focusing on climate change and increased forest fire risks, advocating for adaptive strategies and financial commitments to mitigate risks.

- Climate change is a central theme in EU policy documents addressing forest fires, featuring 285 times across both legally binding (hard, $n = 149$) and non-legally binding (soft, $n = 136$) texts. Forest fires are commonly negatively depicted in these policies because of climate change, enhancing the risk and severity of fires and their socio-economic impacts. Forest fires are typically integrated into a larger narrative on natural hazards, such as droughts, placing them within a disaster management framework that includes firefighting strategies.
- The discussion around forest fires, particularly within this context, emphasises the need for climate action and the potential consequences of failing to act. The narrative often stresses that climate change and increased fire risks are inevitable, advocating for enhanced adaptation ($n = 32$) and resilience ($n = 15$) as key responses. This reflects a broader understanding that these risks are now integral to the natural environment.
- One significant aspect of the legally binding (or hard) policy documents is their significant focus on financial aspects, with 44 instances focusing on funding strategies to mitigate forest fire risks. This emphasises the EU's efforts to financially support adaptive and resilience-building measures to target the increasing threat of forest fires in the context of climate change.

Sustainable Energy Transformation

Linking forest fire risk reduction with sustainable energy practices, suggesting a mutual benefit in managing forest biomass and promoting renewable energy.

- The energy narrative ($n = 29$) presents an interesting perspective on forest fire management by linking renewable energy and forestry practices as methods to reduce forest fire risks. This narrative proposes that active management of forest biomass, such as through bioenergy production, can reduce the fuel available for fires, lowering the risk of forest fires.
- This positive framing uniquely promotes a win-win (or mutually beneficial) approach where it is argued that sustainable energy practices can lead to fewer forest fires, thereby establishing a strategic connection between energy policy and fire risk reduction.
- Forest fires are also linked to the adverse effects of climate change ($n = 10$), underscoring the need for policies that support a transition to renewable energy sources. This dual narrative acts both as a warning about the consequences of climate change and as a call for proactive measures. It strategically advocates for integrated solutions addressing energy sustainability and forest fire prevention.

(Continued)

Table 7.1 (Continued)

Technocratic perspectives on forest fires

Importance of environmental monitoring and data accessibility, highlighting technological approaches like EFFIS for effective forest fire management.

- The thematic areas of environmental monitoring and data accessibility emerged as significant in EU policy discussions on forest fires, highlighting monitoring efforts ($n = 161$) and data accessibility ($n = 85$). This data-oriented frame underscores the important role of real-time information on fires and collaborative firefighting across the EU. While 'data' does not constitute a policy domain per se, it is a cross-sectoral area of work that affects different policy domains, including awareness campaigns, firefighting capabilities, and disaster management strategies. It is, as such, an important part of shaping the overall narrative around forest fires.
- The importance of environmental monitoring and data is evidenced by the frequent emphasis on tools such as EFFIS ($n = 102$), earth observation technologies ($n = 39$), and early warning systems ($n = 44$). This focus reflects a tendency towards a technocratic approach in managing forest fires, closely aligned with advancing new technologies like remote sensing.
- The emphasis on reporting obligations in legally binding (hard) policy documents highlights the significance of environmental indices and indicators. These tools are instrumental in shaping how the natural environment is understood and, more importantly, in influencing perceptions and policymaking related to forest fires. This data-oriented perspective demonstrates the EU's technology-driven approach to forest fire management and prioritising innovative solutions and data analysis associated with forest fires.

Ecological Diversity and Stewardship

Forest fires threaten biodiversity, stressing forest restoration and management as key preventive measures.

- Biodiversity conservation ($n = 60$) is closely interlinked with forest management ($n = 16$), particularly regarding mitigating the impact on biodiversity from forest fires. This collective frame generally emphasises the role of forest restoration as a strategy for fire prevention rather than for enhancing biodiversity. Given the immediate threat posed by fires, this narrative may be understandable, but it also indicates a negative perception of forest fires in policy discussions.
- Rather than acknowledging the possible natural role and ecological benefits of forest fires, this collective frame argues that fires are harmful, often a result of neglect, and a significant obstacle to conservation. In other words, forest fires are consistently set out as a destructive force that limits conservation efforts.
- This perspective implies a belief in the need for forest management and preventive measures as a tool to protect biodiversity. The framing highlights the urgent need for interventions and supports forest management practices that balance fire prevention and ecological preservation.

Agriculture and Rural Development

Strategic forest management within the context of the CAP, emphasising concrete actions and investments for forest fire mitigation.

- The frame on agriculture and rural development ($n = 226$) addresses forest fires directly through the CAP and national RDPs. This narrative emphasises strategic forest management, focusing on the funding and incentives to mitigate forest fire risks, including prevention, protective measures, and control strategies. These strategies ($n = 84$) include wildfire prevention, post-fire restoration, and reforestation.
- Unlike other frames that use risk as a narrative device (such as for the climate and biodiversity), this frame concentrates on forestry measures and initiatives driving efforts to reduce the frequency and severity of forest fires. It also emphasises the importance of prescribed burning, acknowledging the natural fire cycle.
- This pragmatic framing is likely due to the legally binding nature of the CAP, which requires direct action from the European Commission. For instance, the policy documents wherein this frame can be found detail the practices eligible for RDP support, ranging from direct to indirect fire prevention measures. These measures are closely aligned with the goals of the EUs rural development policy, highlighting a focus on practical measures in forest fire management within the agricultural and rural development sector.

(Continued)

Table 7.1 (Continued)

Risk Mitigation and Protective Governance *Centres around disaster management and civil protection, highlighting measures and activities for enhanced fire monitoring and resilience.* • This frame, covering disaster management, natural hazards, and civil protection, is fundamentally interlinked with risk mitigation, identified 208 times across EU policy domains, and especially related to water and climate policy. This frame is action-oriented, focusing on the risks and threats of forest fires to human safety, critical infrastructure, and economic stability. It stresses the importance of active measures and funding ($n = 36$) for preventive strategies. These cover initiatives to improve fire monitoring, national restoration programs to enhance fire resilience, and EU-wide collaborative efforts to strengthen firefighting capabilities. • Central to this frame is the focus on societal vulnerabilities in relation to forest fires, with an emphasis on protecting human life. It consistently argues for increased preparedness and responsive measures, acknowledging the significant impact of forest fires on local communities and the environment. This frame sets out a clear call for action to safeguard against the risks posed by forest fires.

The discourse on forest fires, as covered by EU policy documents, typically highlights increasing risks, with the renewable energy discussions serving as an exception to this trend. Forest fires are employed as a narrative device (or tool) in different EU policy domains to underscore the need for policy action, examples being calls for climate action and biodiversity conservation. The general approach strategically positions forest fires within a broader environmental context, often integrated into larger narratives about risks and resilience or together with other natural hazards (e.g., droughts and storms). This framing process suggests that forest fires are mainly perceived as part of a wider set of environmental challenges that Europe wish to address.

However, when EU policy documents specifically focus on forest fires, the emphasis shifts towards areas where the EU has explicit competencies. One example is EFFIS, which plays an important role in monitoring and reporting forest fires across Europe, helping to inform EU-wide fire prevention, management, and mitigation strategies. Additionally, EU policies highlight the importance of civil protection measures to forest fires. This includes emergency response and firefighting capabilities and strategies like fire prevention programs, awareness campaigns, and integrating fire management into land use and urban planning.

Moreover, the different narratives associated with forest fires and risks highlight a diverse approach to framing across policy domains:

- Forest fires are linked to proactive forest management strategies in renewable energy policies. These policies outline a mutually beneficial relationship between renewable energy and forest fire management by advocating for biomass removal as a concrete measure that reduces forest fire risks. More specifically, this solution-oriented narrative and positive framing suggest that effective forest management, such as using biomass for energy, increases renewable energy production and reduces the fuel load in the forest, emphasising the dual benefits of energy production and fire risk mitigation.
- In comparison, biodiversity and climate-centric policies convey a more 'alarmist' picture where forests are under threat. Here, forest fires describe the increasing risks associated with climate change, leading to the increasing frequency of forest fires. This narrative emphasises the need for urgent and preventive measures to protect biodiversity and forest health. It focuses on the negative impacts of forest fires, including biodiversity loss and ecosystem damages, to argue for mitigating measures.
- Beyond environmental concerns, the framing of forest fires also encompasses broader socio-economic implications. This includes the impact of forest fires on international trade, agriculture, rural development, and the socio-economic stability of countries and regions.

For instance, discussions on the impact of forest fires on rural communities are used to call for strategies beyond fire mitigation, covering economic support and resilience building among rural communities.

- Forest fires are often viewed as anthropogenic effects, not as natural events. In other words, this frames fires as events caused by human actions, such as land-use changes. This policy discourse consequently extends to the drivers underlying climate change.

The varied use of 'risk' narratives in forest fire-related policies suggests a strategic approach in policy framing and agenda setting. For instance, to influence the EU's broader policy agenda, each policy domain uses forest fires to highlight specific concerns and objectives. Whether this is to promote renewable energy, protect biodiversity, manage climate risks, or ensure socio-economic stability, framing forest fires is used to shape policy priorities and direct EU action. All-in-all, the discourse around forest fires exemplifies how an environmental challenge can be used differently to advance specific policy goals and objectives.

> Forest fires are framed through six distinct but interconnected collective frames (Table 7.1), reflecting varied policy approaches to forest fire. These frames highlight the relationship between climate change and increased forest fire risks, link forest fire risk reduction with sustainable energy practices, emphasise the importance of technological solutions, present forest fires as a threat to biodiversity, consider forest fire management within an agricultural context, and focus on enhanced fire monitoring and resilience for disaster management and civil protection. These frames reflect a strategic approach to policy framing, using forest fire narratives to advance specific policy goals and objectives.

3.3. Key frames in the EU forest fire policy discourse

In the analysis of EU policy documents, about 30% of all references to forest fires are covered by four collective frames: (1) climate adaptation and resilience, (2) risk mitigation and protective governance, (3) agriculture and rural development., and (4) technocratic perspectives on forest fires. These frames are consistently represented across both types of policy documents, namely, 34% in non-legally binding (soft) and 32.4% in legally binding (hard) policy documents (Figures 7.3 and 7.4), emphasising how dominant they are in EU policies on forest fires. When accounting for thematic areas like reporting (e.g., monitoring obligations associated with EFFIS), prevention, and funding, these collective frames account for around 55% of all references in the policy documents. This is slightly higher in non-legally binding (soft) policy documents (56.7%) as compared to legally binding (hard) policy documents (54.5%). Nevertheless, this distribution stresses the overall influence of these frames in guiding the discussion around forest fires. More crucially, they are not only central to the narratives in these documents but also play a role in resource allocation and shaping policy actions.

Arguably, the collective frames mirror EU policy priorities and highlight varied strategic approaches to forest fire management within the broader context of different EU policy domains. The alignment with EU competencies also suggests a degree of fragmentation. This argument comes from earlier work, which has stressed the fragmented and incoherent nature of the EU's forest-related policy landscape. Research has pointed to this fragmentation in areas where the EU does not have explicit competencies, such as with forest fires and forests more generally (Edwards and Kleinschmit, 2013, Elomina and Pülzl, 2021, Aggestam and Giurca, 2021). Nevertheless, a

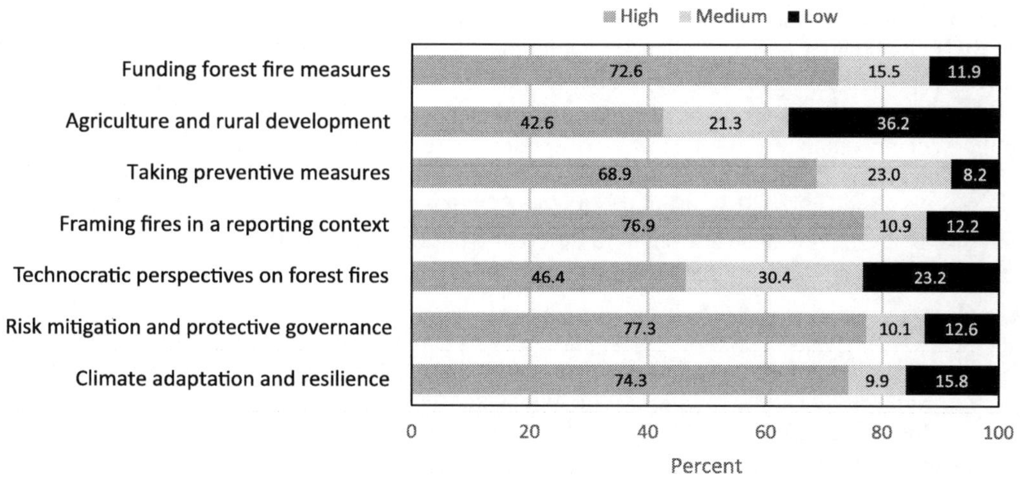

Figure 7.3 Breakdown of collective frames in soft EU policies.

Source: own figure

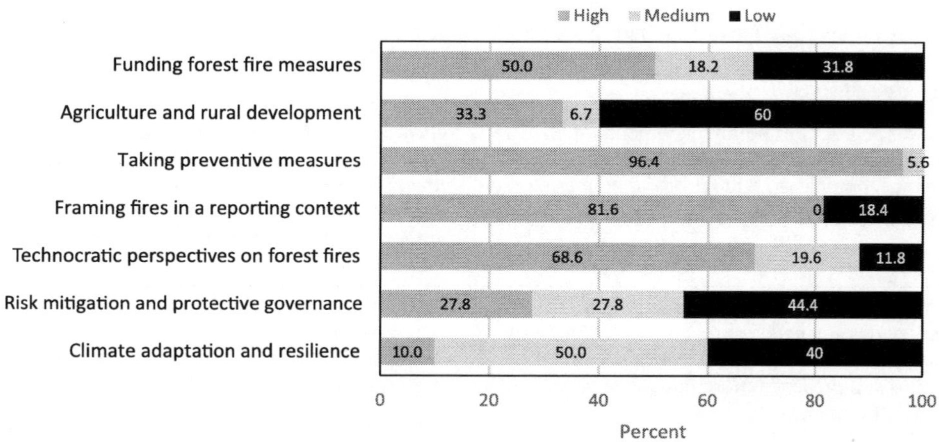

Figure 7.4 Breakdown of collective frames in hard EU policies.

Source: own figure

more nuanced picture emerges when taking a closer look at individual policy domains, such as agriculture and renewable energy. More specifically, each policy domain appears to approach forest fire management in a way that aligns with its specific mandate and objectives, suggesting a level of coherence that might not be apparent when considering the macro level. This would suggest that while the EU's policy approach may seem fragmented, there is still consistency in how the respective policy domains address forest fires, each tailoring its approach to fit its respective mandate. Furthermore, the repeated representation of specific collective frames in soft and hard policy documents indicates a deliberate emphasis on certain aspects associated with forest fires. This consistency across policy documents would suggest a concerted effort to have a coordinated response across the policy domain. This is particularly relevant in the context of the EU's broader policy challenges, including the need to find a balance between different levels of governance and the complexities associated with cross-sectoral coordination.

Interestingly, the analysis of EU policy documents reveals differences in how soft and hard policy documents focus on forest fires while representing unified strategic narratives underpinning both (Table 7.2).

The EU's forest fire policy discourse is shaped not only by the thematic content of the frames but also by the significance attributed to different types of policy documents. This becomes evident when analysing the importance attributed to different collective frames:

- *Variable importance of frames*: 'Agriculture and rural development' is frequently found in documents classified as medium to low in relevance. This suggests that, while acknowledged, this frame may not be at the forefront of the policy agenda on forest fires. In contrast, the 'technocratic perspectives on forest fires' frame is mainly associated with documents of high relevance, suggesting the EU emphasise data and information as a key tool in forest fire management.
- *Soft versus hard policy documents*: The prevalence of frames such as 'climate adaptation and resilience' and 'risk mitigation and protective governance' in soft policy documents considered highly relevant may *reflect* the EU's strategic focus. Soft policy documents provide a format whereby the EU can explore broader and more integrative approaches to forest fire management. In contrast, hard policy documents, which are legally binding and tend to be

Table 7.2 Comparison of soft and hard EU policy documents.

Soft policy documents	*Hard policy documents*
Broader scope: Soft policy documents often address overarching themes like climate change, emphasising the need for adaptive strategies and resilience-building.	***Specific focus***: Legally binding documents prioritise actionable strategies, including the provision of assistance to affected areas, ensuring access to data and information for effective management, and outlining specific development initiatives.
Advisory nature: Being non-legally binding, soft policy documents offer recommendations and guidelines, allowing them to explore diverse topics and present a more holistic view of forest fire management.	***Prescriptive nature***: Hard policies are more directive and action-oriented, often detailing the specific measures and protocols to be followed in forest fire management.
Value-driven narratives: Soft policy documents tend to incorporate more value-driven narratives, focusing on the environmental and socio-economic implications of forest fires and the importance of integrated, cross-sectoral approaches.	***Regulatory compliance***: Hard policy documents emphasise regulatory compliance, detailing the obligations and responsibilities of various stakeholders in managing and mitigating forest fire risks.
Common strategic narratives	
Complementary roles: Soft policies provide the conceptual framework and overarching guidelines, while hard policies translate these into concrete actions and legal requirements. ***Policy coherence***: The underlying consistency within policy domains reflects a deliberate effort to maintain policy coherence, ensuring that forest fire management is consistently addressed. ***Integrating forest fires*** *within* ***broader environmental and social contexts***: The narratives often place them within a wider environmental framework, linking them to biodiversity conservation, rural development, and socio-economic stability.	

more directive, may prioritise different aspects of forest fire management, such as specific regulatory measures or compliance requirements.

- *Implications for policy development*: This prioritisation pattern suggests that different frames are addressed in different ways, depending on the type of policy instrument. By placing different emphases on *various* frames across document types, the EU's approach balances immediate, actionable strategies (often found in hard policy documents) with long-term, strategic visions (typically found in soft policy documents).

> The analysis reveals narratives around forest fires shaped by content and context, demonstrating the influence collective frames have across non-legally binding (soft) and legally binding (hard) policy documents. The results highlight a certain degree of fragmentation, yet they also reveal a degree of coherence within individual policy domains. Each policy domain approaches forest fire management in line with its mandates, suggesting a siloed approach. Furthermore, the consistent representation of these frames across policy instruments points towards a coordinated and domain-specific effort in addressing forest fires.

3.4. Longitudinal changes in forest fire frames

When considering the historical trajectory of EU policymaking, it is relevant to recognise the evolving nature of policy frameworks, particularly in relation to forest management. This progression is evident in policies' changing focus and scope over time. For instance, initial forest-relevant EU policies in the 1980s were primarily focused on protection, specifically protecting forests from threats like fires and atmospheric pollution. This is reflected in regulations such as Council Regulation 3529/86 and 1696/87. During this period, policies were more targeted, displaying a narrower scope. They were largely aiming to safeguard forests through direct and prescriptive measures. Over time, the EU's approach to forest governance has broadened and become more integrative. This shift moves away from protection-oriented strategies whereby newer policies cover a wider range of societal concerns, including climate change adaptation, biodiversity conservation, and SFM. This progression also underscores the importance of adapting analytical frameworks that capture EU policy development's dynamic and complex nature.

The shift in focus on forest fires represents a significant change in how EU policy perceives and manages fires (whether natural or anthropogenic). Moreover, changes over the last decade align with an increased legislative and policymaking activity that directly addresses forest fires and integrates them into broader environmental and socio-economic agendas. Key factors contributing to this shift include the growing emphasis on climate action, which has led to policies that consider forest fires through the lens of climate change and forest stewardship. This has expanded the discourse beyond fire mitigation to include strategies for adaptation and resilience. Another example is the EU's commitment to renewable energy, which has introduced a new dimension to forest fire management. Policies now often discuss forest biomass management to prevent forest fires and as a component of renewable energy strategies. The integration of circular economy principles into EU policy has also broadened the perspective on resource utilisation and conservation, including forest resources. This has also affected how forest fires are factored into managing forest resources. Other examples include biodiversity conservation, recreation, public health, and the provision of non-timber forest products. These developments represent a change in policy focus from traditional objectives like timber production and reforestation to a more multifunctional view of forest ecosystem services. In other words, forest fires are no longer seen in isolation but as part of a complex ecological system.

However, it should also be recognised that there is significant consistency regarding the collective frames found in EU policy documents, particularly regarding forest utility, over time. This aligns with earlier work on the longitudinal stability of frames (Aggestam, 2015), which suggests that dominant or 'master' frames generally remain stable over extended periods (Benford and Snow, 2000). It is presumed that this also affects the interpretation and discourse around forest fires. Alongside this continuity in framing, the changing and expanding policy landscape around forest fires reflects the increasing range of forest-related policy instruments that contain competing and alternative frames. Another aspect of this change is the increasing focus on 'risk', particularly regarding climate change. This shift signifies a broader trend in environmental policy, where framing issues around risk and its management has become more important. This complex interplay between different frames and narratives and an evolving policy landscape reflects a policy environment where stakeholders from different sectors vie to influence policymaking in the EU.

> Initially centred around forest protection, EU policies addressing forest fires have expanded into other domains, such as climate change, biodiversity, and SFM. In parallel, there has also been an expansion in the number of competing frames. This shift marks a significant evolution in EU forest fire policy, illustrating the complex interplay between different perspectives. However, certain dominant frames have remained remarkably stable, especially regarding forest utility.

4. Understanding the role of frames in EU forest fire management

Fire not only plays a role in shaping forest ecosystems, but it also acts as a natural agent for ecological change, stimulating cycles of growth and regeneration that are essential for the health of certain ecosystems. Wildfires are needed in fire-dependent ecosystems for the reproductive processes of numerous plant and animal species (Sturtevant et al., 2009). For instance, several conifer species have evolved so that the heat from a fire is necessary to release their seeds, a mechanism that ensures their propagation and survival. However, Europe has observed a significant increase in wildfires frequency, magnitude, and severity in recent years (Fernandez-Anez et al., 2021, JRC, 2022, Canelas da Silva et al., 2023). This trend emphasises the need for research into policies and strategies that govern and manage forest fires, particularly as forest fire management needs to address the interplay between natural forest dynamics and the anthropogenic influence on fire regimes. This chapter has contributed to this study area by systematically analysing how forest fires are conceptualised and framed within EU policy documents. It has reviewed the narratives and perspectives embedded in non-legally binding (soft) and legally binding (hard) policy documents to enhance our understanding of the framing of forest fires across policy domains.

The analysis of EU policy documents that address forest fires revealed six collective frames, each using 'forest fires' to advance specific narratives. These narratives range from land use planning to health and safety to socio-economic concerns and are adapted to the priorities and objectives of the respective policy domains. Moreover, the frames are not only narrative devices but also act as mechanisms that guide policy developments and implementation, highlighting the critical role of framing in forest fire policy and management (Dewulf and Bouwen, 2012, Linnell et al., 2015, Steinberg, 1998, van Hulst and Yanow, 2014). For instance, in policy domains like climate and civil protection, forest fires are framed as a significant risk, emphasising the need for strategies to mitigate and adapt to the impacts of fires. These frames stress the need for developing and implementing policies that can effectively manage the dual challenges of forest fires and climate change. One interesting variation is observed in the renewable energy sector. While the energy-based narrative revolves around risk, the focus is on leveraging opportunities to reduce risks. For example, policies discuss the strategic management of forest biomass, not only as a method to prevent fires but also as a sustainable energy source, thus mixing forest fire management with renewable energy initiatives.

Different frames also emphasise different impacts, such as effects on biodiversity conservation or forest-based industries. These differences indicate that the framing of forest fires is closely linked with the concerns and expertise of the specific policy domain. For instance, in conservation-focused discussions, forest fires are framed in terms of their impact on wildlife and habitats. In contrast, in industry-related domains, the emphasis is on socio-economic effects and the need to protect natural assets. The call to action across these narratives is multi-dimensional. Some frames call for adopting specific adaptive strategies, like enhancing forest resilience to fires or adapting forest management practices to reduce fire risks. Others emphasise preventive measures, advocating for policies that avoid fires through land management, community engagement, and increased awareness. Another significant aspect of these frames is the push for investments in scientific and technological advancements. This suggests a techno-centric outlook on combatting forest fires, highlighting the need for innovation in fire detection, monitoring, and suppression technologies. This would suggest that the framing process is influenced by the epistemic communities that participate in these policy discussions. These knowledge communities, comprising experts, scientists, policymakers, and other stakeholders, shape the narratives around forest fires based on their collective understanding, perspectives, and priorities.

It can also be noted that forest fire narratives often appear in a reporting context, such as in documents that include scientific findings or provide updates on forest fire incidents across Europe. This also includes reporting on response strategies and preventive measures adopted by EU Member States. Additionally, forest fire reporting commonly includes an analysis of broader environmental trends, linking forest fires with other environmental risks such as heatwaves and droughts. By situating forest fires within this larger environmental and climate narrative, the EU policy documents stress the interconnected nature of these societal challenges. Forest fires are, as such, rarely presented in an isolated context. However, this broad approach reflects the EU's limited competencies in managing forest fires (Edwards and Kleinschmit, 2013, Aggestam and Pülzl, 2018, 2020). The exception to this trend is in policy areas where the EU has explicit legal mandates, such as monitoring and reporting on forest fires. In these cases, EU policy documents provide detailed guidelines and requirements for EU Member States, for example, directing actions related to data collection, analysis, and sharing. This specificity in monitoring and reporting contrasts with other EU policy documents' more general and integrative approach and possibly reflects an effort to extend existing competencies to cover forest fires.

Expanding on the frames used in EU forest fire policy reveals additional insights into how these narratives shape policy development and stakeholder perceptions. For example, the *'climate adaptation and resilience'* frame argues for strategies to live with and adapt to the impacts of climate change, which includes increasing forest fire incidents. It advocates for adjustments to practices, processes, and structures. The emphasis here is on acknowledging forest fires as part of a new reality, which in turn requires adjustments in how forests are managed. In contrast, the *'ecological diversity and stewardship'* frame aims to preserve and enhance the biodiversity of forest ecosystems. Strategies under this frame often include promoting a mix of species and age groups in forests, maintaining natural fire breaks, and implementing sustainable forestry practices. The main focus is on mitigating the intensity and frequency of forest fires. The *'risk mitigation and protective governance'* frame focuses on proactive measures to prevent forest fires or minimise their impact. This includes physical measures like creating firebreaks and controlled burns and regulatory measures like enforcing land-use policies and building codes. These examples demonstrate the interplay between adaptation and prevention as a key feature of forest fire policy framing. While adaptation acknowledges the inevitability of some forest fire activity, especially regarding climate change, prevention focuses on minimising anthropogenic fires and mitigating risks.

Realistically, how these dominant frames are communicated in policy documents affect how stakeholders perceive and manage forest fire risks (Moser, 2010; Howell et al., 2016). For instance, a policy document focusing on adaptation might lead to more investment in emergency response infrastructure and community preparedness programs. In contrast, a frame emphasising prevention could lead to stricter land management policies and public education campaigns. The point is that the dominance of a particular frame can steer the direction of future policies. Consequently, by understanding and analysing these frames, stakeholders can better understand the underlying values and priorities driving forest fire policy, leading to more informed and effective decision-making (Björnehed and Erikson, 2018). However, the relationship between knowledge and policy frames is complex. It is not always clear whether it is the frame that shapes the expert's understanding or if it is the expert's knowledge that reinforces these frames. This study cannot determine the direction of this influence. Nevertheless, it does highlight a critical concern, namely, the tendency towards path dependency in policymaking. This suggests that new policies and measures are commonly influenced by past perspectives and decisions, even when new innovative approaches emerge (Kay, 2003).

The collective frames also provide insights into the interplay between expert knowledge, policy framing, and policy action. The dominant frames 'climate adaptation and resilience', 'ecological diversity and stewardship', 'agriculture and rural development', and 'risk mitigation and protective governance' can be compared to lenses through which policymakers and experts address policymaking (Björnehed and Erikson, 2018, Howell et al., 2016). For instance, a frame focusing on climate adaptation may lead to policies emphasising resilience-building, whereas a frame centred on risk mitigation might prioritise preventive land management practices. However, as argued above, the relationship between frames and expert knowledge could go in two directions. On the one hand, the frames might influence how experts interpret and approach forest fires, potentially limiting their perspective to the boundaries characterised by the frame. On the other hand, the experts' pre-existing knowledge could reinforce the frame. This suggests a feedback loop between framing and expert input. In line with this argument, the EU's reliance on technocratic problem framing (Daviter, 2018), driven by expert committees, emphasises the important role of epistemic communities in defining policy problems and recommending solutions (van Schendelen, 2002, Bergström, 2005, Egeberg et al., 2003). While these expert committees provide valuable technical insights and specialised knowledge, this process can also prevent the identification and interpretation of frames as experts remove political considerations from the discourse (Daviter, 2018, Heinelt and Münch, 2018, van Schendelen, 2002). Moreover, the potential influence of epistemic communities on policy decisions at various levels raises questions about the inclusiveness of the policymaking process.

4.1. Framing the future of EU forest fire policy

This chapter has revealed a variety of collective frames related to forest fires within the EU policy landscape, principally connected to policy domains where the EU has explicit competencies, such as agriculture and rural development, renewable energy and the environment. Among these, six frames stand out, with four being identified as particularly dominant: (1) climate adaptation and resilience; (2) risk mitigation and protective governance, (3) agriculture and rural development., and (4) technocratic perspectives on forest fires. All the frames have been developed in isolation within their respective policy domains, which suggests a potential gap in cross-sectoral communication and collaboration. However, it is also acknowledged that the analysis was limited to already adopted EU policy documents. This implies that the actual policy development and formulation process were not covered. Nevertheless, these frames still mirror the agendas and priorities of different epistemic communities that have tried to influence policymaking through their specialised knowledge and perspectives.

The analysis of EU forest fire policies reveals policy boundaries and variations in framing across sectors and document types (hard/soft), with each sector adjusting the forest fire narrative to specific mandates and objectives. This finding aligns with academic work on the fragmentation and need for coherence in EU forest-related policy. However, when examined individually, each sector appears to exhibit a coherent and consistent strategy towards forest fires. Moreover, a common thread running through nearly all sectors is to emphasise the risks associated with forest fires. Using risks as a narrative device highlights a tendency towards negative framing whereby inaction is strategically equated with negative consequences, such as damage to infrastructure, loss of human life, and impacts on biodiversity. This means the use of 'risk' is more than a communication choice as it has been utilised as a tool to shape the discourse around forest fires.

Furthermore, the results stress how important it is to understand the role of framing in policy communication. The way forest fires are framed within the policy discourse not only influences the direction and content of policy development but also affects how the public perceives and understands forest fires. The framing process consequently has significant implications, extending beyond policymaking into the broader societal understanding and response to environmental challenges. This highlights the need for a nuanced approach to policy analysis that considers the content of policies and the underlying narratives and implications in terms of policy effectiveness and public perception.

Expanding on the conclusion, the results also emphasise the need for more research. Potential areas for future research include a deeper analysis of stakeholder and expert perspectives. For instance, research could focus on the policymaking process to examine how the frames have influenced policy choices and the implementation of specific measures. Another relevant area of work could be to unpack the value systems underpinning the collective frames, such as reviewing the core values and principles that underlie the emphasis on specific frames, such as adaptation versus prevention, to reveal how these values shape policy directions. Understanding values in the policymaking process would be crucial for assessing the compatibility of forest fire policies with broader EU environmental and social goals. Further research into the longitudinal stability of frames could offer insights into the adaptability and responsiveness of EU forest fire policy to emerging challenges and changing societal needs. Finally, conducting a comparative analysis with other regions makes sense. Comparing the EU's framing of forest fires with approaches in other regions could provide insights into different policy strategies and their effectiveness.

Want to find out more?

For those seeking more knowledge on forest fire policy, please see the references provided below or some of the many references throughout the text.

- Towards Integrated Fire Management – Outcomes of the European Project Fire Paradox (Silva et al., 2010).
- Forest Fires: Sparking firesmart policies in the EU (EC, 2018).
- Land-based wildfire prevention: Principles and experiences on managing landscapes, forests and woodlands for safety and resilience in Europe (EC, 2021).
- Current Wildland Fire Patterns and Challenges in Europe: A Synthesis of National Perspectives (Fernandez-Anez et al., 2021).
- Mapping the forest disturbance regimes of Europe (Senf and Seidl, 2021)

The set of questions listed below may also help guide further inquiry into specific areas of interest:

1. How do forest fire policies differ between EU Member States, and what factors contribute to these differences?
2. How might future climate change projections affect the frequency and severity of forest fires in Europe?
3. How can interdisciplinary research contribute to more effective forest fire management?
4. How has the framing of forest fires in EU policy discourse evolved over time?
5. How can EU policies be designed to be more resilient and adaptable in the face of ongoing and future climate change impacts on forest fires?

5. Assignment VII: Developing a webinar or workshop proposal on forest fire policy

Description

This assignment is focused on developing a proposal for a workshop or webinar that addresses the issue of forest fires within the broader context of climate change resilience and evolving policy frameworks in the EU. Using Chapter 7 as a foundation, this exercise will allow the student to explore the challenges of forest fires, including their increased frequency and intensity due to climate change and the policy responses required at the EU level.

Purpose

Students will use insights from Chapter 7 and other materials to design a webinar or workshop that can bring together stakeholders from different sectors, such as policymakers, scientists, forest managers, and the public. The webinar or workshop should focus on challenges associated with forest fire management, exploring the latest research, and identifying policy recommendations.

Outline of task

- *Identify a topic*: Review interesting themes and challenges related to forest fires, climate change resilience, and EU policy frameworks.
- *Design a workshop or webinar*: Elaborate a design and structure (including a timeline) for a workshop or webinar, considering keynote speeches, panel discussions, breakout sessions, and other interactive activities. Each element should aim to address the specific theme(s) identified.
- *Engaging stakeholders*: Identify and propose a strategy for how to engage relevant stakeholders.
- *Define clear goals*: Outline a few goals and outcomes for the workshop or webinar, such as developing policy recommendations or fostering partnerships and dialogue.

Suggested additional reading

- Explore the European Forest Fire Information System (EFFIS).[4]
- Forzieri, G., Dakos, V., McDowell, N.G. et al. (2022) Emerging signals of declining forest resilience under climate change. *Nature*, 608, 534–539.
- EC (2018) Forest Fires: Sparking firesmart policies in the EU. Luxembourg: Publications Office of the European Union.
- Fernandez-Anez N, Krasovskiy A, Müller M, et al. (2021) Current wildland fire patterns and challenges in Europe: A synthesis of national perspectives. *Air, Soil and Water Research*, 14, 1–19.

Deliverable

- Prepare a detailed webinar or workshop proposal that outlines the objectives, structure, topics to be covered, target audience, and expected outcomes. The proposal should serve as a blueprint for organising a real-world webinar or workshop on forest fires and policy in the European context.

6. Notes

1 Earlier iterations of this regulation go back to 1986 (Council Regulation 3529/86).
2 See http://eurlex.europa.eu.
3 The complete list of policy documents is available in Aggestam (2023).
4 See https://effis.jrc.ec.europa.eu/.

7. References

AGGESTAM, F. 2015. Framing the ecosystem concept through a longitudinal study of developments in science and policy. *Conservation Biology*, 29, 1052–1064.

AGGESTAM, F. 2024. Through smoke to policy: Framing the EU forest fire policy landscape. *Land*, 13(9), 1450, DOI: https://doi.org/10.3390/land13091450.

AGGESTAM, F. & GIURCA, A. 2021. The art of the 'green' deal: Policy pathways for the EU Forest Strategy. *Forest Policy and Economics*, 128, 102456.

AGGESTAM, F. & PÜLZL, H. 2018. Coordinating the uncoordinated: the EU Forest Strategy. *Forests*, 9, 125, DOI: https://doi.org/10.3390/f9030125.

AGGESTAM, F. & PÜLZL, H. 2020. Downloading Europe: A regional comparison in the uptake of the EU Forest Action Plan. *Sustainability*, 12.

AGUILAR, S. & MONTIEL, C. 2011. The challenge of applying governance and sustainable development to wildland fire management in Southern Europe. *Journal of Forestry Research*, 22, 627–639.

ANDERSON, D., CHUBB, P. & DJERF-PIERRE, M. 2018. Fanning the blame: Media accountability, climate and crisis on the Australian 'fire continent'. *Environmental Communication*, 12, 928–941.

ASCOLI, D., PLANA, E., OGGIONI, S. D., TOMAO, A., COLONICO, M., CORONA, P., GIANNINO, F., MORENO, M., XANTHOPOULOS, G., KAOUKIS, K., ATHANASIOU, M., COLAÇO, M. C., REGO, F., SEQUEIRA, A. C., ACÁCIO, V., SERRA, M. & BARBATI, A. 2023. Fire-smart solutions for sustainable wildfire risk prevention: Bottom-up initiatives meet top-down policies under EU green deal. *International Journal of Disaster Risk Reduction*, 92, 103715.

BACHMANN, A. & ALLGOWER, B. 2000. The need for a consistent wildfire risk terminology. *In:* NEUENSCHWANDER, L. F., RYAN, K. C. & GOLLBERG, G. E. (eds.) *Crossing the Millennium: Integrating Spatial Technologies and Ecological Principles for a New Age in Fire Management.* Moscow: University of Idaho and the International Association of Wildland Fire.

BARBOSA, P., CAMIA, A., KUCERA, J., LIBERTÀ, G., PALUMBO, I., SAN-MIGUEL-AYANZ, J. & SCHMUCK, G. 2008. Chapter 8 Assessment of forest fire impacts and emissions in the European Union based on the European Forest Fire Information System. *In:* BYTNEROWICZ, A., ARBAUGH, M. J., RIEBAU, A. R. & ANDERSEN, C. (eds.) *Developments in Environmental Science.* Elsevier.

BAUMGARTNER, F. R. & MAHONEY, C. 2008. The two faces of framing: Individual-level framing and collective issue definition in the European Union. *European Union Politics*, 9, 435–449.

BELAND LINDAHL, K., STÉNS, A., SANDSTRÖM, C., JOHANSSON, J., LIDSKOG, R., RANIUS, T. & ROBERGE, J.-M. 2017. The Swedish forestry model: More of everything? *Forest Policy and Economics*, 77, 44–55.

BENFORD, R. D. & SNOW, D. A. 2000. Framing processes and social movements: an overview and assessment. *Annual Review of Sociology*, 26, 611–639.

BERGLEZ, P. & LIDSKOG, R. 2019. Foreign, domestic, and cultural factors in climate change reporting: Swedish media's coverage of wildfires in three continents. *Environmental Communication*, 13, 381–394.

BERGSTRÖM, C. F. 2005. *Comitology: Delegation of Powers in the European Union and the Committee System.* Oxford: Oxford University Press.

BJÖRNEHED, E. & ERIKSON, J. 2018. Making the most of the frame: developing the analytical potential of frame analysis. *Policy Studies*, 39, 109–126.

BOSOMWORTH, K. 2015. Climate change adaptation in public policy: frames, fire management, and frame reflection. *Environment and Planning C: Government and Policy*, 33, 1450–1466.

BRUNO, L. F. C. & LAY, E. G. E. 2008. Personal values and leadership effectiveness. *Journal of Business Research*, 61, 678–683.

CALKIN, D. E., COHEN, J. D., FINNEY, M. A. & THOMPSON, M. P. 2014. How risk management can prevent future wildfire disasters in the wildland-urban interface. *Proc Natl Acad Sci USA*, 111, 746–751.

CANELAS DA SILVA, M., DURÃO, R., RUSSO, A. & GOUVEIA, C. 2023. The 2022 fire season over Europe. Vienna: EGU General Assembly 2023.

CASTELLÓ, E. & MONTAGUT, M. 2019. Framing forest fires and environmental activism: a storytelling contest about human intervention in nature. *Communication and Society*, 32, 291–306.

CONINX, I., DEWULF, A. R. P. J., STEINGROVER, E. G., WAL, V. D. M. F., OPDAM, P. & VOS, C. C. 2015. Framing ecosystem services: Affecting behaviour of actors in collaborative landscape planning? *Land Use Policy*, 46, 223–231.

CORONA, P., ASCOLI, D., BARBATI, A., BOVIO, G., COLANGELO, G., ELIA, M., GARFÌ, V., IOVINO, F., LAFORTEZZA, R., LEONE, V., LOVREGLIO, R., MARCHETTI, M., MARCHI, E., MENGUZZATO, G., NOCENTINI, S., PICCHIO, R., PORTOGHESI, L., PULETTI, N., SANESI, G. & CHIANUCCI, F. 2015. Integrated forest management to prevent wildfires under Mediterranean environments. *Annals of Silvicultural Research*, 39, 1–22. http://dx.doi.org/10.12899/asr-946.

DAMIANIDIS, C., SANTIAGO-FREIJANES, J. J., DEN HERDER, M., BURGESS, P., MOSQUERA-LOSADA, M.-R., GRAVES, A., PAPADOPOULOS, A., PISANELLI, A., CAMILLI, F., ROIS-DÍAZ, M., KAY, S., PALMA, J. H. N. & PANTERA, A. 2021. Agroforestry as a sustainable land use option to reduce wildfires risk in European Mediterranean areas. *Agroforestry Systems*, 95, 919–929.

DAVITER, F. 2007. Policy framing in the European Union. *Journal of European Public Policy*, 14, 654–666.

DAVITER, F. 2018. Chapter 5: The framing of EU policies. *In: Handbook of European Policies*. Cheltenham: Edward Elgar Publishing.

DEWULF, A. & BOUWEN, R. 2012. Issue framing in conversations for change: Discursive interaction strategies for 'doing differences'. *The Journal of Applied Behavioral Science*, 48, 168–193.

DEWULF, A., GRAY, B., PUTNAM, L., LEWICKI, R., AARTS, N., BOUWEN, R. & VAN WOERKUM, C. 2009. Disentangling approaches to framing in conflict and negotiation research: A meta-paradigmatic perspective. *Human Relations*, 62, 155.

DONAHUE, W. A., ROGAN, R. G. & KAUFMAN, S. 2011. *Framing Matters: Perspectives on Negotiation Research and Practice in Communication*. New York: Peter Lang.

DUNLOP, A. C. 2013. Epistemic communities. *In:* ARARAL, E., FRITZEN, S., HOWLETT, M., RAMESH, M. & WU, X. (eds.) *Routledge Handbook of Public Policy.* New York: Routledge.

DUPUY, J.-L., FARGEON, H., MARTIN-STPAUL, N., PIMONT, F., RUFFAULT, J., GUIJARRO, M., HERNANDO, C., MADRIGAL, J. & FERNANDES, P. 2020. Climate change impact on future wildfire danger and activity in southern Europe: a review. *Annals of Forest Science*, 77, 35.

EC 2018. Forest fires: Sparking firesmart policies in the EU. *In:* FAIVRE, N. (ed.) *Research & Innovation: Projects for Policy.* Luxembourg: Publications Office of the European Union,.

EC 2021. Land-based wildfire prevention: Principles and experiences on managing landscapes, forests and woodlands for safety and resilience in Europe. Luxembourg: European Commission.

EDWARDS, P. & KLEINSCHMIT, D. 2013. Towards a European forest policy: conflicting goals. *Forest Policy and Economics*, 33, 87–91.

EGEBERG, M., SCHAEFER, G. & TRONDAL, J. 2003. The many faces of EU committee governance. *West European Politics*, 26, 19–40.

EKAYANI, M., NURROCHMAT, D. R. & DARUSMAN, D. 2016. The role of scientists in forest fire media discourse and its potential influence for policy-agenda setting in Indonesia. *Forest Policy and Economics*, 68, 22–29.

EKENGREN, M., MATZÉN, N., RHINARD, M. & SVANTESSON, M. 2006. Solidarity or sovereignty? EU cooperation in civil protection. *Journal of European Integration*, 28, 457–476.

ELO, S., KÄÄRIÄINEN, M., KANSTE, O., PÖLKKI, T., UTRIAINEN, K. & KYNGÄS, H. 2014. Qualitative content analysis: A focus on trustworthiness. *SAGE Open*, 4, 2158244014522633.

ELOMINA, J. & PÜLZL, H. 2021. How are forests framed? An analysis of EU forest policy. *Forest Policy and Economics*, 127, 102448.

ENTMAN, R. M. 1993. Framing: Toward clarification of a fractured paradigm. *Journal of Communication*, 43, 51–58.

ERLINGSSON, C. & BRYSIEWICZ, P. 2017. A hands-on guide to doing content analysis. *Afr J Emerg Med*, 7, 93–99.

FERNANDEZ-ANEZ, N., KRASOVSKIY, A., MÜLLER, M., VACIK, H., BAETENS, J. & HUKIĆ, E. 2021. Current wildland fire patterns and challenges in Europe: A synthesis of national perspectives. *Air, Soil and Water Research*, 14, 11786221211028185.

FRIESE, S. 2019. *Qualitative Data Analysis With ATLAS.ti.* London: Sage.

GALIANA, L., AGUILAR, S. & LÁZARO, A. 2013. An assessment of the effects of forest-related policies upon wildland fires in the European Union: Applying the subsidiarity principle. *Forest Policy and Economics*, 29, 36–44.

GAMSON, W. A. & MODIGLIANI, A. 1989. Media discourse and public opinion on nuclear power: A constructionist approach. *American Journal of Sociology*, 95, 1–37.

GANTEAUME, A., CAMIA, A., JAPPIOT, M., SAN-MIGUEL-AYANZ, J., LONG-FOURNEL, M. & LAMPIN, C. 2013. A review of the main driving factors of forest fire ignition over Europe. *Environmental Management*, 51, 651–662.

GIGERENZER, G. & TODD, M. P. 1999. *Simple Heuristics that Make us Smart.* New York: Oxford University Press.

GOFFMAN, E. 1974. *Frame Analysis: An Essay on the Organisation of Experience.* New York: Harper & Row.

HEINELT, H. & MÜNCH, S. 2018. *Handbook of European Policies.* Cheltenham: Edward Elgar Publishing.

HOWELL, R. A., CAPSTICK, S. & WHITMARSH, L. 2016. Impacts of adaptation and responsibility framings on attitudes towards climate change mitigation. *Climatic Change*, 136, 445–461.

JRC 2022. Forest Fires in Europe, Middle East and North Africa 2021. *In:* SAN-MIGUEL-AYANZ, J., DURRANT, T., BOCA, R., MAIANTI, P., LIBERTÁ, G., ARTÉS-VIVANCOS, T., OOM, D., BRANCO, A., DE RIGO, D., FERRARI, D., PFEIFFER, H., GRECCHI, R. & ONIDA, M. (eds.) *JRC Technical Report.* Luxembourg: Joint Research Centre.

KALTENBORN, B. P. & BJERKE, T. 2002. Associations between environmental value orientations and landscape preferences. *Landscape and Urban Planning*, 59, 1–11.
KAY, A. 2003. Path dependency and the CAP. *Journal of European Public Policy*, 10, 405–420.
KLEINSCHMIT, D., ARTS, B., GIURCA, A., MUSTALAHTI, I., SERGENT, A. & PÜLZL, H. 2017. Environmental concerns in political bioeconomy discourses. *International Forestry Review*, 19, 41–55.
KRIPPENDORFF, K. 2004. *Content Analysis: An Introduction to its Methodology*, 2nd ed. Thousand Oaks, CA: Sage.
LANGER, E. R. L. & MCGEE, T. K. Wildfire risk awareness and prevention by predominantly Māori rural residents, Karikari Peninsula, Aotearoa New Zealand. *International Journal of Wildland Fire* 26(9): 820–828. https://doi.org/10.1071/WF16133, 26.
LIDSKOG, R., JOHANSSON, J. & SJÖDIN, D. 2019. Wildfires, responsibility and trust: public understanding of Sweden's largest wildfire. *Scandinavian Journal of Forest Research*, 34, 319–328.
LIDSKOG, R. & SJÖDIN, D. 2016. Risk governance through professional expertise. Forestry consultants' handling of uncertainties after a storm disaster. *Journal of Risk Research*, 19, 1275–1290.
LINNELL, J. D. C., KACZENSKY, P., WOTSCHIKOWSKY, U., LESCUREUX, N. & BOITANI, L. 2015. Framing the relationship between people and nature in the context of European conservation. *Conservation Biology*, 29, 978–985.
LÓPEZ-I-GELATS, F., TÀBARA, J. D. & BARTOLOMÉ, J. 2009. The rural in dispute: Discourses of rurality in the Pyrenees. *Geoforum*, 40, 602–612.
LÓPEZ-VICENTE, M., CERDÀ, A., KRAMER, H. & KEESSTRA, S. 2021. Post-fire practices benefits on vegetation recovery and soil conservation in a Mediterranean area. *Land Use Policy*, 111, 105776.
MCINERNEY, D., SAN-MIGUEL-AYANZ, J., CORTI, P., WHITMORE, C., GIOVANDO, C. & CAMIA, A. 2013. Design and function of the European Forest Fire Information System. *Photogrammetric Engineering & Remote Sensing*, 79, 965–973.
MIEZÏTE, L. E., AMEZTEGUI, A., DE CÁCERES, M., COLL, L., MORÁN-ORDÓÑEZ, A., VEGA-GARCÍA, C. & RODRIGUES, M. 2022. Trajectories of wildfire behavior under climate change. Can forest management mitigate the increasing hazard? *Journal of Environmental Management*, 322, 116134.
MONTIEL-MOLINA, C. 2013. Comparative assessment of wildland fire legislation and policies in the European Union: Towards a Fire Framework Directive. *Forest Policy and Economics*, 29, 1–6.
MONTIEL, C. & KRAUS, D. 2010. Best practices of fire use – Prescribed burning and suppression fire programmes in selected case-study regions in Europe. *Research Report 24.* European Forest Institute.
MOSER, S. C. 2010. Communicating climate change: history, challenges, process and future directions. *WIREs Climate Change*, 1, 31–53.
MOURAO, P. R. & MARTINHO, V. D. 2019. Forest fire legislation: Reactive or proactive? *Ecological Indicators*, 104, 137–144.
MÜLLER, M. M., VILÀ-VILARDELL, L. & VACIK, H. 2020. Towards an integrated forest fire danger assessment system for the European Alps. *Ecological Informatics*, 60, 101151.
NEIDERMEIER, A. N., ZAGARIA, C., PAMPANONI, V., WEST, T. A. P. & VERBURG, P. H. 2023. Mapping opportunities for the use of land management strategies to address fire risk in Europe. *Journal of Environmental Management*, 346, 118941.
NILSSON, S. & ENANDER, A. 2020. 'Damned if you do, damned if you don't': Media frames of responsibility and accountability in handling a wildfire. *Journal of Contingencies and Crisis Management*, 28, 69–82.
PAN, Z. & KOSICKI, G. M. 1993. Framing analysis: An approach to news discourse. *Political Communication*, 10, 55–75.
PARKER, C. F., PERSSON, T. & WIDMALM, S. 2019. The effectiveness of national and EU-level civil protection systems: evidence from 17 member states. *Journal of European Public Policy*, 26, 1312–1334.
POLMAN, D. 2018. Learning from practical experience: Implementation epistemic communities in the European Union. *In:* DUNLOP, C. A., RADAELLI, C. M. & TREIN, P. (eds.) *Learning in Public Policy: Analysis, Modes and Outcomes.* Cham: Springer International Publishing.
REESE, S. D., GANDY, J., O.H., & GRANT, A. E. 2001. *Framing Public Life: Perspectives on Media and Our Understanding of the Social World.* Abingdon, Routledge.
REGULATION 2012/2002. Establishing the European Union Solidarity Fund. *In:* EUROPEAN PARLIAMENT (ed.) *L 311/3.* Brussels: Official Journal of the European Communities.
REGULATION 2152/2003. Monitoring of forests and environmental interactions in the Community (Forest Focus). *L 324/1.* Brussels: Official Journal of the European Union.

REGULATION 2158/92. Protection of the Community's forests against fire. *In:* EUROPEAN PARLIAMENT (ed.) *No L 217/3*. Brussels: Official Journal of the European Communities.

REIN, M. & SCHÖN, D. A. 1996. Frame-critical policy analysis and frame-reflective policy practice. *Knowledge and Policy*, 9, 85–104.

RODRIGUES, M., CUNILL CAMPRUBÍ, A., BALAGUER-ROMANO, R., COCO MEGÍA, C. J., CASTAÑARES, F., RUFFAULT, J., FERNANDES, P. M. & RESCO DE DIOS, V. 2023. Drivers and implications of the extreme 2022 wildfire season in Southwest Europe. *Science of The Total Environment*, 859, 160320.

SAN-MIGUEL-AYANZ, J. 2012. Comprehensive monitoring of wildfires in europe: The European Forest Fire Information System (EFFIS). *In:* TIEFENBACHER, J. P. (ed.) *Approaches to Managing Disaster – Assessing Hazards, Emergencies and Disaster Impacts*. IntechOpen.

SAN-MIGUEL-AYANZ, J., MORENO, J. M. & CAMIA, A. 2013a. Analysis of large fires in European Mediterranean landscapes: Lessons learned and perspectives. *Forest Ecology and Management*, 294, 11–22.

SAN-MIGUEL-AYANZ, J., SCHULTE, E., SCHMUCK, G. & CAMIA, A. 2013b. The European Forest Fire Information System in the context of environmental policies of the European Union. *Forest Policy and Economics*, 29, 19–25.

SCHÖN, D. A. & REIN, M. 1994. *Frame Reflection: Toward the Resolution of Intractable Policy Controversies*. New York: Basic Books.

SENF, C. & SEIDL, R. 2021. Mapping the forest disturbance regimes of Europe. *Nature Sustainability*, 4, 63–70.

SHMUELI, D., ELLIOT, M. & KAUFMAN, S. 2006. Stakeholder frames in the mapping of of the Lower Kishon River Basin conflict. *Conflict Resolution Quarterly*, 21, 211–238.

SILVA, J., REGO, F., FERNANDES, P. & RIGOLOT, E. 2010. *Towards Integrated Fire Management – Outcomes of the European Project Fire Paradox*.

SNOW, D. A. & BENFORD, R. D. 1988. Ideology, frame resonance, and participant mobilization. *In:* KLANDERMANS, B., KRIESI, H. & TARROW, S. (eds.) *From Structure to Action: Social Movement Participation Across Cultures*. Greenwich: JAI Press.

SÖDERBERG, C. & ECKERBERG, K. 2013. Rising policy conflicts in Europe over bioenergy and forestry. *Forest Policy and Economics*, 33, 112–119.

STEINBERG, W. M. 1998. Tilting the frame: Considerations on collective action framing from a discursive turn. *Theory and Society*, 27, 845–872.

STERN, C. P., DIETZ, T., ABEL, T., GUAGNANO, A. G. & KALOF, L. 1999. A value-belief-norm theory of support for social movements: The case of environmentalism. *Human Ecology Review*, 6, 81–97.

STRAUSS, A. & CORBIN, J. 1990. *Basics of Qualitative Research: Grounded Theory Procedures and Techniques*. Newbury Park, CA: Sage.

STURTEVANT, B. R., MIRANDA, B. R., YANG, J., HE, H. S., GUSTAFSON, E. J. & SCHELLER, R. M. 2009. Studying fire mitigation strategies in multi-ownership landscapes: Balancing the management of fire-dependent ecosystems and fire risk. *Ecosystems*, 12, 445–461.

TEDIM, F., XANTHOPOULOS, G. & LEONE, V. 2014. Forest fires in Europe: Facts and challenges. *In:* SHRODER, J. F. and PATON, D. (eds.),*Wildfire Hazards, Risks and Disasters*. Elsevier, 77–99.

VALENTE, S., COELHO, C., RIBEIRO, C., LINIGER, H., SCHWILCH, G., FIGUEIREDO, E. & BACHMANN, F. 2015. How much management is enough? Stakeholder views on forest management in fire-prone areas in central Portugal. *Forest Policy and Economics*, 53, 1–11.

VAN HULST, M. & YANOW, D. 2014. From policy 'frames' to 'framing': Theorizing a more dynamic, political approach. *The American Review of Public Administration*.

VAN SCHENDELEN, R. 2002. The in-sourced experts. *The Journal of Legislative Studies*, 8, 27–39.

VELEZ, A.-L. K., DIAZ, J. M. & WALL, T. U. 2017. Public information seeking, place-based risk messaging and wildfire preparedness in southern California. *International Journal of Wildland Fire*, 26, 469–477.

VUGTEVEEN, P., LENDERS, H. J. R., DEVILEE, L. A. J., LEUVEN, S. E. W. R., VAN DER VEEREN, J. H. M. R., WIERING, A. M. & HENDRIKS, A. J. 2010. Stakeholder value orientations in water management. *Society & Natural Resources: An International Journal of the Commons*, 23, 805–821.

ZITO, A. R. 2001. Epistemic communities, collective entrepreneurship and European integration. *Journal of European Public Policy*, 8, 585–603.

8 Nature protection, productivity, and multifunctional forest use

1. Introduction

The increasing demand for different forest ecosystem services, globally and within Europe, has resulted in the development of progressive forest management paradigms over recent decades (Martynova et al., 2021, Morgan et al., 2021, Caicoya et al., 2023, Konczal et al., 2023). These innovative strategies and approaches have challenged traditional forestry's singular focus on timber production, including the notion that the provision of different ecosystem services must be spatially isolated, such as creating one distinct zone for wood production and another for conservation. This fundamental shift towards more inclusive approaches, considering multiple ecosystem services, has been catalysed by the introduction of sustainable forest management (SFM) (MCPFE, 2005, Sotirov et al., 2015, Linser et al., 2018, Lazdinis et al., 2019). The gradual acceptance of sustainability principles has effectively helped to change how people view and manage forests (Spathelf, 2009). This can be exemplified by the now growing role of biodiversity conservation in forest management, with an increased emphasis on balancing the need for timber, a provisioning service, with regulating (e.g., flood regulation), supporting (e.g., nutrient cycling), and cultural (e.g., tourism) services (TEEB, 2010, Maes, 2011, EC, 2015, EC et al., 2020). Underlying this shift is a movement towards integrative strategies and approaches, such as retention forestry, multifunctional forestry, and close-to-nature forestry, all of which aim to balance various forest functions (Hoogstra-Klein et al., 2017, Başkent, 2018, Gustafsson et al., 2019).

Integrated forest management has emerged as a promising strategy in this context. It represents a change towards combined-objective forestry that addresses varied societal demands within a single spatial area (e.g., within a specific forest stand) rather than seeking to maximise individual objectives (e.g., timber production) in isolated spaces across larger landscapes (Kraus and Krumm, 2013, Blattert et al., 2018, Manning et al., 2018). Integrated forest management arguably stands out as a feasible pathway to navigate key trade-offs confronting the forest-based sector, especially the tension between nature conservation and biomass production (Cosyns et al., 2020, Krumm et al., 2020, Morgan et al., 2021, Konczal et al., 2023). The practical challenge for integrated forest management resides in addressing these trade-offs directly at the forest stand level, requiring an equilibrium between biomass extraction and habitat retention. For instance, local decision-making processes can be influenced by a complex array of factors, ranging from socio-economic to ecological determinants that shape the readiness or capacity of forest managers and owners to embrace integrated forest management practices (Borrass et al., 2017, Maier and Winkel, 2017, Sotirov and Arts, 2018, Başkent, 2018).

Integrated forest management is an adaptive approach to forestry that recognises the inherent complexity of forest ecosystems and the varied objectives of forest policy and management.

DOI: 10.4324/9781003198987-8

This approach, while conceptualised differently across different regions (Kreutzwiser and Wright, 1990, Cairns and Meganck, 1994, Corona et al., 2015), is unified in its aim to balance and integrate the needs of timber production with the need for habitat conservation, water and soil protection, wildlife management, disaster prevention, and the enhancement of recreational and health benefits (Borrass et al., 2017, Hoogstra-Klein et al., 2017, Sutherland and Huttunen, 2018). Integrated forest management is responsive to various aspects, including economic fluctuations, technological innovations, societal values, and evolving cultural and political contexts (Sotirov and Arts, 2018, Aggestam et al., 2020, Konczal et al., 2023). It seeks to resolve the problem of competing goals associated with forest lands. Through a lens that captures the totality of these aspects, integrated forest management strives to provide a balanced solution that serves the multifunctional role of forests.

This chapter investigates the factors that might impede or facilitate the adoption of integrated forest management practices in Europe, offering a detailed understanding of how this approach can be furthered. The focus is mainly set on one of the most complex aspects of integrated forest management, namely, how to harmonise the objectives of nature conservation and wood production. In this context, the chapter will also explore how changing climates might influence integrative strategies, considering the increasing frequency of extreme weather events and their impact on forest health and productivity. Additionally, the chapter delves into the socio-political dimensions of integrative strategies, examining how policy and economic incentives can be aligned to support the transition towards more integrated approaches. By exploring these themes, the aim is to provide a comprehensive picture of the current state of integrated forest management in Europe and the pathways toward its successful uptake.

> Understanding and researching integrative approaches is relevant as it offers an alternative and more holistic approach to governing the natural resources that forests provide. In other words, integrated forest management considers the interplay between ecological, economic, and social aspects of forest ecosystems. Examining factors such as biodiversity conservation, carbon sequestration, sustainable timber harvesting, recreational opportunities, and the well-being of local communities can enable forest managers to optimise forest resource use while reconciling critical trade-offs and minimising negative environmental and societal impacts.

1.1. Why do integrative approaches matter in forest management?

The exploration of integrative management strategies, particularly the integration of nature conservation into current forest management practices, has garnered increasing scholarly attention (Phalan et al., 2011, Blattert et al., 2018, Mölder et al., 2019, Ontl et al., 2019). Studies have delved into the benefits and limitations of aligning conservation objectives with forest management, predominantly through an ecological lens (Paillet et al., 2010, Puettmann et al., 2015, Naumov et al., 2018). These investigations have often concentrated on distinct strategies or methodologies, such as retention or adaptive forestry, which involves preserving certain trees or specific habitats during harvesting to maintain ecological diversity (Gustafsson et al., 2012, Gustafsson et al., 2019, Verkerk et al., 2020, Kuuluvainen et al., 2021). These types of studies underscore the critical importance of biodiversity in forest ecosystems. For instance, by preserving key elements of the original forest structure and composition – such as veteran trees, dead wood, or rare habitats – retention forestry ensures that forests continue to support biodiversity, contributing to ecosystem resilience (Kuuluvainen et al., 2019, Koivula and Vanha-Majamaa, 2020). In essence, the exploration of integrative management strategies reflects a broader shift in thinking, from viewing conservation and human use of nature as competing interests to seeing them as complementary components and as an integral part of sustainable land management.

Studies into integrative practices indicate that while forest managers might support the concept of integrative conservation measures, real-world applications on the ground level have faced significant barriers (Borrass et al., 2017, Maier and Winkel, 2017). As documented by social science researchers, these barriers are not one-dimensional but rather stem from economic, social, logistical, and regulatory aspects of forest management, such as the lack of practical knowledge, the complexity of integrative measures, land ownership and rights issues, lack of legal clarity regarding conservation practices, and differing cultural perspectives on forest management. Moreover, research into the drivers and perspectives of forest managers concerning integrative forest management have mainly been conducted within the context of Central European forestry, such as focusing on the multifunctional forestry paradigm in Germany (Borrass et al., 2017, Maier and Winkel, 2017, Krumm et al., 2020, Aggestam et al., 2020, Benz et al., 2020, Konczal et al., 2023). This body of research underscores the complex motivations behind forest managers' decisions, highlighting personal and organisational values, economic considerations, and societal expectations.

Despite the recognition that more research is needed (Schulz et al., 2014, Nagel et al., 2017, Coll et al., 2018), there exists a gap in understanding the specific incentives or barriers forest owners and managers encounter in Europe, particularly when it comes to expanding the integration of conservation measures into forest management practice. Expanding this research to the European scale is crucial because forest ecosystems do not adhere to national boundaries, and ecological challenges are often global in scope. Europe's forests are also diverse, spanning the boreal forests of Scandinavia to the Mediterranean forests in the south, each with unique conservation requirements (EEA, 2016). Understanding the national and regional challenges and perspectives could help inform more effective, region-specific policies and conservation strategies, contributing to the broader goal of SFM (SFC, 2015, Forest Europe, 2020).

2. Approach

The chapter builds on an earlier publication – 'Can nature conservation and wood production be reconciled in managed forests? A review of driving factors for integrated forest management in Europe' (Aggestam et al., 2020). The current text has been extended and updated to reflect recent developments in this area of work.

2.1. Methodological framework

The study design applied a three-stage methodology:

1. *Literature review*: The aim was to survey existing literature on the integration of nature conservation with forest management and to identify prevailing knowledge gaps and the factors influencing integrated forest management (Jalali and Wohlin, 2012).
2. *Expert workshop*: The study conducted an interdisciplinary focus group discussion on factors for integrated forest management. The Social, Technological, Economic, Environmental, and Political (STEEP) framework was used to structure the discussion and classify factors that

either facilitate or impede the integration of nature conservation and wood production in forest management (Bowman, 1998, Kim-Keung Ho, 2014, Bonsu et al., 2017, 2020, Savastano et al., 2023).

3. *Expert-based cooperative analysis*: The cooperative analysis aimed to draft concise narratives (approximately 500 words each) that covered the prioritised factors within each STEEP category – addressing socio-cultural, technological, economic, environmental, and political factors – and have them reviewed by experts (Gibert et al., 2010).

2.1.1. Reviewing the literature

The literature review involved a screening of publications, academic studies, reports, and related projects. Emphasis was placed on materials explicitly discussing integrated forest management, nature conservation, timber production, sustainable forestry practices, and climate change dynamics (Aggestam et al., 2020). This process included snowball sampling/referral sampling (Jalali and Wohlin, 2012).

The review resulted in a synthesis of relevant state-of-the-art knowledge on integrated forest management. It also facilitated the compilation of an initial draft list of 32 potential factors, which formed the basis for initial discussions during the expert workshop discussions (see below).

2.1.2. Expert workshop

The expert workshop was carried out during the launch of the Integrated Forest Management Learning Architecture (INFORMAR) project,[1] which was attended by 34 people. Participants included representatives from European academic circles, private forest owners' associations, public forestry representatives, international organisations, non-governmental organisations, government ministries, and the European Commission. The participants that took part in the workshop were comprised of 16 experts. The group's composition was mixed to involve many complimentary perspectives from science, policy, and forestry (INFORMAR, 2018).

The interactive focus-group discussion was designed to encourage participants to contribute additional factors within each STEEP category. The preliminary list of factors, derived from the literature review, initiated the dialogue. A subsequent roundtable discussion allowed for collective reflection on the importance of each factor. This collaborative approach ensured that the list of factors was not only discussed and validated but also jointly refined and expanded. The discussion finished with the participants identifying the three most important factors for integrated forest management, indicated by allocating three stickers to the consolidated list (Aggestam et al., 2020).

2.1.3. Expert-based cooperative analysis

The final step centred on building on the information and insights gained from the expert workshop to develop draft narratives, each corresponding to a category within the STEEP framework. This means five narratives were developed, which were subsequently disseminated among a panel of experts for further comment and review. These were experts who had not taken part in the original workshop. The primary purpose of this process was to connect expert knowledge with the data obtained from the interdisciplinary focus group, especially concerning novel aspects introduced during the discussions. It principally acted as a step to verify the assumptions made during the workshop.

3. Results and analysis

The main output from the literature review and workshop is a list of factors, collectively prioritised and grouped according to the STEEP categories. The results from the workshop are presented in Table 8.1; however, the full range of the factors considered can be found in Aggestam et al. (2020). All the factors were considered as supporting or preventing the uptake of integrative forest management practices on the ground level. The prioritisation represents the overall relevance allocated by the workshop participants. Further, each entry on the list of factors was considered relevant by at least one participant in the workshop; however, the overall ranking indicates how many experts attributed importance to the respective factors at the end of the workshop.

Table 8.1 reveals that a total of 37 factors were identified and discussed. Among these, a trio emerged as especially critical to most participants. These factors concern the economic competitiveness of integrative forest management, the socio-cultural identity of forest owners or managers, and outreach and education associated with integrative practices. Interestingly, socio-cultural factors were considered significantly more important, representing 48% of the points allocated by the participants. The emphasis on socio-cultural dynamics underscores a critical recognition, namely, that the identity, values, and traditional knowledge systems of those who manage or own forests are decisive in shaping their receptivity towards integrative practices.

Table 8.1 Breakdown of prioritised factors from the expert workshop.

Socio-cultural factors	*Prioritisation*	*Economic factors (continued)*	*Prioritisation*
1. Identity	**7** (14.3%)	*20. Incentives*	**0** (0%)
2. Outreach & education	**5** (10.2%)	*21. Bioeconomy*	**0** (0%)
3. Nature connectedness	**3** (6.1%)	*22. Diversification*	**0** (0%)
4. Social capital	**3** (6.1%)	*23. Certification*	**0** (0%)
5. Skilled labour	**2** (4.1%)	**Environmental factors**	
6. Forest ownership	**2** (4.1%)	*24. Regional variations*	**3** (6.1%)
7. Public opinion	**1** (2.0%)	*25. Climate change*	**1** (2.0%)
8. Entrepreneurship	**0** (0%)	*26. Naturalness*	**0** (0%)
9. Urbanisation	**0** (0%)	*27. Land use change*	**0** (0%)
Technological factors		*28. Degradation*	**0** (0%)
10. Mechanisation	**1** (2.0%)	*29. Nitrogen deposition*	**0** (0%)
11. Product development	**1** (2.0%)	**Political factors**	
12. Research & development (R&D)	**0** (0%)	*30. Leadership*	**4** (8.2%)
13. Information & communication tools	**0** (0%)	*31. Policy coherence*	**3** (6.1%)
14. Data sharing	**0** (0%)	*32. Policy incentives*	**2** (4.1%)
15. Educational tools	**0** (0%)	*33. Regional variations*	**1** (2.0%)
Economic factors		*34. Soft policy instruments*	**1** (2.0%)
16. Competitiveness	**7** (14.3%)	*35. Policy barriers*	**1** (2.0%)
17. Global competition	**1** (2.0%)	*36. Taxes*	**0** (0%)
18. Cross-sectoral competition	**0** (0%)	*37. Regulatory burden*	**0** (0%)
19. Economic viability	**0** (0%)		

3.1. Socio-cultural factors

The workshop participants identified socio-cultural factors as vital in expanding the uptake and application of integrated forest management. Central to this assertion is the evolving and often contrasting viewpoints held by various societal groups – such as forest owners versus tourists, urban populations versus rural communities, and forestry professionals versus conservation advocates. These disparate viewpoints give rise to multiple demands on forest resources, thereby influencing their utilisation (Bonsu et al., 2017, Maier and Winkel, 2017, Deuffic et al., 2018). Moreover, it was noted that the perceptions of natural landscapes play an important role in shaping public sentiment and, by extension, the development of relevant policies (Rametsteiner et al., 2009, Pülzl et al., 2022).

Several trends are instrumental in this dynamic, including demographic shifts like urbanisation, ageing demographics, and evolving family structures, transformations in forest management cultures and traditions, diminished economic dependency on forests, and an increase in nature-centric activities. These societal shifts are changing expectations regarding ecosystem services, subsequently reshaping the prerequisites for forest management (Borrass et al., 2017). A case in point is the escalating demand for forests to serve multiple and parallel purposes, particularly in or close to urban areas. This may range from recreational activities to environmental conservation to public health (Borrass et al., 2017, Başkent, 2018, Ciesielski and Stereńczak, 2018).

The above-noted socio-cultural factors present a unique leverage point. They can be strategically utilised to generate support for nature conservation objectives within the broader forest management framework. This is primarily because these factors can help to amplify societal demands for forests to deliver beyond timber production, emphasising the intangible benefits they offer, such as enhancing mental health, aesthetic values, and as part of our cultural heritage (Bjärstig and Kvastegård, 2016).

A prominent socio-cultural factor, deeply rooted in evolving societal perspectives on natural landscapes, is the emergence of a distinct category of urban forest owners. Characterised by their minimal financial dependence on forest land resources, these individuals represent a growing demographic. Their approach to forest ownership is commonly influenced by a sense of nature connectedness and personal identity, resulting in management objectives prioritising nature connectedness and personal identity over traditional economic gains (Feliciano et al., 2017, Weiss et al., 2019). This shift reflects a broader societal transition towards valuing natural spaces for their intrinsic and personal value, such as enjoying the beauty of a forest (Aggestam, 2015, Irvine et al., 2016, Pülzl et al., 2022).

However, this trend also has its challenges. It has been linked to debatable outcomes, including land abandonment and a decreasing number of viable opportunities for active forest management (Benayas et al., 2007, Ustaoglu and Collier, 2018). Traditional forestry practices, often considered necessary for ecological health, such as controlled thinning or prescribed burns, may be underutilised, leading to overgrown and potentially unhealthy forests. Moreover, the rise in urban forest ownership correlates, whether coincidentally or directly, with a troubling decline in the number of experienced forest workers, which has led to a decline in the availability of specialised forestry skills (Lawrence et al., 2017). This trend has affected rural economies as forests increasingly cease to be a reliable source of employment and income they historically were. The implications of these shifts for integrated forest management are nuanced and double-edged.

Conversely, there is a legitimate concern that the reduced prevalence of experienced forestry professionals and a potential decreasing interest in forest management could result in adverse outcomes. These outcomes range from increased frequencies of land neglect to the commercial acquisition of forest land by larger enterprises less interested in integrated forest management. In other words, the risk is a landscape increasingly vulnerable to unsustainable practices, biodiversity loss, and environmental mismanagement. The increasing preference of urban and new forest owners toward alternative, more integrative forest management objectives nevertheless presents opportunities. Time will tell if urban and new forest owners' readiness to pursue non-commercial practices will increase the uptake of more innovative strategies that broaden the scope for embedding nature conservation into forest management paradigms.

While urban forest owners are particularly important for privately held forest lands, the management of public forest land – such as those owned by the State or local municipalities – is predominantly influenced by the forest managers' perspective. Their attitudes, shaped by diverse management cultures, educational backgrounds, and historical traditions, play a decisive role in forest management strategies (Maier and Winkel, 2017). These varying perspectives result in a spectrum of management practices, some aligned with integrated forest management principles and others that may be at odds with SFM (Deuffic et al., 2018, Verkerk et al., 2011, 2020).

The importance of social capital and reputation came up significantly during the discussions between workshop participants. They emphasised that within rural communities, there is often an embedded respect for foresters, rooted in a longstanding trust in their expertise and the personal relationships that have developed over time (Guillén et al., 2015). This contrast becomes apparent when compared to the doubt or criticism that foresters might face from urban and new forest owners, who may lack a direct connection to or understanding of SFM practices (Primmer and Karppinen, 2010, Buijs and Lawrence, 2013, Guillén et al., 2015). Such disconnects have, at times, escalated into conflicts, with foresters at the receiving end of the blame for the perceived overexploitation of forest resources, biodiversity loss, and an assumed disregard for nature conservation. These types of conflict underscores the urgent need for foresters and forestry professionals to reassess and adapt their communication strategies. A shift towards more transparent, inclusive, and relatable communication may help avoid misunderstanding, facilitate a more effective two-way communication process, and foster trust (Bethmann et al., 2018). Enhancing social capital, particularly by developing trust and mutual understanding between varied stakeholders – such as forestry professionals and rural populations, urban forest owners, and conservation advocates – is crucial. This implies that building social capital offers one pathway forward concerning the uptake of integrated forest management practices.

Moreover, a substantial proportion – nearly half – of the workshop participants emphasised the critical role of outreach and education, highlighting the need for platforms that facilitate dialogue and the sharing of perspectives and values. This focus on enhanced outreach and educational efforts is crucial in expanding the general understanding of the opportunities and barriers with regard to the harmonised delivery of forest ecosystem services (FES), whether in the public or private sector (Coll et al., 2018, Leban et al., 2016). For instance, the participants pointed out the need for forest owners and managers to receive increased social support and educational resources. This need stems from the challenge of effectively implementing integrated forest management in the face of the diverse and, at times, conflicting expectations placed on forest resources. In these instances, comprehensive educational programs and community outreach initiatives, such as workshops, seminars, and interactive online resources, can help increase the knowledge, skills, and support needed to navigate the complexities of integrated forest management. Additionally, creating or strengthening networks between forest owners and managers can facilitate the sharing of best practices and lessons learned. Another example is mentorship programmes, where experienced practitioners can help guide urban and new forest owners by illustrating the trade-offs and synergies involved in forest management decisions. In the context of conflicting demands on forests, it would also be essential to extend these educational initiatives to the broader public. Enhancing public understanding regarding forest ecosystems is integral to the long-term success and viability of integrative strategies.

Socio-cultural factors are fundamental for integrated forest management, as diverse group interests, ranging from forest owners to conservationists, affect forest use and policy. Continued urbanisation and lifestyle changes mean that today's forests must fulfil multiple roles. Moreover, new and urban forest owners, often valuing forests for personal and environmental reasons, diverge from profit-driven forest management, challenging traditional forestry practices. Effective forest management requires improved communication between stakeholders and educational initiatives to build social capital and understanding. Outreach and shared learning is essential to address varied expectations and is integral to promoting the adoption of conservation-focused management strategies.

3.2. Technological factors

Although the workshop participants considered technological factors to be of a lesser priority compared to other factors, it was acknowledged that advancements in technology and the development of new and innovative forest products could create opportunities for integrated forest management, aside from also addressing critical challenges in sustainable forestry, including the integration of nature conservation objectives. For instance, enhanced capabilities for forest mapping and monitoring have transformed how forest resources are managed. Technological innovations, such as geospatial monitoring systems (Mikula et al., 2021, Varghese et al., 2022), have increased the detail and accuracy in identifying and tracking valuable habitats and tree stands. These technological advances can enable forest managers to adapt conservation efforts more effectively. Additionally, non-invasive techniques for assessing tree biomass (Dittmann et al., 2017, Ye et al., 2019) can contribute to a deeper understanding of forest structure and carbon stock without harming the trees or surrounding ecosystems. Such methodologies can facilitate a more nuanced approach to forest management that balances the need for wood production with preserving natural resources.

The integration of new tools and technologies can furthermore result in more environmentally friendly forest management practices. For example, optimising machine traffic on the stand level can reduce soil compaction and disturbance to wildlife habitats, thereby mitigating the environmental impact of forestry operations. Moreover, the strategic use of technology can support a more informed and efficient forest management planning process. By leveraging more detailed environmental data, forest managers could identify and exploit the synergetic relationships between conservation and wood production (Biber et al., 2020). This could not only increase the sustainability of forest operations but also enhance their economic viability.

Another relevant factor concerns the access to and dissemination of data and information relevant to shaping the uptake of integrative practices. The distribution and sharing of high-quality and relevant data with forest managers and owners, policymakers, and the public should be a cornerstone for improving decision-making in forest management. A well-informed approach not only allows for better planning and operations but can facilitate an improved understanding and acceptance of integrative practices. However, a notable challenge in this context is the observed information deficit among forest owners and managers and other relevant stakeholders. For instance, a lack of environmental awareness and understanding of integrated forest management could impede its wider adoption (Aggestam et al., 2020, Konczal et al., 2023). Ensuring that stakeholders have access to relevant data is, for this reason, crucial to overcoming these types of obstacles, which are often interlinked with the underlying infrastructure that supports data and information sharing. Moreover, integrating new technologies that streamline data sharing and translation is becoming increasingly significant. Conversely, when this infrastructure is robust, it can not only enhance the transparency of integrative initiatives but also allow forest owners and managers to observe, understand, and support conservation efforts (Baycheva-Merger et al., 2018, Eriksson and Klapwijk, 2019). These innovations, including relevant platforms that facilitate information access on different market demands, both private and commercial, are central to transforming attitudes and perspectives towards nature conservation and timber production (Bonsu et al., 2017).

The nexus between outreach, education, and the practical application of forest knowledge was discussed during the workshop, particularly concerning the methodologies used to share information with forest management practitioners. One illustrative example is the role of Marteloscopes, as part of the European network of demonstration sites for integrated forest management, which serve as an educational tool. These educational tools provide a platform for users to get hands-on experience balancing economic interests and ecological values within a forest stand (Soucy et al., 2016, Kraus et al., 2018, Krumm et al., 2019). In other words, they are instrumental in illustrating the immediate and long-term impacts of different forest management strategies and approaches. Particularly as they enable practitioners to simulate decision-making processes that reveal the effects of preserving or harvesting trees that contribute to biodiversity, such as through the conservation of microhabitats (Bütler et al., 2013, Cosyns et al., 2018, Santopuoli et al., 2019). These simulations are not just theoretical exercises; they are based on real-life complexities and consequences related to forest management decisions.

These factors highlight that in-depth knowledge and skills about forests are vital for successfully adopting silvicultural practices that underpin integrative approaches. For instance, close-to-nature forestry, representing many integrated forest management principles, requires a comprehensive understanding of forest ecology and management. It is based on silvicultural techniques that require a fundamental understanding of the varied relationships within a forest ecosystem (O'Hara, 2016, Roitsch et al., 2023). Investments in educational platforms like Marteloscopes are therefore not just about knowledge transfer; they are about shaping perspectives and developing skills that enable foresters to manage forests in a way that is not only sustainable and ecologically sound but also economically viable.

The advancement of mechanisation in the forest-based sector presents a nuanced spectrum of benefits and risks that were also considered. It was noted that the progression towards using larger, more powerful machinery has raised concerns interlinked with increased soil compaction and damage to trees not designated for harvesting (Dudáková et al., 2020, Labelle et al., 2022). Nonetheless, modern forestry equipment is often equipped with cutting-edge features designed to reduce environmental impacts and allow for more precise execution of silvicultural activities as compared to older, less robust machinery (Vanclay, 2011, Magagnotti et al., 2012). Moreover, integrating GPS technology into forestry equipment has become a game-changer. By facilitating the careful planning of harvesting pathways, GPS technologies can enable operators to bypass ecologically sensitive areas, such as wetlands or trees of particular ecological or cultural significance, and avoid disturbing historical or archaeological sites, as well as structures like utility lines and pipelines (Mohtashami et al., 2017). Additionally, developments of booms and hydraulic systems enable more deliberate and delicate felling of trees and handling of logs. Such precision plays a critical role in reducing harm to the surrounding vegetation as well as the structural integrity of the forest during the logging process (Spinelli et al., 2014).

Finally, innovation, alongside research and development, was considered fundamentally interconnected with the wider adoption of integrated forest management practices. One pathway through which innovation may facilitate the implementation of integrated forest management is the emergence of new methodologies to evaluate the economic value of forest ecosystem services. For example, the development of Payment for Ecosystem Services (PES) models not only incentivises the preservation and sustainable management of forest resources but helps to quantify the value of these services (Kant et al., 2016, Bösch et al., 2018, Chen et al., 2020). Furthermore, advancements in how forest managers and owners understand and map the provision of ecosystem services are crucial. They can, on the one hand, improve conservation strategies and, on the other hand, pave the way for discovering additional services provided by forests, such as non-timber forest products. These services may, in turn, offer supplementary revenue streams or even serve as an alternative to timber sales, thus enhancing the economic viability of integrative practices. Put another way, bolstering the financial sustainability of forest ecosystems may result in an increased consensus on the importance of multifunctional forestry.

> Technological advancements are critical in advancing sustainable forestry. Improved forest mapping and non-invasive assessments can allow for better conservation and management. Efficient machinery use can minimise environmental damage and enhance operational precision. Critical to this progress is the effective sharing of data and information among forest stakeholders to inform decisions and promote sustainable practices. For instance, education, through tools like Marteloscopes, can equip practitioners with skills to balance economic output with ecological value. Innovations, such as GPS-equipped machinery, can reduce the ecological footprint of forestry operations, while economic models valuing ecosystem services can incentivise forest conservation.

3.3. Economic factors

The competitiveness of the forest-based sector stood out as a key concern for nearly half of the workshop participants, who regarded it as a crucial driver for integrated forest management. In line with the inherent nature of business, the European forest-based sector (as any other sector) constantly seeks to enhance profitability by creating added value for forest products and services. However, the sector faces increasing challenges concerning its financial sustainability, primarily due to heightened global competition, such as from regions with lower labour costs, leading to lower profit margins (Rivera León et al., 2016, Ziolo et al., 2019). Moreover, the pursuit of competitiveness, while essential for the economic viability of the sector, can increase conflicts with regard to nature conservation objectives (Damania et al., 2018, Felton et al., 2020, Eyvindson et al., 2021). A notable example of such a conflict is the balancing act between harvesting forest biomass for economic benefits vis-a-vis retaining it to support biodiversity, which includes the conservation of deadwood and habitat trees within forest ecosystems (Winkel et al., 2015, Bauhus et al., 2017, Sabatini et al., 2019). Furthermore, the forest-based industries' preference for uniform and large-scale timber production can be at odds with the growing demand for variable and small-scale single-tree forest management techniques, such as selective logging, that create artificial structural diversity. Such methods often conflict with industrial requirements for bulk quantities of wood or biomass produced cost-effectively. It can be noted that the literature presents mixed views on the cost-efficiency of different silvicultural practices, particularly when forest risks are considered, suggesting that economic outcomes are heavily dependent on the ecological and socio-economic contexts of the forests in question (Müller et al., 2019, Kożuch, 2022). Thus, the interplay between the economic goals of the forest-based sector and the incorporation of conservation values into forest management presents a complex and not yet fully understood relationship. The challenge lies in navigating this dynamic process to find a sustainable and balanced pathway that can reconcile economic imperatives with the ecological and social values of forests.

During the workshop, participants underscored a growing willingness among those working within the forest-based sector, including forest managers and owners, to prove that they can fulfil conservation requirements. This drive for sustainability is not just about meeting environmental standards but is seen as a strategic advantage to stay competitive in the marketplace. Moreover, the commitment to protecting forest biodiversity can be a critical factor in maintaining political support and securing the social license needed for the sector's operations (Moffat et al., 2016, Gehman et al., 2017). Despite these positive trends towards integrated forest management, there remains a tension between environmental and economic objectives. These conflicting objectives can complicate the forest-based sector's target for long-term sustainability (Eyvindson et al., 2018, Triviño et al., 2017). For example, one of the central challenges to enhancing the role of nature conservation within forestry is to illustrate the added value of biodiversity. This includes identifying and promoting new market opportunities for wood-based products, like biofuels and textiles, and non-wood forest products, such as medicinal herbs and mushrooms, as well as other ecosystem services. Economically, there is also potential in safeguarding forest genetic resources as an integral part of standard production systems, particularly with tree species that yield high-value timber (Potter et al., 2017). From an environmental and economic resilience standpoint, there is a positive link between the diversity of tree species in a forest stand and its overall resilience. Biodiversity within a forest stand can not only provide access to a broader range of products and services but can also offer a degree of security against various threats. This includes buffering impacts from disturbances such as diseases or pests, as well as the variable conditions brought on by climate change. To simplify, a diverse forest is also a resilient forest that can withstand and adapt to a variety of environmental challenges.

Evolving market dynamics are increasingly aligning with nature conservation objectives, facilitating the adoption of integrated forest management. For example, forest certification programs like the Forest Stewardship Council (FSC) have not only opened doors for greater conservation efforts, such as the requirement to preserve habitat trees or maintain deadwood within managed forests (Gustafsson et al., 2019) but have also provided economic incentives for forest owners (Kalonga et al., 2016, Gutierrez Garzon et al., 2020, Dieterle and Karsenty, 2020). These certification schemes can enhance market access that benefits forest owners financially. However, in contexts where markets for non-wood forest products and ecosystem services are not fully developed, incentives become crucial for enabling integrative practices. Incentives offer a pragmatic solution for forest owners whose operations are primarily profit-driven. Through incentives, owners are encouraged to engage in conservation even when these might conflict with economic gains (Tian et al., 2018, Gundersen et al., 2019, Tian and Pelkki, 2021). Incentive mechanisms could also be broadened beyond the basic payment schemes to include tax-related incentives, such as deductions or credits for engaging in conservation, and social benefits, such as healthcare provisions. These types of innovative approaches could not only compensate for the direct costs associated with conservation but serve to support the multifunctional contributions of forest ecosystems.

The bioeconomy concept, which emerged as a discussion point during the workshop, underscores the complex balance required between being sustainable and being competitive. The bioeconomy refers, in this instance, to an economic model that leverages natural resources, such as forests, with regard to their capacity to provide multiple ecosystem services, thereby promoting

SFM (Winkel et al., 2022). While the bioeconomy can create new economic opportunities, for example, by providing biomass for energy or other ecosystem services (Landis et al., 2018, Gregg et al., 2020, Hetemäki et al., 2022), it also presents a unique set of challenges that requires reconciling economic and conservation objectives within forest management practices. Expanding on the potential of the bioeconomy, there is a recognition that the provision of multiple ecosystem services offers an opportunity for discovering and capitalising on conservation synergies (a win-win scenario). For instance, innovative business models are beginning to emerge, such as funeral forests, where the conservation of cultural forest ecosystem services intersects with economic viability. More specifically, these models allow for the monetisation of services that were previously non-market, such as providing spaces for natural burials in forests, which in turn also supports the retention of old-growth trees within high habitat values (Roux et al., 2022). However, it should be recognised that the rising demand for timber could constrain the broader adoption of integrative practices. The pressure to harvest more timber, often driven by financial incentives, might be at odds with integrative principles, which strive for a balance between economic returns and maintaining forest ecosystem health. The critical determinant for the success of integrative forest management within the context of a growing bioeconomy will, as such, be the willingness and ability of forest owners, both public and private, to shift towards a focus on ecosystem services. Fundamentally, this entails recognising the benefits that forests provide – beyond timber – and integrating such values into forest management decisions.

> The European forest-based sector faces the challenge of remaining competitive while addressing biodiversity conservation. Economic pressures from global competition are set against the need for biodiversity and sustainable management practices. While there is a push for profitability through large-scale timber production, there is also a growing demand for diverse, conservation-oriented practices like selective logging. However, the tension between economic gains and ecological health remains despite market incentives from certification programs and a recognition of the multifunctional role of forests. The bioeconomy may offer new opportunities but requires integrating the full spectrum of ecosystem services into forest management.

3.4. Environmental factors

The workshop discussions delved into pressing environmental issues such as climate and land use change and environmental degradation. A significant outcome from these discussions was the consensus on the importance of regional variation in forest ecosystems as a factor influencing the uptake of integrated forest management. The participants highlighted that the success of integrated forest management is contingent on local and regional bio-geographical conditions, which include variables such as the diversity of tree species, soil types, climate patterns (including precipitation and temperature), elevation, and geological features (Kraus et al., 2018, Krumm et al., 2020, Konczal et al., 2023). It was essentially argued that integrative practices are not one-size-fits-all and vary distinctly across different biomes, implying that what might work in the boreal forests with their vast expanses and colder climates will not necessarily be as effective in the more varied and often warmer temperate forests or the dry Mediterranean regions (Puettmann et al., 2015, Halbe et al., 2018). Each of these biomes requires tailored management approaches that address the region's unique environmental conditions.

Furthermore, the interplay between various drivers was seen as essential to understanding effective integrated forest management. Bio-geographical factors, such as the native tree species and the terrain they grow on, are significantly intertwined with socio-cultural factors, including historical land management practices. This entanglement influences the practical implementation of biodiversity conservation (Borrass et al., 2017, Maier and Winkel, 2017, Caicoya et al., 2023). For instance, in the boreal forests, small-scale clear-cutting might be employed to mimic natural disturbances, whereas in other regions, such as those following close-to-nature forestry principles or retention forestry practices – where certain trees or groups of trees are left standing to preserve biodiversity – might be preferred (Gustafsson et al., 2013, 2019, Kuuluvainen et al., 2019). All-in-all, the implications are that integrative practices must be fine-tuned to address the specific ecological and socio-cultural realities of the region in question. This customisation ensures that conservation measures are more effective and harmonised with the ecological processes and services inherent to the local forest landscapes.

Climate change acts as a significant determinant in shaping integrative practices. However, the implications of climate change are different as they vary greatly from region to region, resulting in a wide range of challenges for forest ecosystems. These challenges can include varied temperature increases, alterations in precipitation patterns, and more frequent and intense disturbances, such as wildfires and pest outbreaks. Disturbance regimes, notably those involving forest fires and pest infestations like bark beetles, are particularly influential. It is widely recognised that mono-cultured or even-aged forests composed of limited tree species are especially vulnerable to such disturbances (Triviño et al., 2017, Pohjanmies et al., 2021). This vulnerability highlights the benefits of structurally diverse forests, such as those managed under continuous cover forestry systems. These typically exhibit greater resistance to biotic and abiotic threats due to their complexity and species richness (Aszalós et al., 2022). One strategy to bolster forests against these challenges is enhancing forest diversity across multiple dimensions. This can include improving genetic diversity, cultivating a variety of tree species, increasing the structural complexity, and considering the landscape context as critical steps in building the adaptive capacity of forests. The main argument is that diversity serves as a foundation for resilience in the face of climate change, advancing conservation objectives by supporting a range of habitats and species (Nikinmaa et al., 2020, Messier et al., 2022).

While increasing the structural and species diversity of forests tends to have positive outcomes for nature conservation, other climate adaptation strategies may have mixed effects (Jandl et al., 2019). For instance, reducing the amount of forest biomass by employing shorter rotation periods is used to lower the risk of large-scale disturbances. This approach can lead to more frequent harvesting, which may, in turn, reduce the conservation value of forests by limiting the growth of older trees and diminishing the availability of habitats typically found in mature stands (Ontl et al., 2019, Felton et al., 2020). Integrated forest management must, therefore, navigate these strategies with an understanding of their implicit trade-offs. Balancing the need to mitigate the risks posed by climate change with the imperative to maintain and enhance biodiversity requires a context-sensitive approach.

Forest naturalness was identified as a factor influencing forest management strategies. 'Naturalness' encompasses a range of elements, including the natural processes within the forest ecosystem, its physical structure, and the composition of species inhabiting it. The central concern was the degree to which the species composition remains native to a given area (Bončina et al., 2017, Lazzaro et al., 2020). The impact of introducing non-native or exotic tree species was criticised due to its potential to alter ecological functions. For example, when exotic species are introduced, they can disrupt the soil chemistry, particularly by changing nitrogen levels. Such alterations can extend beyond the soil, potentially undermining the nutrient cycle of the forest and ultimately leading to detrimental effects concerning indigenous biodiversity. The presence of non-native species may also affect the forests' microclimate as well as the diversity of fauna and flora (Paillet et al., 2010, Langmaier and Lapin, 2020, Milanović et al., 2020, Rose et al., 2020). In addition, the workshop participants noted the scarcity of forests with a high degree of naturalness, such as primary or ancient forests, in Europe (Sabatini et al., 2018, 2021). These ecosystems are valuable for their complexity, resilience, and the unique habitats they offer. Preserving and promoting native species is essential not only for the conservation of biodiversity but also for stabilising soil conditions, which in turn support a multitude of ecosystem services. Moreover, maintaining native species helps ensure that the habitat requirements for a more comprehensive array of native wildlife are met, contributing to the overall health of the forest ecosystem. In this context, the maintenance of naturalness is arguably not only about conserving what is currently present in the forest but also about deliberate forest management choices aimed at protecting native species, restoring natural forest structures and functions, and resisting the pressures to introduce invasive species that could compromise the ecological integrity of the forest landscape (IPBES, 2023).

Land use patterns, which emerge from a complex interplay between environmental, socio-cultural, and economic elements, is another critical factor that shapes forest use. These patterns are not merely the outcome of present-day decision-making; instead, they are rooted in historical contexts that continue to influence forest management practices (Jakovac et al., 2021, Long et al., 2021). Historically, certain socio-cultural factors have led to environmental changes that affect forest health and productivity. A prime example is the acidification of soils, which can be traced back to the establishment of monoculture spruce plantations in areas ill-suited for such species (Hopf et al., 2023). These historical decisions have left a legacy of environmental challenges that must now be addressed through forest management. Moreover, certain types of forestry initiated for their economic benefits, such as coppice forestry or the cultivation of economically valuable species like chestnuts, have now become integral to both the ecology and the economy of their regions. These practices have significantly impacted nature conservation, demonstrating the intricate relationship between economic imperatives and environmental stewardship. The game management issue is another dimension where historical land use decisions continue to have effects today. Overpopulation of game species can lead to excessive browsing, impeding the regeneration of certain tree species and causing a cascade of negative economic and ecological effects. The consequences include not only financial loss due to poor forest regeneration but also a decrease in biodiversity (Boulanger et al., 2018, Neumann et al., 2022).

The historical trends associated with land use have resulted in a patchwork of local and regional landscape patterns that set out the contextual framework for implementing integrated forest management. These patterns highlight the influence of past land use decisions, underscoring the importance of considering historical land use factors when planning for SFM. The challenge for current and future forest management is reconciling these historical influences with contemporary conservation objectives (Martin et al., 2020, Felton et al., 2020). In other words, understanding the historical land use matrix is essential for informing the uptake of integrative approaches that align economic objectives with conservation priorities and societal values.

The success of integrated forest management depend on regional environmental conditions such as biodiversity, soil types, climate, and geological features, requiring region-specific strategies. Moreover, socio-cultural factors like historical land use intertwine with bio-geographical factors, affecting the implementation of conservation efforts. For instance, land use patterns shaped by historical decisions continue to influence forest management, highlighting the significance of past practices for conservation. Climate change further complicates these dynamics by differing regionally, highlighting the need for diversity-based resilience and adaptive capacity in forests. Other factors, such as naturalness in forests, play a critical role in managing forest ecosystems. Introducing non-native species can disrupt these processes, affecting soil chemistry and biodiversity.

3.5. Political factors

The successful implementation of integrated forest management is contingent upon having a supportive political and legislative framework that encourages integration. Workshop participants stressed the critical nature of policy coherence in this context, highlighting that with it, the adoption of integrative practices would be substantially improved. This implies that for integrated forest management to be more widely taken up, it is crucial that policies set clear and coherent targets that align with multifunctional objectives. These objectives can range from timber production to conservation and carbon sequestration, each with a set of priorities and metrics for success (Maier and Winkel, 2017, Pülzl et al., 2018). Policy coherence could, in this case, facilitate the integration of varied policy goals and targets into a unified management strategy (Aggestam and Giurca, 2021, Aggestam et al., 2020, Aggestam and Pülzl, 2020). The complexities of soft and hard policy frameworks are nevertheless evident in the various legislative structures and related policy instruments that exist across different international, regional, and national contexts. These frameworks can introduce incentives that, while beneficial in one context, may lead to conflicting policy targets in another. For example, in Europe, forestry measures aimed at promoting timber production, often encouraged through EU Rural Development Programmes, may conflict with objectives for nature conservation and bioenergy set at both the EU and national level. This type of conflict reflects the challenging balancing act between promoting industry and protecting the environment through policy (Aggestam and Giurca, 2021, Aggestam and Pülzl, 2018, 2020).

To navigate these complexities and implement integrated forest management, it is essential that policy frameworks not only recognise inherent trade-offs but also seek solutions that can mitigate the competition between different land uses. Even more, in cases where policy frameworks are slow to resolve these conflicts, it becomes imperative for practitioners to address these trade-offs through forest management decisions made on the stand level. This can entail forest managers making informed choices that reflect integrative principles and site realities, even if policy support still needs to be fully aligned with the principles of integrated forest management. However, ultimately, the pathway to integrated forest management resides in having a policy environment that is dynamic and able to transcend traditional silos to encompass economic, ecological, and social dimensions.

During the workshop, it was underscored that adopting integrated forest management is fundamentally connected to effective leadership. Such leadership is not merely a function of authority or power but covers a range of capabilities and associated skills critical for advocating and achieving established conservation objectives (Mann et al., 2021, Webb et al., 2022). In other words, effective leadership can enable a shift from conventional, production-oriented forest management to more integrative approaches. The literature reinforces the importance of leadership, emphasising that individuals in leadership roles must possess not only a vision for sustainability but also the competence to navigate complex social, economic, and environmental challenges. Leaders must have various skills, including strategic planning, stakeholder engagement, conflict resolution, and change management (Mattijssen et al., 2020, Mann et al., 2022). For example, a policymaker with a leadership mindset can create and reinforce frameworks that promote the integration of conservation objectives into forest management practices. Likewise, forest managers with strong leadership skills are crucial for translating policy into practice, ensuring conservation becomes an integral part of forest operations. In these contexts, leaders must bridge the gap between policy and practice to effectively communicate the value of integrated forest management to different stakeholders, from policymakers to local communities and industry.

Alongside the need for effective leadership and coherent policy frameworks, strategies that encourage integrated forest management adoption are diverse. Examples include financial measures, such as compensation for income foregone due to conservation practices like leaving deadwood or protecting habitats (Mann et al., 2021, Sarvašová et al., 2019a) or offering tax relief for those who invest in nature conservation, such as personal income tax credits, have proven to be effective tools for embedding biodiversity considerations within forest management, particularly for privately owned forest lands (Besacier et al., 2021, Juerges et al., 2021, Sarvašová et al., 2019b). However, while the use of economic incentives is increasingly recognised as an effective mechanism for encouraging biodiversity and ecosystem services conservation, these types of incentives are just one piece of a more extensive policy toolbox. Other policy instruments that may not directly involve financial incentives can also be highly impactful. These include participatory approaches, which leverage the power of community engagement and bottom-up policy development to advance conservation objectives (Reed et al., 2019, Marques et al., 2020). Such strategies often encompass collaborative efforts where various stakeholders, ranging from local communities to forestry experts, come together to devise and implement conservation-friendly practices. Establishing vertical networks, linking different levels of governance, and horizontal, connecting various stakeholders within the forest-based sector – is another effective strategy. Networks can facilitate the sharing of best practices, encourage goal alignment, and support the overall integration of conservation objectives across different policy domains (Elliott and Schlaepfer, 2001, Sotirov and Winkel, 2016, Kleinschmit et al., 2018, Lazdinis et al., 2019). It should also be stressed that the varying success and implementation of these types of policy-driven incentives underscore the influence that policy has on integrated forest management, especially in regions of Europe where integrative practices are politically prominent (Winkel and Sotirov, 2016, Borrass et al., 2017, Pülzl et al., 2018). In such cases, political will can be the primary driver for adopting integrated forest management by promoting policies that incentivise, value and legitimise conservation efforts.

The complex interplay between policy, leadership, and strategy forms one crucial cornerstone for advancing integrative practices. Political support and legislative frameworks are pivotal in this context. They need to provide clear, coherent targets that align with the multifunctional nature of forests, encompassing timber production, conservation, and carbon sequestration. However, policy coherence is often challenged by competing priorities across different levels of governance, such as the EU's push for timber production potentially conflicting with national conservation goals. Strategic incentives also play a role in promoting IFM. Financial measures, like compensations or tax reliefs, have been effective, particularly for private lands, by encouraging biodiversity conservation. Moreover, participatory approaches and the establishment of networks among stakeholders help to align goals and integrate conservation into forest management.

4. Driving forces behind integrated forest management

This chapter has aimed to introduce some of the essential framework conditions that shape the adoption of integrated forest management practices. Specifically, it has explored factors that can facilitate the harmonisation of biodiversity conservation with wood production, which are central elements of integrated forest management. To achieve this goal, the chapter introduced results from a comprehensive review of driving factors undertaken during an expert workshop and cooperative analysis (Aggestam et al., 2020), where a total of 37 factors were identified (Table 8.1). It is noteworthy that each of the listed factors was recognised as impactful by at least one expert participant, underscoring the varied nature of factors that can encourage or impede the uptake of integrative practices. More specifically, factors influencing the adoption of integrated forest management range from being more dynamic and interactive to being more static or slow to evolve. On the dynamic end of the spectrum, there are factors like economic competitiveness and the ever-shifting landscape of policy frameworks, as well as the uncertain impacts associated with climate change. In contrast, on the static end of the spectrum, bio-geographical conditions and forest ownership structures change at a more gradual pace. All factors are nevertheless important in shaping integrated forest management outcomes.

The array of factors outlined throughout this chapter aligns with recent scholarly definitions of integrated forest management, theoretical frameworks proposed in the literature, and empirical research findings that shed light on the complex interplay between forest management and biodiversity conservation (Lazdinis et al., 2019, Storch et al., 2020, Asbeck et al., 2021). It is particularly notable that the factors most frequently prioritised by the workshop participants – when asked to select their top three – echo the key enablers for integrative practices identified in the existing literature. These include societal expectations and pressures, which reflect the growing awareness and demand for SFM, and economic pressures that influence the competitiveness within forest product value chains. Previous research underscores the significance of several factors in enabling integrated forest management, including political factors like policies, laws, and regulations, socio-cultural factors such as forest ownership patterns, economic factors exemplified by the global timber market, and environmental factors, most notably climate change. These have been highlighted as vital considerations for adopting integrated forest management. Moreover, these factors are not only significant in themselves but also in how they interact with each other, presenting a complex landscape for integrated forest management implementation (Konczal et al., h2023, Kraus and Krumm, 2013, Krumm et al., 2020).

The synthesis of expert inputs reveals a critical consensus, namely that socio-cultural factors profoundly influence the management of forests. It is the mosaic of individual and family values, local customs, and social networks that principally shape forest use. This intricate web of human interactions and cultural dynamics is fundamental in determining the success or failure of integrative approaches. Just as pivotal are the economic dimensions of the forest-based sector. The sector's competitiveness emerged as a vital economic driver, essential not only for its viability but also for the effective implementation of integrative practices. Regarding environmental factors, the experts highlighted the significance of local environmental conditions, such as climate, soil, and biodiversity, that are crucial in tailoring integrated forest management strategies to specific countries or regions. Anticipated effects of climate change were also underlined as a central environmental consideration, emphasising the need for integrated forest management to be adaptable and resilient in the face of evolving ecological scenarios. From a technological standpoint, although not ranked as the top priority, the range of tools identified by experts as enablers for integrated forest management spans from advanced, low-impact forestry equipment to educational resources that enhance management capabilities. These innovations have the potential to refine forest management practices substantially, enabling more responsible stewardship of forest resources while also improving the economic outcomes of forestry activities. Lastly, the political factors capture the role of governance in shaping the adoption of integrative practices. However, policy frameworks are double-edged swords. They can either facilitate the pathway towards adopting integrated forest management or create barriers that hinder its progress. Furthermore, the expert inputs underlined the role of leadership in championing integrated forest management. Effective leadership is necessary at all levels, from the formulation of policies to the hands-on practices associated with forest management.

Interestingly, insights from the workshop participant's prioritisation spotlighted the impact of social identity and societal perceptions of forests. For instance, the changing landscape of forest ownership reflects a shift in forest planning and integrative management beyond traditional economic considerations. The motivations of new or urban forest owners are more multifaceted, ranging from conservation to economic utilisation, challenging the conventional dichotomy between economic interests and environmental stewardship (Quiroga et al., 2019, Takala et al., 2017, Torralba et al., 2020). Forest owners and managers bring a spectrum of goals, capabilities, and decision-making processes to the table, each responding differently to the respective political, economic, and socio-cultural factors (Weiss et al., 2019, ECE, 2020, Torralba et al., 2020). This diversity may present an opportunity to advance the adoption of integrated forest management as it can help to bridge the gap between the emerging emphasis on nature protection and the longstanding economic use of forest resources. However, an apparent and critical barrier to the broader adoption of integrated forest management concerns the prevalent gap in forest management knowledge among forest owners and the public. To overcome this, outreach and education naturally emerged as an essential factor for fostering integration concerning the interface between forestry and biodiversity. Empowering stakeholders, such as forest owners, forestry professionals and conservationists, with knowledge about how to practically balance nature conservation with timber production is crucial. It is this informed and educated approach that could unlock the successful implementation of integrative strategies.

Competitiveness within the forest-based sector is fundamentally tied in with generating economic returns through forest management and, more broadly, the wider spectrum of forest-based value chains. This critical relationship suggests that the interaction between integrated forest management and competitiveness deserves in-depth exploration and understanding (Rivera León et al., 2016, Biber et al., 2020). Competitiveness can, on the one hand, pose a challenge to the integration of non-economic considerations, particularly those that diverge from the primary goal of maximising biomass production and minimising costs. This is especially true for biodiversity conservation efforts, where there is often a tension between the goals of increased harvest yields and the retention of biological diversity, such as balancing biomass extraction with the preservation of deadwood or reconciling the efficiency of large-scale clear-cutting with the maintenance of structurally diverse forests. The extent to which this tension impacts forest management practices is contingent on various contextual factors. These include the types of forest ownership, the motivations of the forest owners, and the demands from the forest product markets. For instance, a privately owned forest with an owner invested in long-term sustainability may operate differently from a publicly owned forest under pressure to meet short-term economic goals. On the other hand, competitiveness is also increasingly redefined to include sustainable and credible management practices. Credibility can, as such, become a catalyst for the integration of biodiversity conservation into forest management, essentially becoming a driving force for integrated forest management. In this emerging paradigm, demonstrating sustainable practices can provide a license to operate, particularly for European forests where environmental credentials are increasingly valued (Toppinen et al., 2017). In this scenario, competitiveness could become synonymous with sustainable practices that align economic interests with forest stewardship. It suggests that the market could play an important role in incentivising the adoption of integrated forest management by favouring products that are certified as sustainably managed, thus encouraging forest owners and managers to align their practices with broader conservation objectives. This could create a marketplace where the value is placed not only on the quantity and cost-efficiency of forest products but also on the ecological and social values maintained through forest management.

Lastly, the STEEP typology – focusing on social, technological, economic, environmental, and political factors – served as a practical framework for organising and evaluating the driving forces behind forest management, both during the workshop and subsequent analysis (Bowman, 1998, Kim-Keung Ho, 2014, Bonsu et al., 2017, 2020, Savastano et al., 2023). This approach provided a structured method for analysing factors influencing integrated forest management. However, it is important to acknowledge that while this compartmentalised methodology helps to clarify the varied landscape underlying integrated forest management, it also simplifies a reality where these factors are deeply intertwined. For example, socio-cultural drivers do not operate in a vacuum; they lay the groundwork for the economic conditions, shape environmental attitudes, and influence the political climate surrounding forest management. The values and norms of a society can dictate how forests are perceived and used, which in turn impacts economic strategies, environmental policies, and technological adoption within the forest-based sector. These interdependencies underscore the limitations of viewing such factors in isolation. It is clear that a more cross-cutting view is necessary. An integrated perspective is essential for understanding all the factors affecting forest management (Kraus and Krumm, 2013, Aggestam et al., 2020, Krumm et al., 2020, Konczal et al., 2023). For example, it would be essential to reflect on how socio-cultural expectations can drive technological innovations tailored for sustainable forestry or how policy instruments might be developed in response to environmental concerns and economic pressures.

4.1. Paving the way for integrated forest management

The results from this work suggest that the commitment and actions of key individuals at the operational level are crucial for adopting integrated forest management. These individuals often include early adopters, innovative entrepreneurs, and influential community leaders who can demonstrate the tangible benefits of integrative practices. These frontrunners not only endorse the integration of new forest management practices, but they also become instrumental in merging policy with practice (Deuffic et al., 2018, Eriksson and Klapwijk, 2019, Gundersen et al., 2019, Dieterle and Karsenty, 2020). In other words, the initial success and performance of any forest management approach, integrated forest management included, depend significantly on these pioneers. These leaders possess the foresight and the ability to translate integrative principles into actionable strategies. By doing so, they lay the groundwork for a wider shift towards SFM practices. At the policy-making level, effective leadership is just as critical. Leaders with a clear vision for the future of forest management can develop and implement policies that reflect the multifunctional role of forests. They can provide an enabling environment for integrated forest management. The importance of leadership was also a point of consensus among workshop participants, indicating that strong governance structures and capable leadership are crucial.

It is furthermore necessary to consider the most effective governance approaches. This includes addressing complex questions such as whether forest owners should be mandated or incentivised to preserve features like deadwood or habitat trees and determining which stakeholders, whether it is a conservationist, forest experts, landowners, or the public, should have the right to make decisions regarding conservation objectives. These considerations are significantly connected with the policy and institutional frameworks governing forestry and nature conservation. For example, command-and-control strategies may offer clarity and immediate impact by prescribing specific actions and creating a requirement for compliance, potentially leading to biodiversity gains if enforced (Aggestam and Pülzl, 2018, Weiss et al., 2019, ECE, 2020). However, these types of strategies can also face resistance due to their prescriptive nature, which essentially reduces the autonomy of the target groups. In contrast, self-regulation can grant greater flexibility, enabling forest owners and managers to apply their judgment and make decisions that align with their perspectives and motivations. This approach relies strongly on the provision of adequate resources, such as funding and education, and depends on the commitment and voluntary action of stakeholders. The lack of enforcement mechanisms would further mean that outcomes may vary from being simple symbolic actions to comprehensive application of integrative principles (Blattert et al., 2018, Chen et al., 2020, Mattijssen et al., 2020, Besacier et al., 2021). This variance in governance approaches and potential outcomes highlights the need for tailored, adaptive, and innovative policy designs to effectively integrate nature conservation into forest management. More specifically, policy instruments need to account for specific contextual factors that drive the integration of biodiversity considerations into integrated forest management practices.

As climate change progresses, it should also be noted that the strategies for biodiversity conservation, as part of integrated forest management, will encounter complex challenges. Current integrative practices emphasise the use of native tree species, the promotion of natural forest regeneration, the cultivation of structurally diverse forests, and the conservation of both deadwood and habitat trees. These practices are fundamental for maintaining forest biodiversity (Kraus and Krumm, 2013, Krumm et al., 2020). Yet, as environmental conditions slowly change, defining clear objectives and measures for forest biodiversity, such as the concept of naturalness, becomes more problematic. A warming climate may require adaptations in these strategies to maintain their effectiveness. In line with this argument, there exists a divergence in the perspectives of public, political, and professional stakeholders concerning forest management and conservation strategies related to the new climate reality. More specifically, the forest-based sector and actors working on nature conservation exhibit varying opinions on adapting to these changes (Maier and Winkel, 2017, Naumov et al., 2018, Santopuoli et al., 2019). One of the most significant questions is reconciling climate change mitigation efforts with the twin objectives of nature conservation and timber production. Finding a balance that ensures the health and productivity of forests while protecting biodiversity is a complex task that will continue to challenge Europe in the coming decades. It is not only a matter of scientific and technical adaptation but also consensus-building among diverse stakeholder groups with conflicting interests (Gustafsson et al., 2012, Buijs and Lawrence, 2013, Bonsu et al., 2017). Future-oriented integrated forest management must, therefore, be dynamic and responsive, accommodating changes not only in the ecosystem but also in societal values and policy frameworks.

Paving the way for integrated forest management will ultimately depend on a nuanced understanding of the relationships between different factors within the social, environmental, economic, technological, and political landscapes. It is essential to recognise that these factors are shaped by the specificities of each locale where integrative practices are being applied. There is, as such, a compelling need for detailed, context-specific insights that can guide forest managers and owners toward more informed implementation strategies. While it is recognised that the work underlying this chapter is limited in scope, it could help to further the development of integrated forest management. It also establishes a framework for research focused on a better understanding of individual factors, such as the role of market competitiveness or the collective impact these factors may have under specific regional or socio-economic conditions. By encouraging collaborative discourse and targeted research on integrated forest management there is an opportunity to respond to the challenges and opportunities of the twenty-first century and to ensure that European forests continue to provide multiple ecosystem services.

Want to find out more?

For those seeking more knowledge on integrated forest management, please see the references provided below or some of the many references throughout the text.

- How to balance forestry and biodiversity conservation – A view across Europe (Krumm et al., 2020).
- Integrative approaches as an opportunity for the conservation of forest biodiversity (Kraus and Krumm, 2013).
- Integrating forest management across the landscape: a three pillar framework (Morgan et al., 2021).
- The use of tree-related microhabitats as forest biodiversity indicators and to guide integrated forest management (Asbeck et al., 2021).
- Resolving the trade-off between production and biodiversity conservation in integrated forest management: comparing tree selection practices of foresters and conservationists (Cosyns et al., 2020).
- Forest Management Based on the Principles of Multifunctional Forest Use (Martynova et al., 2021).

The set of questions listed below may also help guide further inquiry into specific areas of interest:

1. How can societal expectations and pressures influence the priorities and practices of forest management in different European countries?
2. What specific policies, laws, or regulations do you think are most influential in adopting integrated forest management?
3. How does economic competitiveness within the forest-based sector impact the balance between timber production and biodiversity conservation?
4. How might integrative practices need to evolve to address the effects of climate change on European forests?
5. What challenges and opportunities arise from the changing landscape of forest ownership, particularly with regard to new or urban forest owners?
6. How do social, economic, and environmental factors interact to either support or prevent the adoption of integrated forest management?
7. What are the limitations of using the STEEP typology to analyse integrated forest management, and how can a more holistic approach be developed?

5. Assignment VIII: Developing a poster on multifunctional forest management

Description

Focusing on visual communication, this assignment is about designing a poster that addresses the complexities surrounding multifunctional forest use, focusing on how nature conservation can be integrated with forest productivity while addressing some of the socio-political and environmental challenges outlined in Chapter 8.

Purpose

Students will be asked to translate information from Chapter 8 and other materials into a visually engaging poster. The poster should demonstrate key multifunctional forest management and use aspects, including conservation, productivity goals, and socio-political and environmental considerations. It should educate readers about the importance of balancing FES to meet different types of demands and needs.

Outline of task

- *Identifying core concepts*: Using Chapter 8 (and additional readings), review and summarise the concept of multifunctional (or integrative) forest management.
- *Poster design*: Develop a poster representing these concepts, using visual elements such as graphs and images to demonstrate how multifunctional forest management can balance conservation, productivity, and societal demands. Ensure the poster conveys a clear and coherent message about the challenges and opportunities surrounding multifunctional forest management.
- *Presentation*: Prepare a brief (verbal) presentation that explains the poster's content and how it addresses multifunctional forest management.

Suggested additional reading

- Kraus, D. & Krumm, F. (2013) Integrative approaches as an opportunity for the conservation of forest biodiversity. European Forest Institute (EFI).
- Krumm, F., Schuck, A. & Rigling, A. (2020) How to balance forestry and biodiversity conservation – A view across Europe. European Forest Institute (EFI).
- Martynova, M., Sultanova, R., Khanov, D., Talipov, E. & Sazgutdinova, R. (2021) Forest management based on the principles of multifunctional forest use. *Journal of Sustainable Forestry*, 40(1): 32–46.
- Santos, J.P.B., Romanelli, J.P., Gardon, F.R., Krainovic, P.M., de Resende, A.F., Souza, L.R., Piotto, D. & Rodrigues, R. R. (2023) Multifunctional forest restoration in Brazil: A critical analysis of the trends and knowledge gaps in the scientific literature. *Sustainability*, 15(22): 15782.

Deliverable

- Prepare an A2 or A1 poster, in electronic format, that effectively communicates the complexities and approaches surrounding multifunctional forest management.

6. Note

1 See https://informar.eu/.

7. References

AGGESTAM, F. 2015. Framing the ecosystem concept through a longitudinal study of developments in science and policy. *Conservation Biology*, 29, 1052–1064.

AGGESTAM, F. & GIURCA, A. 2021. The art of the 'green' deal: Policy pathways for the EU Forest Strategy. *Forest Policy and Economics*, 128, 102456.

AGGESTAM, F., KONCZAL, A., SOTIROV, M., WALLIN, I., PAILLET, Y., SPINELLI, R., LINDNER, M., DERKS, J., HANEWINKEL, M. & WINKEL, G. 2020. Can nature conservation and wood production be reconciled in managed forests? A review of driving factors for integrated forest management in Europe. *Journal of Environmental Management*, 268, 110670.

AGGESTAM, F. & PÜLZL, H. 2018. Coordinating the uncoordinated: The EU Forest Strategy. *Forests*, 9, 125. https://doi.org/10.3390/f9030125.

AGGESTAM, F. & PÜLZL, H. 2020. Downloading Europe: A regional comparison in the uptake of the EU Forest Action Plan. *Sustainability*, 12.

ASBECK, T., GROßMANN, J., PAILLET, Y., WINIGER, N. & BAUHUS, J. 2021. The use of tree-related microhabitats as forest biodiversity indicators and to guide integrated forest management. *Current Forestry Reports*, 7, 59–68.

ASZALÓS, R., THOM, D., AAKALA, T., ANGELSTAM, P., BRŪMELIS, G., GÁLHIDY, L., GRATZER, G., HLÁSNY, T., KATZENSTEINER, K., KOVÁCS, B., KNOKE, T., LARRIEU, L., MOTTA, R., MÜLLER, J., ÓDOR, P., ROŽENBERGAR, D., PAILLET, Y., PITAR, D., STANDOVÁR, T., SVOBODA, M., SZWAGRZYK, J., TOSCANI, P. & KEETON, W. S. 2022. Natural disturbance regimes as a guide for sustainable forest management in Europe. *Ecological Applications*, 32, e2596.

BAŞKENT, E. Z. 2018. A review of the development of the multiple use forest management planning concept. *International Forestry Review*, 20, 296–313.

BAUHUS, J., KOUKI, J., PAILLET, Y., ASBECK, T. & MARCHETTI, M. 2017. How does the forest-based bioeconomy impact forest biodiversity? *In:* WINKEL, G. (ed.) *Towards a Sustainable European Forest-Based Bioeconomy – Assessment and the Way Forward. What Science Can Tell Us 8.* European Forest Institute.

BAYCHEVA-MERGER, T., SOTIROV, M., HOLMGREN, S. & SELTER, A. 2018. Institutional and actor-oriented factors constraining expert-based forest information exchange in Europe: A policy analysis from an actor-centred institutionalist approach. *Forests* 9, 129.

BENAYAS, J. M. R., MARTINS, A., NICOLAU, J. M. & SCHULZ, J. J. 2007. *Abandonment of Agricultural Land: An Overview of Drivers and Consequences*. Wallingford: CABI.

BENZ, J. P., CHEN, S., DANG, S., DIETER, M., LABELLE, E. R., LIU, G., HOU, L., MOSANDL, R. M., PRETZSCH, H., PUKALL, K., RICHTER, K., RIDDER, R., SUN, S., SONG, X., WANG, Y., XIAN, H., YAN, L., YUAN, J., ZHANG, S. & FISCHER, A. 2020. Multifunctionality of forests: A white paper on challenges and opportunities in China and Germany. *Forests*, 11, 266.

BESACIER, C., GARRETT, L., IWEINS, M. & SHAMES, S. 2021. Local financing mechanisms for forest and landscape restoration – A review of local level investment mechanisms. *Forestry Working Paper No. 21.* Rome: Food and Agriculture Organization (FAO).

BETHMANN, S., SIMMINGER, E., BALDY, J. & SCHRAML, U. 2018. Forestry in interaction. Shedding light on dynamics of public opinion with a praxeological methodology. *Forest Policy and Economics*, 96, 93–101.

BIBER, P., FELTON, A., NIEUWENHUIS, M., LINDBLADH, M., BLACK, K., BAHÝL, J., BINGÖL, O., BORGES, J. G., BOTEQUIM, B., BRUKAS, V., BUGALHO, M. N., CORRADINI, G., ERIKSSON, L. O., FORSELL, N., HENGEVELD, G. M., HOOGSTRA-KLEIN, M. A., KADıOĞULLARı, A. I., KARAHALIL, U., LODIN, I., LUNDHOLM, A., MAKRICKIENĖ, E., MASIERO, M., MOZGERIS, G., PIVORIŪNAS, N., POSCHENRIEDER, W., PRETZSCH, H., SEDMÁK, R. & TUČEK, J. 2020. Forest biodiversity, carbon sequestration, and wood production: Modeling synergies and trade-offs for ten forest landscapes across Europe. *Frontiers in Ecology and Evolution*, 8.

BJÄRSTIG, T. & KVASTEGÅRD, E. 2016. Forest social values in a Swedish rural context: The private forest owners' perspective. *Forest Policy Econ.*, 65, 17–24. https://doi.org/10.1016/j.forpol.2016.01.007.

BLATTERT, C., LEMM, R., THEES, O., HANSEN, J., LEXER, M. J. & HANEWINKEL, M. 2018. Segregated versus integrated biodiversity conservation: Value-based ecosystem service assessment under varying forest management strategies in a Swiss case study. *Ecol. Indic.*, 95, 751–764. https://doi.org/10.1016/j.ecolind.2018.08.016.

BONČINA, A., KLOPČIČ, M., SIMONČIČ, T., DAKSKOBLER, I., FICKO, A. & ROZMAN, A. 2017. A general framework to describe the alteration of natural tree species composition as an indicator of forest naturalness. *Ecol. Indic.*, 77, 194–204. https://doi.org/10.1016/j.ecolind.2017.01.039.

BONSU, N. O., DHUBHÁIN, A. N. & O'CONNOR, D. 2017. Evaluating the use of an integrated forest land-use planning approach in addressing forest ecosystem services conflicting demands: Experience within an Irish forest landscape. *Futures*, 86, 1–17. https://doi.org/10.1016/j.futures.2016.08.004.

BONSU, N. O., TYREEHAGEMAN, J. & KELE, J. 2020. Beyond Agenda 2030: Future-oriented mechanisms in localising the Sustainable Development Goals (SDGs). *Sustainability*, 12, 9797.

BORRASS, L., KLEINSCHMIT, D. & WINKEL, G. 2017. The 'German model' of integrative multifunctional forest management – Analysing the emergence and political evolution of a forest management concept. *Forest Policy Econ.*, 77, 16–23. https://doi.org/10.1016/j.forpol.2016.06.028.

BÖSCH, M., ELSASSER, P., FRANZ, K., LORENZ, M., MONING, C., OLSCHEWSKI, R., RÖDL, A., SCHNEIDER, H., SCHRÖPPEL, B. & WELLER, P. 2018. Forest ecosystem services in rural areas of Germany: Insights from the national TEEB study. *Ecosyst. Serv.*, 31, 77–83. https://doi.org/10.1016/j.ecoser.2018.03.014.

BOULANGER, V., DUPOUEY, J.-L., ARCHAUX, F., BADEAU, V., BALTZINGER, C., CHEVALIER, R., CORCKET, E., DUMAS, Y., FORGEARD, F., MÅRELL, A., MONTPIED, P., PAILLET, Y., PICARD, J.-F., SAÏD, S. & ULRICH, E. 2018. Ungulates increase forest plant species richness to the benefit of non-forest specialists. *Global Change Biology*, 24, e485–e495.

BOWMAN, C. 1998. *Strategy in practice*, Prentice Hall Europe.

BUIJS, A. & LAWRENCE, A. 2013. Emotional conflicts in rational forestry: Towards a research agenda for understanding emotions in environmental conflicts. *Forest Policy Econ.*, 33, 104–111. https://doi.org/10.1016/j.forpol.2012.09.002.

BÜTLER, R., LACHAT, T., LARRIEU, L. & PAILLET, Y. 2013. *Habitat Trees: Key Elements for Forest Biodiversity*. European Forest Institute.

CAICOYA, A. T., POSCHENRIEDER, W., BLATTERT, C., EYVINDSON, K., HARTIKAINEN, M., BURGAS, D., MÖNKKÖNEN, M., UHL, E., VERGARECHEA, M. & PRETZSCH, H. 2023. Sectoral policies as drivers of forest management and ecosystems services: A case study in Bavaria, Germany. *Land Use Policy*, 130, 106673.

CAIRNS, M. A. & MEGANCK, R. A. 1994. Carbon sequestration, biological diversity, and sustainable development: Integrated forest management. *Environ. Manage.*, 18, 13–22. https://doi.org/10.1007/BF02393746.

CHEN, H. L., LEWISON, R. L., AN, L., TSAI, Y. H., STOW, D., SHI, L. & YANG, S. 2020. Assessing the effects of payments for ecosystem services programs on forest structure and species biodiversity. *Biodiversity and Conservation*, 29, 2123–2140.

CIESIELSKI, M. & STEREŃCZAK, K. 2018. What do we expect from forests? The European view of public demands. *J. Environ. Manage.*, 209, 139–151. https://doi.org/10.1016/j.jenvman.2017.12.032.

COLL, L., AMEZTEGUI, A., COLLET, C., LÖF, M., MASON, B., PACH, M., VERHEYEN, K., ABRUDAN, I., BARBATI, A., BARREIRO, S., BIELAK, K., BRAVO-OVIEDO, A., FERRARI, B., GOVEDAR, Z., KULHAVY, J., LAZDINA, D., METSLAID, M., MOHREN, F., PEREIRA, M., PERIC, S., RASZTOVITS, E., SHORT, I., SPATHELF, P., STERBA, H., STOJANOVIC, D., VALSTA, L., ZLATANOV, T. & PONETTE, Q. 2018. Knowledge gaps about mixed forests: What do European forest managers want to know and what answers can science provide? *Forest Ecol. Manag.*, 407, 106–115. https://doi.org/10.1016/j.foreco.2017.10.055.

CORONA, P., ASCOLI, D., BARBATI, A., BOVIO, G., COLANGELO, G., ELIA, M., GARFÌ, V., IOVINO, F., LAFORTEZZA, R., LEONE, V., LOVREGLIO, R., MARCHETTI, M., MARCHI, E., MENGUZZATO, G., NOCENTINI, S., PICCHIO, R., PORTOGHESI, L., PULETTI, N., SANESI, G. & CHIANUCCI, F. 2015. Integrated forest management to prevent wildfires under Mediterranean environments. *Annals of Silvicultural Research*, 39, 1–22. http://dx.doi.org/10.12899/asr-946.

COSYNS, H., JOA, B., MIKOLEIT, R., KRUMM, F., SCHUCK, A., WINKEL, G. & SCHULZ, T. 2020. Resolving the trade-off between production and biodiversity conservation in integrated forest management: comparing tree selection practices of foresters and conservationists. *Biodiversity and Conservation*, 29, 3717–3737.

COSYNS, H., KRAUS, D., KRUMM, F., SCHULZ, T. & PYTTEL, P. 2018. Reconciling the tradeoff between economic and ecological objectives in habitat-tree selection: A comparison between students, foresters, and forestry trainers. *Forest Science*, 65, 223–234.
DAMANIA, R., RUSS, J., WHEELER, D. & BARRA, A. F. 2018. The road to growth: Measuring the tradeoffs between economic growth and ecological destruction. *World Dev.*, 101, 351–376. https://doi.org/10.1016/j.worlddev.2017.06.001.
DEUFFIC, P., SOTIROV, M. & ARTS, B. 2018. 'Your policy, my rationale'. How individual and structural drivers influence European forest owners' decisions. *Land Use Policy*, 79, 1024–1038. https://doi.org/10.1016/j.landusepol.2016.09.021.
DIETERLE, G. & KARSENTY, A. 2020. 'Wood Security': the importance of incentives and economic valorisation in conserving and expanding forests. *International Forestry Review*, 22, 81–92.
DITTMANN, S., THIESSEN, E. & HARTUNG, E. 2017. Applicability of different non-invasive methods for tree mass estimation: A review. *Forest Ecol. Manag.*, 398, 208–215. https://doi.org/10.1016/j.foreco.2017.05.013.
DUDÁKOVÁ, Z., ALLMAN, M., MERGANIČ, J. & MERGANIČOVÁ, K. 2020. Machinery-induced damage to soil and remaining forest stands – Case study from Slovakia. *Forests*, 11, 1289.
EC 2015. Mapping and assessment of forest ecosystems and their services – Applications and guidance for decision making in the framework of MAES *JRC Science for Policy Report.* Brussels: Joint Research Section (JRC).
EC, JOINT RESEARCH CENTRE, MAES, J., TELLER, A., ERHARD, M., CONDÉ, S., VALLECILLO, S., BARREDO, J., PARACCHINI, M., ABDUL MALAK, D., TROMBETTI, M., VIGIAK, O., ZULIAN, G., ADDAMO, A., GRIZZETTI, B., SOMMA, F., HAGYO, A., VOGT, P., POLCE, C., JONES, A., MARIN, A., IVITS, E., MAURI, A., REGA, C., CZÚCZ, B., CECCHERINI, G., PISONI, E., CEGLAR, A., DE PALMA, P., CERRANI, I., MERONI, M., CAUDULLO, G., LUGATO, E., VOGT, J., SPINONI, J., CAMMALLERI, C., BASTRUP-BIRK, A., SAN MIGUEL, J., SAN ROMÁN, S., KRISTENSEN, P., CHRISTIANSEN, T., ZAL, N., DE ROO, A., CARDOSO, A., PISTOCCHI, A., DEL BARRIO ALVARELLOS, I., TSIAMIS, K., GERVASINI, E., DERIU, I., LA NOTTE, A., ABAD VIÑAS, R., VIZZARRI, M., CAMIA, A., ROBERT, N., KAKOULAKI, G., GARCIA BENDITO, E., PANAGOS, P., BALLABIO, C., SCARPA, S., MONTANARELLA, L., ORGIAZZI, A., FERNANDEZ UGALDE, O. & SANTOS-MARTÍN, F. 2020. *Mapping and Assessment of Ecosystems and their Services – An EU Wide Ecosystem Assessment in Support of the EU Biodiversity Strategy*. Publications Office.
ECE 2020. Who owns our forests? Forest ownership in the ECE Region. Geneva, Switzerland: United Nations Economic Commission for Europe.
EEA 2016. European forest ecosystems: State and trends. Denmark: Copenhagen: European Environment Agency.
ELLIOTT, C. & SCHLAEPFER, R. 2001. Understanding forest certification using the Advocacy Coalition Framework. *Forest Policy and Economics*, 2, 257–266.
ERIKSSON, L. & KLAPWIJK, M. J. 2019. Attitudes towards biodiversity conservation and carbon substitution in forestry: a study of stakeholders in Sweden. *Forestry: An International Journal of Forest Research*, 92, 219–229.
EYVINDSON, K., DUFLOT, R., TRIVIÑO, M., BLATTERT, C., POTTERF, M. & MÖNKKÖNEN, M. 2021. High boreal forest multifunctionality requires continuous cover forestry as a dominant management. *Land Use Policy*, 100, 104918.
EYVINDSON, K., REPO, A. & MÖNKKÖNEN, M. 2018. Mitigating forest biodiversity and ecosystem service losses in the era of bio-based economy. *Forest Policy and Economics*, 92, 119–127.
FELICIANO, D., BOURIAUD, L., BRAHIC, E., DEUFFIC, P., DOBSINSKA, Z., JARSKY, V., LAWRENCE, A., NYBAKK, E., QUIROGA, S., SUAREZ, C. & FICKO, A. 2017. Understanding private forest owners' conceptualisation of forest management: Evidence from a survey in seven European countries. *J. Rural Stud.*, 54, 162–176. https://doi.org/10.1016/j.jrurstud.2017.06.016.
FELTON, A., LÖFROTH, T., ANGELSTAM, P., GUSTAFSSON, L., HJÄLTÉN, J., FELTON, A. M., SIMONSSON, P., DAHLBERG, A., LINDBLADH, M., SVENSSON, J., NILSSON, U., LODIN, I., HEDWALL, P. O., STÉNS, A., LÄMÅS, T., BRUNET, J., KALÉN, C., KRISTRÖM, B., GEMMEL, P. & RANIUS, T. 2020. Keeping pace with forestry: Multi-scale conservation in a changing production forest matrix. *Ambio*, 49, 1050–1064.
FOREST EUROPE 2020. State of europe's forests 2020. Bratislava: Ministerial Conference on the Protection of Forests in Europe.

GEHMAN, J., LEFSRUD, L. M. & FAST, S. 2017. Social license to operate: Legitimacy by another name? *Canadian Public Administration*, 60, 293–317.

GIBERT, K., GARCÍA-ALONSO, C. & SALVADOR-CARULLA, L. 2010. Integrating clinicians, knowledge and data: expert-based cooperative analysis in healthcare decision support. *Health Res. Policy Sy.*, 28, 1–16. https://doi.org/10.1186/1478-4505-8-28.

GREGG, J. S., JÜRGENS, J., HAPPEL, M. K., STRØM-ANDERSEN, N., TANNER, A. N., BOLWIG, S. & KLITKOU, A. 2020. Valorization of bio-residuals in the food and forestry sectors in support of a circular bioeconomy: A review. *Journal of Cleaner Production*, 267, 122093.

GUILLÉN, L. A., WALLIN, I. & BRUKAS, V. 2015. Social capital in small-scale forestry: A local case study in Southern Sweden. *Forest Policy Econ.*, 53, 21–28. https://doi.org/10.1016/j.forpol.2014.12.006.

GUNDERSEN, V., KÖHLER, B. & MYRVOLD, K. M. 2019. Seeing the forest for the trees: A review-based framework for better harmonization of timber production, biodiversity, and recreation in boreal urban forests. *Urban Science*, 3, 113.

GUSTAFSSON, L., BAKER, S. C., BAUHUS, J., BEESE, W. J., BRODIE, A., SVERDRUP-THYGESON, A., LÕHMUS, A., PALIK, B., MESSIER, C., LINDENMAYER, D. B., PASTUR, G. M., KOUKI, J., FRANKLIN, J. F., NEYLAND, M., VOLNEY, W. J. A. & WAYNE, A. 2012. Retention forestry to maintain multifunctional forests: A world perspective. *BioScience*, 62, 633–645. https://doi.org/10.1525/bio.2012.62.7.6.

GUSTAFSSON, L., BAUHUS, J., ASBECK, T., AUGUSTYNCZIK, A. L. D., BASILE, M., FREY, J., GUTZAT, F., HANEWINKEL, M., HELBACH, J., JONKER, M., KNUFF, A., MESSIER, C., PENNER, J., PYTTEL, P., REIF, A., STORCH, F., WINIGER, N., WINKEL, G., YOUSEFPOUR, R. & STORCH, I. 2019. Retention as an integrated biodiversity conservation approach for continuous-cover forestry in Europe. *Ambio*, 1–13. https://doi.org/10.1007/s13280-019-01190-1.

GUSTAFSSON, L., BAUHUS, J., KOUKI, J., LÕHMUS, A. & SVERDRUP-THYGESON, A. 2013. Retention Forestry – an integrated approach in practical use. *In:* KRAUS, D. & KRUMM, F. (eds.) *Integrative Approaches as an Opportunity for the Conservation of Forest Biodiversity.* Freiburg: European Forest Institute.

GUTIERREZ GARZON, A. R., BETTINGER, P., SIRY, J., ABRAMS, J., CIESZEWSKI, C., BOSTON, K., MEI, B., ZENGIN, H. & YEŞIL, A. 2020. A comparative analysis of five forest certification programs. *Forests*, 11, 863.

HALBE, J., KNÜPPE, K., KNIEPER, C. & PAHL-WOSTL, C. 2018. Towards an integrated flood management approach to address trade-offs between ecosystem services: Insights from the Dutch and German Rhine, Hungarian Tisza, and Chinese Yangtze basins. *J. Hydrol.*, 559, 984–994. https://doi.org/10.1016/j.jhydrol.2018.02.001.

HETEMÄKI, L., KANGAS, J. & PELTOLA, H. 2022. *Forest Bioeconomy and Climate Change*. Cham: Springer Nature.

HOOGSTRA-KLEIN, M. A., BRUKAS, V. & WALLIN, I. 2017. Multiple-use forestry as a boundary object: From a shared ideal to multiple realities. *Land Use Policy*, 69, 247–258. https://doi.org/10.1016/j.landusepol.2017.08.029.

HOPF, S.-E., TRESCH, S., BELYAZID, S., SVERDRUP, H., AUGUSTIN, S., KURZ, D., RIHM, B. & BRAUN, S. 2023. Dendrochemical indicators of tree rings reveal historical soil acidification in Swiss forest stands. *Dendrochronologia*, 81, 126099.

INFORMAR. 2018. Meeting Report 1: Integrated Forest Management Learning Architecture. *In:* AGGESTAM, F., KONCZAL, A., LINDNER, M., SCHUCK, A., WINKEL, G. & SCHIFFERDECKER, G. (eds.). European Forest Institute. https://informar.eu/.

IPBES 2023. Thematic assessment of invasive alien species and their control. *In:* ROY, H. E., PAUCHARD, A., STOETT, P. & TANARA, R. T. (eds.). Bonn: Intergovernmental Science-Policy Platform on Biodiversity and Ecosystem Services (IPBES).

IRVINE, K. N., O'BRIEN, L., RAVENSCROFT, N., COOPER, N., EVERARD, M., FAZEY, I., REED, M. S. & KENTER, J. O. 2016. Ecosystem services and the idea of shared values. *Ecosystem Services*, 21, 184–193.

JAKOVAC, C. C., JUNQUEIRA, A. B., CROUZEILLES, R., PEÑA-CLAROS, M., MESQUITA, R. C. G. & BONGERS, F. 2021. The role of land-use history in driving successional pathways and its implications for the restoration of tropical forests. *Biological Reviews*, 96, 1114–1134.

JALALI, S. & WOHLIN, C. Systematic literature studies: Database searches vs. backward snowballing. Proceedings of the 2012 ACM-IEEE International Symposium on Empirical Software Engineering and Measurement, 20–21 Sept. 2012, 29–38.

JANDL, R., SPATHELF, P., BOLTE, A. & PRESCOTT, C. E. 2019. Forest adaptation to climate change – is non-management an option? *Annals of Forest Science*, 76, 48.
JUERGES, N., ARTS, B., MASIERO, M., HOOGSTRA-KLEIN, M., BORGES, J. G., BRODRECHTOVA, Y., BRUKAS, V., CANADAS, M. J., CARVALHO, P. O., CORRADINI, G., CORRIGAN, E., FELTON, A., KARAHALIL, U., KARAKOC, U., KROTT, M., VAN LAAR, J., LODIN, I., LUNDHOLM, A., MAKRICKIENĖ, E., MARQUES, M., MENDES, A., MOZGERIS, G., NOVAIS, A., PETTENELLA, D., PIVORIŪNAS, N. & SARı, B. 2021. Power analysis as a tool to analyse trade-offs between ecosystem services in forest management: A case study from nine European countries. *Ecosystem Services*, 49, 101290.
KALONGA, S. K., MIDTGAARD, F. & KLANDERUD, K. 2016. Forest certification as a policy option in conserving biodiversity: An empirical study of forest management in Tanzania. *Forest Ecol. Manag.*, 361, 1–12. https://doi.org/10.1016/j.foreco.2015.10.034.
KANT, S., DEEGEN, P., HOSTETTLER, M., WANG, S., XU, J. & VALATIN, G. 2016. New frontiers of forest economics: Forest economics beyond the perfectly competitive commodity markets. *Forest Policy Econ.*, 72, 1–6. https://doi.org/10.1016/j.forpol.2016.06.009.
KIM-KEUNG HO, J. 2014. Formulation of a systemic PEST analysis for strategic analysis. *European Academic Research*, 2, 6478–6492.
KLEINSCHMIT, D., PÜLZL, H., SECCO, L., SERGENT, A. & WALLIN, I. 2018. Orchestration in political processes: Involvement of experts, citizens, and participatory professionals in forest policy making. *Forest Policy Econ.*, 89, 4–15. https://doi.org/10.1016/j.forpol.2017.12.011.
KOIVULA, M. & VANHA-MAJAMAA, I. 2020. Experimental evidence on biodiversity impacts of variable retention forestry, prescribed burning, and deadwood manipulation in Fennoscandia. *Ecological Processes*, 9, 11.
KONCZAL, A. A., DERKS, J., DE KONING, J. H. C. & WINKEL, G. 2023. Integrating nature conservation measures in european forest management – An exploratory study of barriers and drivers in 9 european countries. *Journal of Environmental Management*, 325, 116619.
KOŻUCH, A. 2022. How unit costs of conducting forestry vary with the natural and economic conditions of forests in eastern Poland. *Sylwan*, 166, 551–565.
KRAUS, D. & KRUMM, F. 2013. Integrative approaches as an opportunity for the conservation of forest biodiversity. European Forest Institute. www.integrateplus.org.
KRAUS, D., SCHUCK, A., KRUMM, F., BÜTLER SAUVAIN, R., COSYNS, H., COURBAUD, B. & WILHELM, G. 2018. Seeing is building better understanding: The Integrate+ Marteloscopes. *Integrate+ Technical Paper.* European Forest Institute. https://informar.eu/marteloscope-sites
KREUTZWISER, R. D. & WRIGHT, C. S. 1990. Factors influencing integrated forest management on private industrial forest land. *J. Environ. Manage.*, 30, 31–46. https://doi.org/10.1016/0301-4797(90)90035-U.
KRUMM, F., LACHAT, T., SCHUCK, A., BÜTLER, R. & KRAUS, D. 2019. Marteloscopes as training tools for the retention and conservation of habitat trees in forests. *Schweizerische Zeitschrift für Forstwesen*, 170, 86–93.
KRUMM, F., SCHUCK, A. & RIGLING, A. 2020. *How to Balance Forestry and Biodiversity Conservation: A View across Europe*. Birmensdorf, European Forest Institute (EFI) & Swiss Federal Institute for Forest, Snow and Landscape Research (WSL).
KUULUVAINEN, T., ANGELSTAM, P., FRELICH, L., JÕGISTE, K., KOIVULA, M., KUBOTA, Y., LAFLEUR, B. & MACDONALD, E. 2021. Natural disturbance-based forest management: Moving beyond retention and continuous-cover forestry. *Frontiers in Forests and Global Change*, 4.
KUULUVAINEN, T., LINDBERG, H., VANHA-MAJAMAA, I., KETO-TOKOI, P. & PUNTTILA, P. 2019. Low-level retention forestry, certification, and biodiversity: case Finland. *Ecological Processes*, 8, 47.
LABELLE, E. R., HANSSON, L., HÖGBOM, L., JOURGHOLAMI, M. & LASCHI, A. 2022. Strategies to mitigate the effects of soil physical disturbances caused by forest machinery: A comprehensive review. *Current Forestry Reports*, 8, 20–37.
LANDIS, D. A., GRATTON, C., JACKSON, R. D., GROSS, K. L., DUNCAN, D. S., LIANG, C., MEEHAN, T. D., ROBERTSON, B. A., SCHMIDT, T. M., STAHLHEBER, K. A., TIEDJE, J. M. & WERLING, B. P. 2018. Biomass and biofuel crop effects on biodiversity and ecosystem services in the North Central US. *Biomass Bioenerg.*, 114, 18–29. https://doi.org/10.1016/j.biombioe.2017.02.003.
LANGMAIER, M. & LAPIN, K. 2020. A systematic review of the impact of invasive alien plants on forest regeneration in European temperate forests. *Front Plant Sci*, 11, 524969.

LAWRENCE, A., SPINELLI, R., TOPPINEN, A. & SALO, E. 2017. What are the implications of the bioeconomy for forest-related jobs? *In:* WINKEL, G. (ed.) *Towards a Sustainable European Forest-Based Bioeconomy – Assessment and the Way Forward. What Science Can Tell Us 8*. European Forest Institute.

LAZDINIS, M., ANGELSTAM, P. & PÜLZL, H. 2019. Towards sustainable forest management in the European Union through polycentric forest governance and an integrated landscape approach. *Landscape Ecology*, 34, 1737–1749.

LAZZARO, L., BOLPAGNI, R., BUFFA, G., GENTILI, R., LONATI, M., STINCA, A., ACOSTA, A. T. R., ADORNI, M., ALEFFI, M., ALLEGREZZA, M., ANGIOLINI, C., ASSINI, S., BAGELLA, S., BONARI, G., BOVIO, M., BRACCO, F., BRUNDU, G., CACCIANIGA, M., CARNEVALI, L., DI CECCO, V., CESCHIN, S., CIASCHETTI, G., COGONI, A., FOGGI, B., FRATTAROLI, A. R., GENOVESI, P., GIGANTE, D., LUCCHESE, F., MAINETTI, A., MARIOTTI, M., MINISSALE, P., PAURA, B., PELLIZZARI, M., PERRINO, E. V., PIRONE, G., POGGIO, L., POLDINI, L., POPONESSI, S., PRISCO, I., PROSSER, F., PUGLISI, M., ROSATI, L., SELVAGGI, A., SOTTOVIA, L., SPAMPINATO, G., STANISCI, A., VENANZONI, R., VICIANI, D., VIDALI, M., VILLANI, M. & LASTRUCCI, L. 2020. Impact of invasive alien plants on native plant communities and Natura 2000 habitats: State of the art, gap analysis and perspectives in Italy. *Journal of Environmental Management*, 274, 111140.

LEBAN, V., PEZDEVŠEK MALOVRH, S., ZADNIK STIRN, L. & KRČ, J. 2016. Forest biomass for energy in multi-functional forest management: Insight into the perceptions of forest-related professionals. *Forest Policy Econ.*, 71, 87–93. https://doi.org/10.1016/j.forpol.2015.07.005.

LINSER, S., WOLFSLEHNER, B., BRIDGE, R. J. S., GRITTEN, D., JOHNSON, S., PAYN, T., PRINS, K., RAŠI, R. & ROBERTSON, G. 2018. 25 years of criteria and indicators for sustainable forest management: How intergovernmental C& I processes have made a difference. *Forests*, 9.

LONG, H., ZHANG, Y., MA, L. & TU, S. 2021. Land use transitions: Progress, challenges and prospects. *Land*, 10, 903.

MAES, J. 2011. A spatial assessment of ecosystem services in Europe: methods, case studies and policy analysis – phase 1. Ispra: Partnership for European Environmental Research.

MAGAGNOTTI, N., SPINELLI, R., GÜLDNER, O. & ERLER, J. 2012. Site impact after motor-manual and mechanised thinning in Mediterranean pine plantations. *Biosystems Engineering*, 113, 140–147.

MAIER, C. & WINKEL, G. 2017. Implementing nature conservation through integrated forest management: A street-level bureaucracy perspective on the German public forest sector. *Forest Policy Econ.*, 82, 14–29. https://doi.org/10.1016/j.forpol.2016.12.015.

MANN, C., LOFT, L. & HERNÁNDEZ-MORCILLO, M. 2021. Assessing forest governance innovations in Europe: Needs, challenges and ways forward for sustainable forest ecosystem service provision. *Ecosystem Services*, 52, 101384.

MANN, C., LOFT, L., HERNÁNDEZ-MORCILLO, M., PRIMMER, E., BUSSOLA, F., FALCO, E., GENELETTI, D., DOBROWOLSKA, E., GROSSMANN, C. M., BOTTARO, G., SCHLEYER, C., KLUVANKOVA, T., GARCIA, G., LOVRIĆ, M., TORRALBA, M., PLIENINGER, T. & WINKEL, G. 2022. Governance Innovations for forest ecosystem service provision – Insights from an EU-wide survey. *Environmental Science & Policy*, 132, 282–295.

MANNING, P., VAN DER PLAS, F., SOLIVERES, S., ALLAN, E., MAESTRE, F. T., MACE, G., WHITTINGHAM, M. J. & FISCHER, M. 2018. Redefining ecosystem multifunctionality. *Nature Ecology & Evolution*, 2, 427–436.

MARQUES, M., OLIVEIRA, M. & BORGES, J. G. 2020. An approach to assess actors' preferences and social learning to enhance participatory forest management planning. *Trees, Forests and People*, 2, 100026.

MARTIN, D. A., OSEN, K., GRASS, I., HÖLSCHER, D., TSCHARNTKE, T., WURZ, A. & KREFT, H. 2020. Land-use history determines ecosystem services and conservation value in tropical agroforestry. *Conservation Letters*, 13, e12740.

MARTYNOVA, M., SULTANOVA, R., KHANOV, D., TALIPOV, E. & SAZGUTDINOVA, R. 2021. Forest management based on the principles of multifunctional forest use. *Journal of Sustainable Forestry*, 40, 32–46.

MATTIJSSEN, T. J. M., GANZEVOORT, W., VAN DEN BORN, R. J. G., ARTS, B., BREMAN, B. C., BUIJS, A. E., VAN DAM, R. I., ELANDS, B. H. M., DE GROOT, W. T. & KNIPPENBERG, L. W. J. 2020. Relational values of nature: leverage points for nature policy in Europe. *Ecosystems and People*, 16, 402–410.

MCPFE 2005. *Sustainable Forest Management and Ecosystem Approach.* Ministerial Conference on the Protection of Forests in Europe Laison Unit Warsaw. Outcome of the MCPFE and EfE/PEBLDS ad hoc Working Group on Development of the Pan-European Understanding of the Linkages of Ecosystem Approach and Sustainable Forest Management. Session held 19–21 April, 2005, in Krakow, Poland.

MESSIER, C., BAUHUS, J., SOUSA-SILVA, R., AUGE, H., BAETEN, L., BARSOUM, N., BRUELHEIDE, H., CALDWELL, B., CAVENDER-BARES, J., DHIEDT, E., EISENHAUER, N., GANADE, G., GRAVEL, D., GUILLEMOT, J., HALL, J. S., HECTOR, A., HÉRAULT, B., JACTEL, H., KORICHEVA, J., KREFT, H., MEREU, S., MUYS, B., NOCK, C. A., PAQUETTE, A., PARKER, J. D., PERRING, M. P., PONETTE, Q., POTVIN, C., REICH, P. B., SCHERER-LORENZEN, M., SCHNABEL, F., VERHEYEN, K., WEIH, M., WOLLNI, M. & ZEMP, D. C. 2022. For the sake of resilience and multifunctionality, let's diversify planted forests! *Conservation Letters*, 15, e12829.

MIKULA, K., ŠIBÍKOVÁ, M., AMBROZ, M., KOLLÁR, M., OŽVAT, A. A., URBÁN, J., JAROLÍMEK, I. & ŠIBÍK, J. 2021. NaturaSat – A software tool for identification, monitoring and evaluation of habitats by remote sensing techniques. *Remote Sensing* [Online], 13.

MILANOVIĆ, M., KNAPP, S., PYŠEK, P. & KÜHN, I. 2020. Linking traits of invasive plants with ecosystem services and disservices. *Ecosystem Services*, 42, 101072.

MOFFAT, K., LACEY, J., ZHANG, A. & LEIPOLD, S. 2016. The social licence to operate: a critical review. *Forestry: An International Journal of Forest Research*, 89, 477–488.

MOHTASHAMI, S., ELIASSON, L., JANSSON, G. & SONESSON, J. 2017. Influence of soil type, cartographic depth-to-water, road reinforcement and traffic intensity on rut formation in logging operations: a survey study in Sweden. *Silva Fennica*, 51.

MÖLDER, A., MEYER, P. & NAGEL, R.-V. 2019. Integrative management to sustain biodiversity and ecological continuity in Central European temperate oak (*Quercus robur*, *Q. petraea*) forests: An overview. *Forest Ecology and Management*, 437, 324–339.

MORGAN, E. A., CADMAN, T. & MACKEY, B. 2021. Integrating forest management across the landscape: a three pillar framework. *Journal of Environmental Planning and Management*, 64, 1735–1769.

MÜLLER, F., AUGUSTYNCZIK, A. L. D. & HANEWINKEL, M. 2019. Quantifying the risk mitigation efficiency of changing silvicultural systems under storm risk throughout history. *Annals of Forest Science*, 76, 116.

NAGEL, T. A., FIRM, D., PISEK, R., MIHELIC, T., HLADNIK, D., DE GROOT, M. & ROZENBERGAR, D. 2017. Evaluating the influence of integrative forest management on old-growth habitat structures in a temperate forest region. *Biological Conservation*, 216, 101–107.

NAUMOV, V., MANTON, M., ELBAKIDZE, M., RENDENIEKS, Z., PRIEDNIEKS, J., UHLIANETS, S., YAMELYNETS, T., ZHIVOTOV, A. & ANGELSTAM, P. 2018. How to reconcile wood production and biodiversity conservation? The Pan-European boreal forest history gradient as an 'experiment'. *J. Environ. Manage.*, 218, 1–13. https://doi.org/10.1016/j.jenvman.2018.03.095.

NEUMANN, W., LEVERS, C., WIDEMO, F., SINGH, N. J., CROMSIGT, J. P. G. M. & KUEMMERLE, T. 2022. Hunting as land use: Understanding the spatial associations among hunting, agriculture, and forestry. *Ecology and Society*, 27.

NIKINMAA, L., LINDNER, M., CANTARELLO, E., JUMP, A. S., SEIDL, R., WINKEL, G. & MUYS, B. 2020. Reviewing the use of resilience concepts in forest sciences. *Current Forestry Reports*, 6, 61–80.

O'HARA, K. L. 2016. What is close-to-nature silviculture in a changing world? *Forestry*, 89, 1–6. https://doi.org/10.1093/forestry/cpv043.

ONTL, T. A., JANOWIAK, M. K., SWANSTON, C. W., DALEY, J., HANDLER, S., CORNETT, M., HAGENBUCH, S., HANDRICK, C., MCCARTHY, L. & PATCH, N. 2019. Forest management for carbon sequestration and climate adaptation. *Journal of Forestry*, 118, 86–101.

PAILLET, Y., BERGÈS, L., HJÄLTEN, J., ODOR, P., AVON, C., BERNHARDT-ROMERMANN, M., BIJLSMA, R. J., DE BRUYN, L., FUHR, M., GRANDIN, U., KANKA, R., LUNDIN, L., LUQUE, S., MAGURA, T., MATESANZ, S., MESZAROS, I., SEBASTIA, M. T., SCHMIDT, W., STANDOVAR, T., TOTHMERESZ, B., UOTILA, A., VALLADARES, F., VELLAK, K. & VIRTANEN, R. 2010. Biodiversity differences between managed and unmanaged forests: Meta-analysis of species richness in Europe. *Conserv. Biol.*, 24, 101–112. https://doi.org/10.1111/j.1523-1739.2009.01399.x.

PHALAN, B., ONIAL, M., BALMFORD, A. & GREEN, R. E. 2011. Reconciling food production and biodiversity conservation: Land sharing and land sparing compared. *Science*, 333, 1289–1291. https://doi.org/10.1126/science.1208742.

POHJANMIES, T., EYVINDSON, K., TRIVIÑO, M., BENGTSSON, J. & MÖNKKÖNEN, M. 2021. Forest multifunctionality is not resilient to intensive forestry. *European Journal of Forest Research*, 140, 537–549.

POTTER, K. M., JETTON, R. M., BOWER, A., JACOBS, D. F., MAN, G., HIPKINS, V. D. & WESTWOOD, M. 2017. Banking on the future: progress, challenges and opportunities for the genetic conservation of forest trees. *New Forests*, 48, 153–180.

PRIMMER, E. & KARPPINEN, H. 2010. Professional judgment in non-industrial private forestry: Forester attitudes and social norms influencing biodiversity conservation. *Forest Policy Econ.*, 12, 136–146. https://doi.org/10.1016/j.forpol.2009.09.007.

PUETTMANN, K. J., WILSON, S., BAKER, S. C., DONOSO, P. J., DRÖSSLER, L., AMENTE, G., HARVEY, B. D., KNOKE, T., LU, Y., NOCENTINI, S., PUTZ, F. E., YOSHIDA, T. & BAUHUS, J. 2015. Silvicultural alternatives to conventional even-aged forest management – what limits global adoption? *Forest Ecosyst.*, 2, 8. https://doi.org/10.1186/s40663-015-0031-x.

PÜLZL, H., AGGESTAM, F., PROKOFIEVA, I., LUKINA, V. N., SOTIROV, M., PECURUL-BOTINES, M., TEBENKOVA, N. D., WIDMARK, C. & ROSINGER, C. 2022. Re-imagining nature communication: The role of societal values in forest ecosystem services. https://papers.ssrn.com/sol3/papers.cfm?abstract_id=3972031

PÜLZL, H., WYDRA, D. & HOGL, K. 2018. Piecemeal integration: Explaining and understanding 60 years of European Union forest policy-making. *Forests*, 9, 719.

QUIROGA, S., SUAREZ, C., FICKO, A., FELICIANO, D., BOURIAUD, L., BRAHIC, E., DEUFFIC, P., DOBSINSKA, Z., JARSKY, V., LAWRENCE, A. & NYBAKK, E. 2019. What influences European private forest owners' affinity for subsidies? *Forest Policy and Economics*, 99, 136–144.

RAMETSTEINER, E., EICHLER, L., BERG, J., AGGESTAM, F., BINDA ZANE, E. & PLUMET, C. 2009. Shaping forest communication in the European Union: public perceptions of forests and forestry. *ECORYS*. Final Report ed.

REED, J., BARLOW, J., CARMENTA, R., VAN VIANEN, J. & SUNDERLAND, T. 2019. Engaging multiple stakeholders to reconcile climate, conservation and development objectives in tropical landscapes. *Biological Conservation*, 238, 108229.

RIVERA LEÓN, L., BOUGAS, K., AGGESTAM, F., PÜLZL, H., ZOBOLI, E., RAVET, J., GRINIECE, E., VERMEER, J., MAROULIS, N., ETTWEIN, F., VAN BRUSSELEN J. & GREEN, T. 2016. An assessment of the cumulative cost impact of specified EU legislation and policies on the EU forest-based industries. Brussels: DG GROW.

ROITSCH, D., ABRUSCATO, S., LOVRIĆ, M., LINDNER, M., ORAZIO, C. & WINKEL, G. 2023. Close-to-nature forestry and intensive forestry – Two response patterns of forestry professionals towards climate change adaptation. *Forest Policy and Economics*, 154, 103035.

ROSE, K. M. E., FRIDAY, J. B., OLIET, J. A. & JACOBS, D. F. 2020. Canopy openness affects microclimate and performance of underplanted trees in restoration of high-elevation tropical pasturelands. *Agricultural and Forest Meteorology*, 292–293, 108105.

ROUX, J.-L., KONCZAL, A., BERNASCONI, A., BHAGWAT, S. A., DE VREESE, R., DOIMO, I., MARINI GOVIGLI, V., KAŠPAR, J., KOHSAKA, R., PETTENELLA, D., PLIENINGER, T., SHAKERI, Z., SHIBATA, S., STARA, K., TAKAHASHI, T., TORRALBA, M., TYRVÄINEN, L., WEISS, G. & WINKEL, G. 2022. Exploring evolving spiritual values of forests in Europe and Asia: a transition hypothesis toward re-spiritualizing forests. *Ecology and Society*, 27.

SABATINI, F. M., BLUHM, H., KUN, Z., AKSENOV, D., ATAURI, J. A., BUCHWALD, E., BURRASCANO, S., CATEAU, E., DIKU, A., DUARTE, I. M., FERNÁNDEZ LÓPEZ, A. B., GARBARINO, M., GRIGORIADIS, N., HORVÁTH, F., KEREN, S., KITENBERGA, M., KIŠ, A., KRAUT, A., IBISCH, P. L., LARRIEU, L., LOMBARDI, F., MATOVIC, B., MELU, R. N., MEYER, P., MIDTENG, R., MIKAC, S., MIKOLÁŠ, M., MOZGERIS, G., PANAYOTOV, M., PISEK, R., NUNES, L., RUETE, A., SCHICKHOFER, M., SIMOVSKI, B., STILLHARD, J., STOJANOVIC, D., SZWAGRZYK, J., TIKKANEN, O.-P., TOROMANI, E., VOLOSYANCHUK, R., VRŠKA, T., WALDHERR, M., YERMOKHIN, M., ZLATANOV, T., ZAGIDULLINA, A. & KUEMMERLE, T. 2021. European primary forest database v2.0. *Scientific Data*, 8, 220.

SABATINI, F. M., BURRASCANO, S., KEETON WILLIAM, S., LEVERS, C., LINDNER, M., PÖTZSCHNER, F., VERKERK PIETER, J., BAUHUS, J., BUCHWALD, E., CHASKOVSKY, O., DEBAIVE, N., HORVÁTH, F., GARBARINO, M., GRIGORIADIS, N., LOMBARDI, F., MARQUES DUARTE, I., MEYER, P., MIDTENG, R., MIKAC, S., MIKOLÁŠ, M., MOTTA, R., MOZGERIS, G., NUNES, L., PANAYOTOV, M., ÓDOR, P., RUETE, A., SIMOVSKI, B., STILLHARD, J., SVOBODA, M., SZWAGRZYK, J., TIKKANEN, O.-P., VOLOSYANCHUK, R., VRSKA, T., ZLATANOV, T. & KUEMMERLE, T. 2018. Where are Europe's last primary forests? *Divers. Distrib.*, 24, 1426–1439. https://doi.org/10.1111/ddi.12778.

SABATINI, F. M., DE ANDRADE, R. B., PAILLET, Y., ÓDOR, P., BOUGET, C., CAMPAGNARO, T., GOSSELIN, F., JANSSEN, P., MATTIOLI, W., NASCIMBENE, J., SITZIA, T., KUEMMERLE, T. & BURRASCANO, S. 2019. Trade-offs between carbon stocks and biodiversity in European temperate forests. *Global Change Biol.*, 25, 536–548. https://doi.org/10.1111/gcb.14503.

SANTOPUOLI, G., DI CRISTOFARO, M., KRAUS, D., SCHUCK, A., LASSERRE, B. & MARCHETTI, M. 2019. Biodiversity conservation and wood production in a Natura 2000 Mediterranean forest. A trade-off evaluation focused on the occurrence of microhabitats. *iForest – Biogeosciences and Forestry*, 12, 76–84.

SARVAŠOVÁ, Z., ALI, T., ĐORĐEVIĆ, I., LUKMINE, D., QUIROGA, S., SUÁREZ, C., HRIB, M., RONDEUX, J., MANTZANAS, K. T. & FRANZ, K. 2019a. Natura 2000 payments for private forest owners in Rural Development Programmes 2007–2013: A comparative view. *Forest Policy and Economics*, 99, 123–135.

SARVAŠOVÁ, Z., BÁLIKOVÁ, K., DOBŠINSKÁ, Z., ŠTĚRBOVÁ, M. & ŠÁLKA, J. 2019b. Payments for forest ecosystem services across Europe – Main approaches and examples from Slovakia. *Ekológia (Bratislava)*, 38, 154–165.

SAVASTANO, M., SUCIU, M.-C., GORELOVA, I. & STATIVĂ, G.-A. 2023. How smart is mobility in smart cities? An analysis of citizens' value perceptions through ICT applications. *Cities*, 132, 104071.

SCHULZ, T., KRUMM, F., BÜCKING, W., FRANK, G., LIER, M., LOVRIĆ, M., VAN DER MAATEN-THEUNISSEN, M., PAILLET, Y., PARVIAINEN, J., VACCHIANO, G. & VANDEKERKHOVE, K. 2014. Institutional determinants of integrative nature protection in forest policy in Europe: a comparative analysis. *Biodivers. Conserv.*, 23, 3425–3450. https://doi.org/10.1007/s10531-014-0817-0.

SFC 2015. Sustainable forest management: Criteria and indicators. *Final Report.* Brussels: Standing Forestry Committee Ad Hoc Working Group on Sustainable Forest Management.

SOTIROV, M. & ARTS, B. 2018. Integrated forest governance in Europe: An introduction to the special issue on forest policy integration and integrated forest management. *Land Use Policy*, 79, 960–967. https://doi.org/10.1016/j.landusepol.2018.03.042.

SOTIROV, M., HOOGSTRA-KLEIN, M., TRUBINS, R., SCHÜLL, E., RIEMER, A., ORAZIO, C., CORDERO, R., SALNÄSS, O. & ERIKSSON, O. 2015. Shaping the future of sustainable forest management. *EFI News*, 23, 10–11.

SOTIROV, M. & WINKEL, G. 2016. Toward a cognitive theory of shifting coalitions and policy change: linking the advocacy coalition framework and cultural theory. *Policy Sciences*, 49, 125–154.

SOUCY, M., ADÉGBIDI, H. G., SPINELLI, R. & BÉLAND, M. 2016. Increasing the effectiveness of knowledge transfer activities and training of the forestry workforce with marteloscopes. *The Forestry Chronicle*, 92, 418–427.

SPATHELF, P. 2009. *Sustainable Forest Management in a Changing World: A European Perspective.* Dordrecht: Springer.

SPINELLI, R., LOMBARDINI, C. & MAGAGNOTTI, N. 2014. The effect of mechanization level and harvesting system on the thinning cost of Mediterranean softwood plantations. *Silva Fennica*, 48.

STORCH, I., PENNER, J., ASBECK, T., BASILE, M., BAUHUS, J., BRAUNISCH, V., DORMANN, C. F., FREY, J., GÄRTNER, S., HANEWINKEL, M., KOCH, B., KLEIN, A.-M., KUSS, T., PREGERNIG, M., PYTTEL, P., REIF, A., SCHERER-LORENZEN, M., SEGELBACHER, G., SCHRAML, U., STAAB, M., WINKEL, G. & YOUSEFPOUR, R. 2020. Evaluating the effectiveness of retention forestry to enhance biodiversity in production forests of Central Europe using an interdisciplinary, multi-scale approach. *Ecology and Evolution*, 10, 1489–1509.

SUTHERLAND, L.-A. & HUTTUNEN, S. 2018. Linking practices of multifunctional forestry to policy objectives: Case studies in Finland and the UK. *Forest Policy Econ.*, 86, 35–44. https://doi.org/10.1016/j.forpol.2017.10.019.

TAKALA, T., HUJALA, T., TANSKANEN, M. & TIKKANEN, J. 2017. Forest owners' discourses of forests: Ideological origins of ownership objectives. *Journal of Rural Studies*, 51, 1–14.

TEEB 2010. *The Economics of Ecosystems and Biodiversity: Mainstreaming the Economics of Nature: A Synthesis of the Approach, Conclusions and Recommendations of TEEB*. Malta: Progress Press.

TIAN, N. & PELKKI, M. 2021. Nonindustrial private forest landowner perspectives on forest certification: A look at awareness and barriers. *Forest Policy and Economics*, 131, 102552.

TIAN, N., POUDYAL, N. C. & LU, F. 2018. Understanding landowners' interest and willingness to participate in forest certification program in China. *Land Use Policy*, 71, 271–280.

TOPPINEN, A., KORHONEN, J., HURMEKOSKI, E. & HANSEN, E. 2017. What makes a European forest-based bioeconomy competitive? *In:* WINKEL, G. (ed.) *Towards a sustainable European forest-based bioeconomy – assessment and the way forward. What Science Can Tell Us 8.* European Forest Institute.

TORRALBA, M., LOVRI, M., ROUX, J.-L., BUDNIOK, M.-A., MULIER, A.-S., WINKEL, G. & PLIENINGER, T. 2020. Examining the relevance of cultural ecosystem services in forest management in Europe. *Ecology and Society*, 25.

TRIVIÑO, M., POHJANMIES, T., MAZZIOTTA, A., JUUTINEN, A., PODKOPAEV, D., LE TORTOREC, E. & MÖNKKÖNEN, M. 2017. Optimizing management to enhance multifunctionality in a boreal forest landscape. *Journal of Applied Ecology*, 54, 61–70.

USTAOGLU, E. & COLLIER, M. J. 2018. Farmland abandonment in Europe: an overview of drivers, consequences, and assessment of the sustainability implications. *Environmental Reviews*, 26, 396–416.

VANCLAY, J. K. 2011. Future harvest: what might forest harvesting entail 25 years hence? *Scandinavian Journal of Forest Research*, 26, 183–186.

VARGHESE, A. O., MANI, J. K. & JHA, C. S. 2022. Applications of geospatial technology in forest resource assessment, management, and monitoring. *In:* JHA, C. S., PANDEY, A., CHOWDARY, V. M. & SINGH, V. (eds.) *Geospatial Technologies for Resources Planning and Management.* Cham: Springer International Publishing.

VERKERK, P. J., ANTTILA, P., EGGERS, J., LINDNER, M. & ASIKAINEN, A. 2011. The realisable potential supply of woody biomass from forests in the European Union. *Forest Ecol. Manag.*, 261, 2007–2015. https://doi.org/10.1016/j.foreco.2011.02.027.

VERKERK, P. J., COSTANZA, R., HETEMÄKI, L., KUBISZEWSKI, I., LESKINEN, P., NABUURS, G. J., POTOČNIK, J. & PALAHÍ, M. 2020. Climate-smart forestry: The missing link. *Forest Policy and Economics*, 115, 102164.

WEBB, S. A., BRUYERE, B., HALLADAY, M. & WALKER, S. 2022. A framework for conceptualizing leadership in conservation. *Oryx*, 56, 664–670.

WEISS, G., LAWRENCE, A., HUJALA, T., LIDESTAV, G., NICHIFOREL, L., NYBAKK, E., QUIROGA, S., SARVAŠOVÁ, Z., SUAREZ, C. & ŽIVOJINOVIĆ, I. 2019. Forest ownership changes in Europe: State of knowledge and conceptual foundations. *Forest Policy Econ.*, 99, 9–20. https://doi.org/10.1016/j.forpol.2018.03.003.

WINKEL, G., BLONDET, M., BORRAS, L., FREI, T., GEITZENAUER, M., GRUPPE, A., JUMP, A., DE KONING, J., SOTIROV, M., WEISS, G., WINTER, S. & TURNHOUT, E. 2015. The implementation of Natura 2000 in forests: A trans- and interdisciplinary assessment of challenges and choices. *Environmental Science and Policy*, 52, 23–32.

WINKEL, G., LOVRIĆ, M., MUYS, B., KATILA, P., LUNDHEDE, T., PECURUL, M., PETTENELLA, D., PIPART, N., PLIENINGER, T., PROKOFIEVA, I., PARRA, C., PÜLZL, H., ROITSCH, D., ROUX, J.-L., THORSEN, B. J., TYRVÄINEN, L., TORRALBA, M., VACIK, H., WEISS, G. & WUNDER, S. 2022. Governing Europe's forests for multiple ecosystem services: Opportunities, challenges, and policy options. *Forest Policy and Economics*, 145, 102849.

WINKEL, G. & SOTIROV, M. 2016. Whose integration is this? European forest policy between the gospel of coordination, institutional competition, and a new spirit of integration. *Environ. Plann. C.*, 34, 496–514. https://doi.org/10.1068/c1356j.

YE, N., VAN LEEUWEN, L. & NYKTAS, P. 2019. Analysing the potential of UAV point cloud as input in quantitative structure modelling for assessment of woody biomass of single trees. *International Journal of Applied Earth Observation and Geoinformation*, 81, 47–57.

ZIOLO, M., FILIPIAK, B. Z., BĄK, I., CHEBA, K., TÎRCA, D. M. & NOVO-CORTI, I. 2019. Finance, sustainability and negative externalities. An overview of the European context. *Sustainability* [Online], 11.

9 A framework for integrative forest policy analysis

1. The integrative path towards a sustainable future

The adoption of sustainable forest management (SFM) as a guiding principle for European forestry has resulted in considerable discussion among academics and practitioners in recent years (Wolfslehner and Vacik, 2008, Newton et al., 2009, Linser et al., 2018, Lazdinis et al., 2019). Researchers have been challenged to extend their analytical lenses and theoretical approaches for a broader understanding of the socio-economic and environmental interactions that shape forest management. More specifically, the paradigm shift away from linear and siloed thinking towards integration and sustainability requires re-evaluating existing methodologies and analytical frameworks, especially in assessing and monitoring the effects of policy interventions (Aggestam et al., 2023). One debated question is how best to assess the impact of relevant interventions (Linser et al., 2018, Christopoulos et al., 2012). In other words, as European countries increasingly align their forest management practices with sustainability principles, there is also an increasing demand for integrative policy analysis frameworks (Box 9.1). Such frameworks would need to capture the multifunctional aspects of forest ecosystems and address the varied socio-economic and environmental dimensions inherent to SFM. Therefore, to open more possibilities for a balanced understanding of SFM and forest policy analysis, there is a need to integrate different theoretical perspectives.

Traditional forest policy analysis (see Chapter 3) often assumes policy systems that are well-defined and manageable, presuming a certain degree of controllability or manageability. In contrast, policy problems are commonly shaped by uncertainties regarding future outcomes, overseen by policymakers with varying and sometimes limited capacities, legally and technically (Emery, 1998, Christopoulos et al., 2012). These policy environments are fundamentally complex, suggesting that researchers might not use the most suitable analytical approaches to examine the increasingly cross-sectoral policy frameworks and governance structures affecting forests (Schmithüsen, 2003). For example, many methods used for policy analysis struggle to address complexities and policy systems experiencing substantial organisational and structural transformations (Emery, 1998, Daviter, 2019, Levin et al., 2012). Such systems are inherently unpredictable, resisting direct control and may instead be more susceptible to influence or pressure.

The European Union's (EU) forest-based sector exemplifies these challenges. Within this sector, EU institutions face limited regulatory capacities due to the absence of a common policy on forests, which means that the European Commission depend on non-binding (soft) policy instruments when addressing forests. Even more, the EU forest-based sector is characterised by policy domains like energy and conservation competing for influence (see Chapters 2 and 4). This interplay between different sectors and interests means that aspects of the forest-based sector is an ideal candidate for integrative policy analysis (Box 9.1).

DOI: 10.4324/9781003198987-9

Box 9.1 What is integrative forest policy analysis?

Integrative policy analysis is a multidisciplinary (or mixed methods) approach that combines different theoretical and methodological perspectives to understand and address policy problems (Dunn, 2017). In other words, an integrative policy analysis combines and integrates different approaches and tools, considering different elements of the policymaking process, how they are connected, and only then deciding on the appropriate analytical approach. This approach acknowledges the complexity of a policy problem, which often requires insights from multiple disciplines to be effectively understood and addressed (Greene, 2007, Muskat et al., 2012). Reducing a policy analysis to a single or segmented analytical perspective would not capture the multidimensional aspect of policy challenges and solutions.

Policy analysis is complex. Building on Dunn's (2017) perspective, who asserted that the 'methodology of policy analysis rests or should rest on epistemological foundations that differ from those of the disciplines of which policy analysis is composed' (p. xxvii), the approach to policy analysis requires a unique epistemological basis (Box 9.2).[1] Integrative analysis should be differentiated from conventional monodisciplinary assumptions that generally form the basis for a forest policy analysis. For instance, the positivism approach, when viewed through a single methodological and epistemological perspective, argues that a policy must be verifiable, logically and empirically. This perspective highlights the importance of observable and measurable evidence as well as logical reasoning in policy analysis. However, relying only on positivism can be limiting as it overlooks the subjective nature of a policy issue (e.g., policies are not only shaped by empirical data but also by socio-economic, cultural, and political factors). This limitation implies that policy researchers would benefit from exploring alternative perspectives and approaches. The underlying assumption, in terms of benefits, is that using different analytical approaches enriches our understanding of complex (or wicked) policy problems, particularly in cases involving multiple policy domains (Head and Alford, 2013, Jax et al., 2018). This may lead to more informed, contextual, and effective policy decisions.

It should also be noted that a policy analysis typically covers descriptive and normative dimensions (Patton et al., 2016). On the descriptive front, it uses methodologies from traditional social science disciplines to outline and explain why a policy instrument may have certain causative variables and effects (e.g., problem structuring). This typically involves an empirical investigation into the factual and observable phenomena surrounding a policy issue. Policy analysis is also normative as it involves making value judgments about what a successful end state (or policy outcome) means rather than just explaining the policy environment. This normative aspect of a policy analysis involves an ethical evaluation, including making choices about desired outcomes (e.g., preferred or expected policy outcomes) and deciding on the most appropriate way to reach this end state. These decisions often involve trade-offs between competing values, requiring the analyst to consider the ethical implications and societal impact of policies. In addressing these trade-offs, the analyst makes value judgements regarding societal contexts and stakeholder interests. This means the researcher should be aware of the theoretical basis for a policy and the practical and ethical implications surrounding policy implementation.

Box 9.2 System of integrated environmental and economic accounts

The System of Integrated Environmental and Economic Accounts (SEEA) is an accounting framework that integrates environmental data with national economic accounts, providing an overview of the interactions between economic and environmental factors (Bartelmus, 2014, UN et al., 2014, Vardon et al., 2018). The SEEA forest accounts, which focus on forest ecosystems, is designed to cover all forest resources, including contributions to the economy, such as timber and non-timber forest products (NWFPs), forest ecosystem services in physical and monetary terms, and the role of forests in carbon sequestration and biodiversity conservation (Haripriya, 2001, La Notte and Rhodes, 2020). By combining data on forest assets, forest accounts provide a more integrative perspective of forest resources than, for example, a system of national accounts (SNA).[2]

SEEA demonstrates that merging different disciplines and approaches is possible under one integrative methodological framework. However, this integration is not without its challenges, ontological[3] and epistemological:

- Ontologically, the challenge concerns the diverse nature of forest resources, which covers biophysical entities and economic assets under a single accounting framework. This requires rediscovering what 'value' means, which includes both market and non-market aspects.
- Epistemologically, integrating varied data sources, such as national forest resource assessments and forest indicator frameworks like the criteria and indicators (C&I) approach, means using data collected and interpreted using different methods. This can lead to inconsistencies when merged under one framework.

These challenges highlight the need for care when merging scientific domains and methodologies and when developing integrated frameworks like SEEA, as they must be able to reconcile ontological and epistemological differences.

1.1. Aim of the chapter

This chapter introduces a heuristic framework (or blueprint) to facilitate the selection of analytical approaches for an integrative policy analysis. The primary aim is to present a simple, dynamic, and adaptable strategy for integrating different methodological approaches to examine policymaking processes. The aim is also to demonstrate its application across policy levels and domains, from the micro to the macro level. The core element of this framework is its integrative nature. More specifically, it is meant to complement existing approaches in that it sets out to identify different methodologies for studying policymaking and governance.

It is crucial to note that the proposed blueprint is not an analytical method. Rather, it can be considered a navigational tool, directing researchers toward the most appropriate methodologies and approaches for a given context. The focus is not on developing new methodologies, as numerous approaches already exist (see Chapter 3); rather, the emphasis is on enabling forest policy analysts to review and select the most appropriate approach. This requires the analyst to move beyond ontological and epistemological biases and personal perspectives. Key considerations include determining what analytical tools are available and their appropriate use, understanding the analysis suitable for different situations, and assessing the added value of an integrative approach. The blueprint could play a role in the planning stage of a project but also throughout implementation so as to ensure that the approaches applied are integrative and suitable for the policy analysis.

2. Introducing a blueprint for an integrative policy analysis

The value of an integrative approach depends on the context and objectives of the research. Some policy problems may not need an integrative approach, particularly in situations where policy problems might be more effectively addressed using a single methodology, such as when the resource demands of an integrative approach may not add proportional value. In some cases, a policy instrument with clear and quantifiable outcomes might be more efficiently analysed and understood using a quantitative method, rather than integrating multiple disciplinary perspectives. Moreover, certain policy questions may be easier to address through the lens of a single discipline to avoid dilution of the core issue being investigated. The scope and scale of the policy analysis also play an important role. For instance, a focused, non-integrative approach might be more practical and feasible for small-scale studies or those with limited resources. Therefore, while integrative policy analysis can offer valuable insights, its utility should be evaluated based on the specific requirements of each research project.

For those seeking more knowledge, resources, and guidance on methods and approaches for policy analysis, the references below can provide a starting point:

- Methods for Policy Research: Taking Socially Responsible Action (Majchrzak and Markus, 2013).
- Handbook of Research Methods and Applications in Comparative Policy Analysis (Peters and Fontaine, 2020).
- A guide to policy analysis as a research method (Browne et al., 2019).
- A Practical Guide for Policy Analysis (Bardach and Patashnik, 2019).

2.1. How to implement an integrated policy analysis – a blueprint

Rather than proposing a prescriptive approach, the blueprint is meant to offer an adaptive framework to develop an integrative study approach. It emphasises the importance of understanding the policy environment, reviewing and selecting appropriate methodologies, and integrating different perspectives before applying analytical approaches, methods, and tools. Following the principles of simple heuristics (Kahneman et al., 1982, Gigerenzer and Todd, 1999, Hertwig et al., 2012), the blueprint outlines a series of steps designed to help analysts break down the policy environment and identify relevant tools for a particular project. This involves an assessment of the policy landscape, an evaluation of available analytical methodologies, and strategically selecting tools that align with the specific demands of the policy issue being investigated. In other words, the framework seeks to simplify the selection process while aiming for the integration of different theoretical and methodological perspectives. The framework comprises seven steps as outlined below and in Table 9.1:

1. *Problem identification and contextual analysis*: This initial exploratory phase involves an in-depth review of the policy problem, including contextual factors. Tools such as PESTEL, SWOT, and stakeholder analysis can be used to collect and organise information on internal

Table 9.1 Blueprint for an integrative policy analysis.

Step	*Objective*	*Process*	*Possible tools*
1. Problem identification and contextual analysis	Understanding and characterising the policy problem.	Collect and analyse information about the policy issue(s) to be studied, including information on relevant stakeholders and the policy environment.	Literature review, PESTEL or SWOT analysis, stakeholder mapping.
2. Defining the scope	Defining the boundaries and scale of the analysis.	Determine at what level the analysis will be carried out.	Micro-meso-macro scale frameworks, systems thinking approaches.
3. Methodological selection	Identifying and selecting methodologies that align with the policy problem and the boundaries and scale of the analysis	Based on the level of analysis, select suitable methodologies.	Decision matrices, flowcharts, comparative analysis, Delphi method.
4. Ensuring an integrative analysis	Applying multiple perspectives to ensure an integrative approach.	Identify how the study will combine insights from different disciplinary perspectives and integrate data sources.	Mixed-methods approaches, interdisciplinary research teams, concept mapping, and data triangulation.
5. Ontological and epistemological concerns	Acknowledging and addressing any ontological and epistemological concerns underpinning the proposed approach.	Reflect on how the selected methodologies and disciplines understand the nature of reality (ontology) and knowledge (epistemology) and how this may affect the study.	Reflexive practices, peer discussions.
6. Evaluating and adapting the approach	Assessing the effectiveness of the chosen approach and adjusting as necessary.	Iteratively evaluate the approach and outcomes of the analysis.	Feedback loops, pilot studies, periodic reviews, outcome mapping
7. Communication and policy recommendations (*alternative step*)	Translating the results into communication materials and policy recommendations considering the perspectives of relevant stakeholders.	Identify target audiences for communication materials and policy recommendations.	Policy briefs, reports, presentations, and stakeholder consultations.

and external factors, covering political, socio-economic, technological, and environmental aspects. For instance, a stakeholder analysis or mapping exercise can reveal key concerns and relationships to the policy issue by reviewing publicly available policy statements. The target is to gain a foundational understanding of the policy landscape and, more importantly, to allow the analyst to approach the policy problem with an open mind.

2. *Defining the scope*: Based on the collected background information, this step focuses on outlining the boundaries and scale of the analysis. The objective here is to clarify the overall purpose of the analysis, considering factors like the level of analysis, available expertise, resources and tools. Whether focusing on individuals, organisations, policy systems, or a combination thereof, using frameworks such as the micro-meso-macro scale and systems thinking can assist in identifying the appropriate level of analysis. It can also help in understanding interconnections within the policy landscape. The key point here is to avoid theoretical tunnel vision[4] by starting to think about the policy problem more broadly (step 1), similar to a grounded approach,[5] before narrowing or refining the scope in terms of what is realistic and feasible (step 2).
3. *Methodological selection*: Leveraging insights from the contextual analysis and scoping exercise, the third step involves choosing methodologies that are suitable to address the policy problem, considering the defined scope and constraints (e.g., financial limitations). The selection could range from qualitative methods, like interviews and case studies, to quantitative approaches, like experiments and statistical analyses. Decision matrices, flowcharts, comparative analysis, or the Delphi method can be used to evaluate and finalise the study approach, ensuring alignment with the parameters established under step 2.
4. *Ensuring an integrative analysis*: This step principally focuses on validating the integrative aspects of the study approach. It involves planning and recording how the different disciplinary perspectives and data types will be combined (see also step 5). This is the final step before getting to implementation, and it can be carried out while drafting the methodological approach for the study. Mixed methods approaches, interdisciplinary research teams, concept mapping, and data triangulation are some of the tools that can be used for this step. The purpose is to ensure that your selection of approaches and tools would work in practice.
5. *Ontological and epistemological concerns*: Running parallel to the policy analysis or as an integral part of step 4, this stage addresses philosophical concerns related to the selected methodologies. Reflecting on ontological and epistemological concerns helps in understanding how the different methods might affect the study outcome and analysis. It can also strengthen the overall methodological approach. Reflexive practices and peer discussions could be used to examine and reflect on the data and information that will be collected for the analysis.
6. *Evaluating and adapting the approach*: As an iterative and reflective phase, this step centres on assessing the appropriateness and effectiveness of the integrative approach, setting the stage for any adjustments that may be needed. This step is context specific as it is only possible for some studies to continuously evaluate their approach, including adjusting their data collection and analytical work. Like steps 4 and 5, it is a reflective process where approaches like feedback loops, pilot studies, periodic reviews, and outcome mapping could be applied to ensure that the study remains reactive to emerging data and findings.
7. *Communication and policy recommendations*: The final step, considered an alternative, involves translating the research findings into policy recommendations and communication materials. Although not applicable to all research projects, developing materials like policy briefs, detailed reports, and presentations for different audiences can significantly enhance the impact and dissemination of the research. This is considered as part of the science-policy interface, making findings accessible across disciplines and policy domains.

Central to the design of the blueprint is its adaptability. In other words, it is meant to allow for flexibility in modifying the approach according to a research project's specific demands and objectives. The blueprint has been deliberately structured to be a non-prescriptive toolkit, offering guidance while encouraging a reflexive process. This flexibility extends to accommodating different levels of analysis, ensuring the blueprint's applicability across policy domains and disciplines. In using the blueprint, analysts are encouraged to consider each step of the blueprint and decide which is most applicable to the specific context of their study (Table 9.1).

This approach recognises that certain steps may be more important than others, depending on factors such as the particularities of the policy problem, the availability of data, and the expected outcomes of the research. For instance, a study with a well-defined policy problem and lots of available data might focus on methodological selection and integrative analysis (steps 3 and 4). Conversely, a more complex project could benefit more from an in-depth review of ontological and epistemological concerns (step 5). Moreover, specific constraints, such as time limitations, should also be considered. These factors influence the feasibility of incorporating certain aspects of the blueprint, such as feedback loops and iterative adjustments, as part of the research process.

3. Integrated framing using the landscape concept

For the illustration of an integrative analysis, the focus is on the landscape concept. The landscape includes natural, artificial, and managed elements. It serves as an interface through which people and society interact with the natural world. Landscapes are as such not only physical but also dynamic spaces where economic activities, such as forestry, intersect with individual and societal priorities, like public health and biodiversity conservation. However, more importantly, the landscape concept connects individuals with the natural environment, whether directly or indirectly, and links policy systems to human communities and forest ecosystems (de Groot et al., 2010, Chazdon et al., 2016, Fischer, 2018).

From an academic perspective, the landscape can be viewed as a product of natural processes shaped by human value systems, societal priorities, and policymaking decisions. These relationships represent both historical and contemporary processes. For example, discussions embodied by the conservation movement, such as the best strategies to preserve nature, represent a complex interplay between human value systems, perspectives, and the landscape. For this example, interests from different stakeholder groups, ranging from climate change advocates to industrial lobbyists, reflect varied socio-economic, cultural, and professional preferences and perspectives (Kaltenborn and Bjerke, 2002, Howley et al., 2011, Morgan et al., 2021). These differences can result in conflicts within or about the landscape based on disagreements around the framing and use of nature. Adopting a landscape approach can offer a naturally integrative perspective, like taking a bird's eye view, that enable the analyst to consider multiple levels and interactions in the same space.

3.1. Problem identification – taking a micro, meso, and macro perspective

The case study is based on an analysis of three distinct yet thematically connected publications (Aggestam, 2014a, 2014b, 2015), each offering a different perspective on framing effects at various levels of implementation. Unlike traditional approaches, where such case studies tend to be analysed in isolation, this chapter will adopt an integrative approach (Figure 9.1). The focus is not only on micro-level framing effects, as seen through the lens of individuals, but also on broader effects at the meso and macro levels, covering institutional and policymaking contexts.

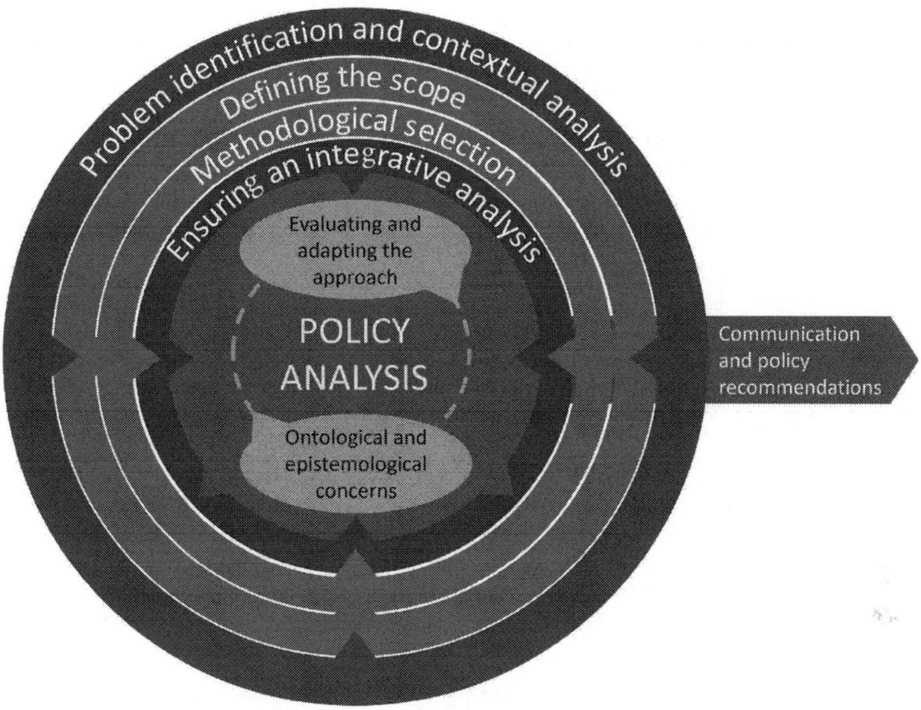

Figure 9.1 Taking an integrative approach – the blueprint.

Source: own figure

The first case is concerned with wetland restoration, examining the influence of landscape frames on restoration efforts (Aggestam, 2014b). The second case reviews how stakeholder participation has been operationalised by individuals and institutions, exploring how framing shapes participatory processes within the landscape (Aggestam, 2014a). The third case considers how the ecosystem concept has been framed over time, as reflected in scientific literature and policy documents (Aggestam, 2015).

Moving from the individual to the policy level, this vertical review emphasises the rationale behind selecting the landscape concept as an analytical framework. The landscape concept offers a framework that can link these case studies and position them within the context of framing research. In other words, it provides a platform to integrate the case studies against a wider spectrum of framing studies, acknowledging varied definitions and applications of frames (Cornelissen and Werner, 2014). Moreover, this chapter will use the landscape concept innovatively, leveraging it to analyse and integrate framing effects across different dimensions. By investigating both vertical and horizontal relationships between frames, the aim is to consider how frames have real-world implications. The challenge resides in unravelling how frames are embedded at different levels and integrating these varied perspectives while considering different disciplinary perspectives and approaches. This approach will seek to improve our understanding of framing effects across levels of implementation and interpretation.

The present case study builds on an earlier publication – Integrated Framing: A Micro to Macro Case for the Landscape (Aggestam, 2017). The case studies has been revised based on the proposed blueprint (section 2) and expanded to reflect developments since its publication.

3.2. Defining the scope – crossing the theoretical divide

Framing has been a research focus in fields as varied as media and communication, policy studies, and psychology, with notable contributions from researchers like Kahneman and Tversky (2000), Druckman (2001), Daviter (2007), Hanggli and Kriesi (2012), and Van Buuren et al. (2014). These varied strands of research offer a wide range of perspectives on how frames and framing (see Chapter 7) can influence perceptions and policymaking across levels. While a comprehensive review of all these areas is outside the scope of this chapter, it is crucial to acknowledge the scale of framing research, in particular, as an integrated approach to framing would require that all these areas of research are considered, recognising the connection between frames at different levels of analysis, from the individual (micro) level to institutional and societal (meso and macro) level (Cornelissen and Werner, 2014). Therefore, this section aims to define the scope by offering an overview of framing across these proposed levels of operationalisation. This will provide the groundwork for understanding the varied applications and implications of framing. Further, the overview will provide the boundary for how framing effects will be analysed and pave the way for an integrated review of its impacts.

3.2.1. Micro level

Framing research at the micro or individual level primarily concentrates on cognitive theories, exploring how frames or 'schemas' function as cognitive structures. These structures, encompassing preferences and values, are shaped by mental functions and personal experiences, as outlined by scholars like Kahneman and Tversky (2000) and Shen and Edwards (2005). This level of framing analysis zeroes in on the individual, examining how a person interprets their surroundings, makes social judgments, and engages in reflective thinking. The core assumption is that cognitive frames are dynamic yet stable 'meaning-making' systems (Markanday and Galarraga, 2021). These systems, grounded in cognitive structures, play an important role in how we learn and retain knowledge as well as how we process and interpret new information, building upon it cumulatively (Nelson et al., 1997). Using cognitive frames is similar to the evaluation process for simple heuristics, as Gigerenzer and Todd (1999) described, providing simple rules for how people process information. These processes are influenced by how the brain responds to external stimuli, which guides behaviour and decision-making. For instance, emotionally charged words can impact decision-making processes (Devignemont and Singer, 2006, Lakoff, 2010), affecting how individuals perceive information and make decisions.[6]

The distinction between the micro, meso, and macro levels of framing research becomes particularly evident when focusing on an individual's cognition. At the micro level, the focus is commonly on the internal cognitive mechanisms of individuals, how personal experiences and environments influence these, and how they shape perceptions. This contrasts with the societal and institutional focus at the meso and macro levels, where the analysis extends to collective frames and their implications in wider contexts. Understanding framing at the micro level offers valuable insights into the cognitive processes that underpin how individuals make sense of the world.

3.2.2. Meso level

At the meso or organisational level, framing is typically understood as the collective construction of shared assumptions, knowledge, and value systems by communities or social groups. This aspect of framing is commonly explored in research related to social movements or political actions, where groups strategically employ frames to persuade policymakers or generate public support (Hanggli and Kriesi, 2012). Snow et al. (1986) describe this as a frame alignment process, wherein a specific frame facilitates individuals to experience a change in their understanding or perspectives on a policy issue, often referred to as re-framing. For instance, within this level, strategic messaging, which could be characterised as socially constructed realities, can change how groups or individuals perceive policy problems. When successful, such messaging can significantly alter public opinions or behaviours (Jerneck and Olsson, 2011). This transformation is not limited to public perceptions but extends to organisational frame constellations. These constellations can include technological or strategic frames, influencing how an organisation communicates its perspectives and values to its staff and wider audience. Such framing might employ rhetorical devices to convey these values effectively (Schmidt, 2008).

Framing at the meso level is predominantly regarded as an interactive meaning-making process rooted in social construction. It involves the interplay of collective understandings, shared values, and strategic communication, all converging to shape or reshape the perceptions and behaviours of individuals, groups, or organisations. Meso-level framing underscores the powerful role of collective belief systems and strategic messaging in influencing and directing group dynamics, actions and policymaking.

3.2.3. Macro level

At the macro or policy level, framing is often conceptualised through neo-institutional theory (see Chapter 3 and the online Support Material). This perspective focuses on phenomena such as emerging new markets, developing shared conceptions, or disseminating innovative ideas and practices, often covered within cultural framing. As scholars like Bateson (1954) and Goffman (1974) highlight, these processes influence social meaning-making, stabilise power arrangements, and affect interactional patterns within societies. The potential of framing at this level resides in its dual function. On the one hand, it serves as a macro framework that can unite individual frames under a common theme, such as shared values or cultural norms. On the other hand, it acts as a temporally stable construct, enabling researchers to analyse and understand how institutions, and individuals within these institutions, perceive and value aspects of their environment (Snow and Benford, 1988, Benford and Snow, 2000). This approach can also provide insights into organisational behaviour and societal responses to changes.

One example of this process is the evolution of attitudes towards landscape management. A new approach to landscape management might encounter resistance or be perceived negatively. However, as it becomes conventionalised and institutionally embedded over time, it can undergo a cognitive shift, leading to broader acceptance and integration into the public consciousness. This transition typically begins with the emergence of a new topic or issue, its establishment as an institutional frame, and eventually ends with a transformative shift in public understanding. While this is a generalisation, it illustrates the role of taking a macro perspective when considering the framing narrative. The macro level provides a critical link in the framing process, defining structures of meaning that shape collective interactions and understanding of reality. It offers a comprehensive view, explaining how societal constructs, institutional norms, and cultural narratives influence and define perceptions of the natural environment.

It is important to acknowledge that the concept of an integrated approach to framing research is not new, as suggested by Dewulf et al. (2009) and Donahue et al. (2011). Matthes (2012b) argued for an integrated research approach in framing studies in a special issue dedicated to framing in politics. This call for integration encourages researchers to move beyond separated case studies to connect findings across multiple cases within larger research projects. Matthes (2012a) and de Vreese (2012) further suggested a comprehensive examination of the entire framing process, from the micro to macro level, rather than focusing on isolated segments.

While this work aligns with Matthes' argument for integration, linking different levels of frames or steps in the framing process to broader models is only one aspect of an integrated approach. Another crucial aspect that requires attention is the theoretical limitations or tunnel vision often encountered in frame research. This refers to the tendency for research on framing to be constrained by different theoretical approaches underpinning the framing concept. For this chapter, integrated framing research is conceptualised as a synthesis of Dewulf et al. (2009) '*meta-paradigmatic perspective*', which argues for delineating and combining approaches to framing from varied ontological and epistemological backgrounds, with Matthes (2012a) call for an overarching integrated model. This blended approach aims to create a more integrative understanding of framing, looking beyond theoretical and methodological boundaries to provide a more nuanced exploration of how frames influence perceptions and policymaking.

Comprehensive reviews of frame theory can be found in the works of Coninx et al. (2015), Cornelissen and Werner (2014), van Hulst and Yanow (2014), and Chong and Druckman (2007). These publications provide in-depth reviews of various aspects and applications of frame theory, making them useful for those seeking to increase their understanding of the subject.

3.3. Methodological selection and ensuring an integrative approach

Frame theory can be categorised into two distinct yet interrelated strands. The first strand concerns stable meaning-making systems, such as cognitive and institutional frames, rooted in cognitive structures and organisational norms that tend to be stable over time (George et al., 2006, Manfredo et al., 2017). The second strand concerns socially generated meaning-making, characterised by temporally constructed frames that emerge during social interactions (Benford and Snow, 2000, Ellingsen et al., 2012). These strands cover a spectrum of interactions and variations across different levels (micro, meso, and macro) and theoretical frameworks (Cornelissen and Werner, 2014).

To facilitate an integrated approach, it is proposed that the framing process be separated into three categories: (1) cognitive framing, (2) contextual framing, and (3) purpose-driven framing. These categories function as 'meta frames' that merge different theoretical perspectives to provide an integrative perspective of the framing process, as illustrated in Figure 9.2. Providing a cross-cutting breakdown, each category offers a different lens through which the framing process can be analysed and understood.

3.3.1. Cognitive framing

Human cognition plays an important role in how individuals perceive and interact with their physical and social environment, particularly at the micro level. This process involves the

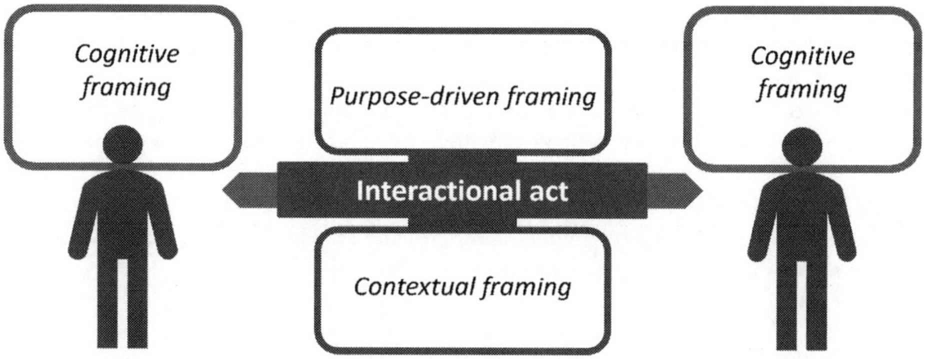

Figure 9.2 Contextual and temporal framing.
Source: own figure

brain evaluating external information, such as media content, which activates specific neural responses that may influence decision-making. For example, certain content can trigger an emotional response that affects how a person interprets and acts on information (Gross and D'Ambrosio, 2004, Hanggli and Kriesi, 2012). In practical terms, this cognitive process means the brain acts as an information processor, filtering and interpreting external information. For instance, as noted by Nickerson (1998), individuals commonly exhibit a bias when searching for, interpreting, and remembering information in ways that reinforce pre-existing beliefs. This is known as a confirmation bias.

Expanding this concept to the meso- and macro-level, cognitive framing consists of collective or institutional frames similarly influencing cognitive processing. These broader frames shape collective perceptions and responses, affecting how groups or institutions interpret information (George et al., 2006). The emphasis shifts from individual cognition to collective or institutional mindsets, where shared beliefs and values influence the interpretation and response to external information. This implies that cognition is central to the framing process at all levels. It dictates individual responses to information and how groups and institutions make sense of and react to the environment they engage with. Further, it underscores the importance of mental structures and biases as part of the cognitive structures underpinning interactions and framing processes.

3.3.2. Contextual framing

The framing process is intrinsically linked to temporally and spatially limited interactions, with contextual framing playing an important role, alongside cognitive framing, in shaping communication. Contextual framing is intertwined with cognitive framing, as it collectively establishes the background for interpersonal communication. The distinction resides in that contextual framing depends on temporary and situational factors, such as social cues, power dynamics, and the specific setting of the interaction at the time of the interaction (Entman, 2007, 2008). These elements give the framing process a distinct shape and character that varies across each interaction. In essence, both contextual and cognitive framing are integral components of a social process that influence how people communicate. This aligns partly with constructivist arguments, which argue that people understand and interpret reality based on their interactions and experiences (Kreuter, 2021). However, Eder (1996) has suggested that while social constructivists offer valuable insights, it is important to recognise the interplay between cognitive framing (innate or acquired) and the collaborative nature of a dialogue. For instance, cognitive framing influences how individuals process information, affecting the interaction dynamics.

Contextual framing, therefore, is unique to each interaction, being time-specific and characterised by the physical and social context in which the interaction occurs. It covers the environmental and situational factors that frame the interaction, adding complexity to how information is communicated and interpreted. Recognising this dual influence of cognitive and contextual framing provides a more complete understanding of the framing process, acknowledging both internal cognitive mechanisms and external social constructs.

3.3.3. Purpose-driven framing

The underlying intent or purpose behind the interaction also influences the framing process. This implies that each interaction is tailored towards a specific goal or objective. For instance, media outlets may frame information to shape public opinion (Goodman, 2006), negotiators might use framing techniques to resolve conflicts (Dewulf and Bouwen, 2012), and policy documents frame narratives to encourage the adoption of collectively negotiated perspectives (Jacoby, 2000). Purpose-driven framing uses both cognitive and contextual framing strategically, but it is distinguished here based on the purposeful manipulation of the framing process. In other words, the focus is on achieving a desired framing effect, such as influencing how information is perceived (Gross, 2008) or shaping power dynamics during issue framing (Entman, 2008). The intent behind purpose-driven framing can be either conscious or subconscious.

The purpose shapes the interactional act, determining how information is communicated. In media, for example, the choice of words, images, and the context in which news is presented has an effect. In negotiations and policymaking, highlighting certain aspects of an issue while downplaying others affects the direction of a discussion. This intentional aspect of framing is a critical factor in the framing process. Thus, purpose-driven framing is not only about the content of the message but also about the strategic intent behind its presentation. It is a nuanced distinction, focusing on how information is packaged and received. Separating this aspect of framing can offer additional insights into the dynamics of the framing process.

3.3.4. Conceptual and analytical approach

The meta-frames (or categories) demonstrate how different theoretical approaches, with varying ontological and epistemological backgrounds, can be merged to provide a more comprehensive and integrative picture of a framing process. Having this in mind, the analytical approach is structured into two interconnected steps aimed at breaking down the framing processes across the case studies:

1. *Describing comparable elements across case studies*: The first step involves delineating comparable elements between and across the case studies. This will be achieved by characterising each case as a distinct interaction, using the conceptual structure in Figure 9.2. The 'interactional act' is broadly defined to cover the analysis, transfer, or exchange of information. This happens through different channels, such as body language, verbal communication, or written texts, whether these exchanges are intentional or not. For each case, three aspects will be examined: the **background** (covering the 'what' and 'where' of each case), the nature of the **interactional act** (covering 'how' the information is exchanged or framed), and **framing effects** (covering impacts on perceptions or behaviours). Additionally, the influence of these

interactions on the landscape, either directly or indirectly, will be explored at each analytical level. It is important to note that the level of detail for each case will be limited. Readers seeking more comprehensive information on the results and methodologies of each case are directed to the associated publications (Aggestam, 2014b, 2014a, 2015).

2. *Defining and applying framing across cases*: The subsequent step delves into the definition of framing and its application across the cases. This involves segmenting the framing process into three functional categories: cognition, context, and purpose. The first step involves identifying and analysing cognitive frames, examining how individual perceptions and mental models influence the interpretation of information. This step focuses on the psychological foundation of framing, concerned with how personal beliefs, experiences, and cognitive biases shape perceptions and policymaking. The second step focuses on contextual and purpose-driven framing. This involves analysing how frames are influenced by and interact within broader social, cultural, and institutional contexts. It also considers the strategic use of framing, examining how frames are employed to achieve specific objectives, such as influencing public opinion or guiding policy decisions.

3.4. Micro case study: Restoring wetlands in an agricultural landscape

The micro case study centres around two wetland restoration programs in southern Sweden, focusing on the Kävlinge and Höje river catchment areas. The Kävlinge River restoration program was initiated in 1995, while the Höje River program started slightly earlier, in 1991. The main objective of both programs was to mitigate nutrient runoff from the agricultural landscape into aquatic ecosystems and the Baltic Sea, a significant environmental concern in the region. The programs employed a variety of strategies to achieve their objectives. Foremost among these were the restoration and rehabilitation of existing wetlands and the construction of artificial wetlands, ponds, and riparian zones. The initiatives were specifically designed to reduce the inflow of nutrients into rivers and lakes within the respective catchment areas in the landscape, thereby reducing the nutrient pollution problem. The success and methodologies of these programs have been well-documented, offering valuable insights into effective practices for wetland restoration and nutrient runoff reduction in agricultural landscapes.

3.4.1. Background

The case considers the dynamics of individual frames and their interplay within the context of the two wetland restoration programs. Central to these programs was the principle of voluntary participation, which inherently made the success of the initiatives dependent on the cooperation and willingness of individual landowners to contribute land for restoration purposes. This implies that the landowners occupied a position of power as compared to other stakeholder groups, such as researchers and non-governmental organisations (NGOs) that were involved. This difference in influence created a power imbalance between the stakeholder groups. More importantly, the stakeholder configuration offered an opportunity to examine how individual frames, specifically those of landowners, influenced the restoration process. This influence is critical to understanding the overall dynamics of the programs and how frames affected implementation.

The power held by the landowners meant that their perceptions, priorities, and decisions had a direct and measurable impact on the landscape. Their frames, encompassing their perspectives, attitudes, and beliefs about land use and conservation, were crucial in shaping the restoration programs. Even more, the landowner's frames had quantifiable framing effects on the landscape as they affected how restoration activities were planned, executed, and maintained. Considering this, the case provides an interesting illustration of the impact of individual frames. Further, it demonstrates the significance of understanding and engaging with these frames, especially in voluntary, collaborative programs where stakeholder buy-in is crucial for success. The case highlights the need for recognising and addressing the power dynamics and varied perspectives among stakeholder groups.

3.4.2. Interactional act

For the wetland restoration programs, the role of a single negotiator in convincing landowners to participate was a key determining factor for their success. The negotiator's engagement with landowners represented the most significant interaction in the restoration process, characterising an interactional act at the micro-level. The outcome of these negotiations directly influenced the restoration outcomes, as the frames associated with wetland restoration became fundamentally important in shaping the restoration strategies. More specifically, the effectiveness of the negotiation process depended on addressing the landowners' frames, which included accounting for their perceptions and attitudes towards wetlands within the landscape. Several factors influenced these frames:

- *Prevailing negative views*: Many landowners had negative perceptions of wetlands, such as concerns about raised water tables impacting their land.
- *Professional interests*: Economic considerations were crucial, with landowners weighing the potential effects of wetland restoration on their income.
- *Past experiences*: Negative interactions with municipal authorities and the public, including issues like property damage, shaped the landowners' willingness to engage.
- *Knowledge gaps*: A lack of understanding and knowledge about wetland ecosystems, including their benefits, shaped the landowner's attitude.

The negotiator's ability to effectively address and reshape these individual frames was central to gaining landowner trust and participation. This involved presenting the benefits of wetland restoration and addressing the landowners' concerns and interests. The process highlights the importance of understanding and directly engaging with different individual perspectives and experiences. These interactions ultimately resulted in a more effective and collaborative restoration process.

3.4.3. Framing effects

The landowners' position of power in the restoration process allowed them to enforce their individual, purpose-driven frames, thereby shaping the operational scope of the programs. Their influence substantially affected the wetlands' physical characteristics and, by extension, the wider landscape. The landowners' purpose-driven framing led to structural adjustments, such as reducing water inflow into the wetlands. These changes diminished the wetlands' capacity to retain nutrients, affecting their efficiency. This framing process was closely interlinked with cognitive and value-based differences between the landowners. Specifically, three basic value orientations emerged as influential in the framing process:

1. Anthropocentric frames: These ranged from weak anthropocentrism,[7] focusing primarily on human (or personal) interests, to environmental pragmatism, where ecological concerns were balanced with human needs.

2. Ecocentric frames:[8] These nature-centred frames prioritised ecological values over human interests, emphasising the intrinsic value of natural ecosystems.
3. Multi-value frames: Representing a blend of ecological, social, and economic values, these frames reflected a more integrative approach to land use considerations.

At the individual level, these orientations can be viewed as a spectrum, ranging from anthropocentric to ecocentric perspectives. Individuals along this spectrum vary in how they perceive the landscape and the importance they assign to different wetland functions and agricultural systems. This diversity in values and perceptions raises questions about the effects of values on land use decisions. For instance, many landowners' negative views of wetlands were often connected to concerns about agricultural productivity. This anthropocentric perspective significantly influenced the direction and outcomes of the restoration efforts. Consequently, this case exemplifies how individual frames, driven by different values and perceptions, affect decision-making processes and the landscape by imposing operational constraints on the programs. It highlights the complex interplay between individual values, cognitive frames, and environmental action, underscoring the need to consider these diverse perspectives when planning and implementing a restoration project.

3.5. Meso case study: institutional framing of stakeholder participation

The meso case study examines nine projects implemented by three international organisations: the United Nations Development Programme (UNDP), the International Commission for the Protection of the Danube River (ICPDR), and the International Institute for Applied Systems Analysis (IIASA). In the original study, these projects were categorised into three separate case studies, based on the implementing organisation and approaches to stakeholder participation. The aim was to review how project managers in these projects framed stakeholder participation and to reveal the strategies and perspectives the managers used to engage and involve stakeholders. The focus, however, goes beyond the managerial framing of participation. More specifically, this case study placed a particular emphasis on exploring the effects of institutional frames. In other words, it investigated how the overarching structures and cultures of the UNDP, ICPDR, and IIASA, as institutions, influenced how stakeholder participation was perceived, approached, and implemented. By examining these institutional frames, the case study aimed to explore how institutional beliefs, values, and practices shape the dynamics of participation, potentially affecting project outcomes and effectiveness.

3.5.1. Background

The emphasis is on institutional framing at the meso level, which provides valuable insights into the practical implementation of stakeholder participation. Furthermore, this case considered the interplay between individual frames, characterised by the perspectives of project managers, and institutional frames, exemplified by international organisations. The focus was on understanding how these frames influenced the operationalisation of stakeholder participation, including the integration of collective frames, as represented by stakeholder groups, within projects that impact the landscape. In other words, the case study investigated the mechanisms through which individual and institutional frames shaped how stakeholder participation was facilitated, structured, and valued. For instance, it examined how project managers, driven by their individual frames, engaged with the broader institutional frames of their organisations. These interactions determined how effectively collective frames were incorporated into project planning and execution.

While the connection to the landscape is indirect, the significance of institutional framing becomes apparent when considering the environmental objectives of these projects. Although the influence of institutional frames on the landscape is not directly quantifiable, it is presumed that the framing processes within these organisations shaped environmental outcomes, as the projects they implement had tangible effects on the landscape. This case study highlights the importance of understanding individual and institutional perspectives when considering the operationalisation and success of landscape-oriented projects.

3.5.2. Interactional act

Most of the projects were characterised by two distinct forms of interactions that played an important role in shaping the implementation of participatory processes:

- *Internal interactions*: Forming what is referred to as the institutional frame, the first interaction occurred between the project manager and the organisations they worked for.
- *External interactions*: Involving the organisations, including project managers and consultants, the second interaction was between relevant stakeholder groups, representing collective frames, and the implementing agencies.

These interactions, within the organisation (internal) and between the implementing agencies and stakeholders (external), demonstrate the significant role of institutional frames in guiding project implementation. They reveal how the values, policies, and cultures of organisations shape project implementation and the adoption of varied approaches to stakeholder participation. Additionally, these interactions highlight the complexities of integrating new value orientations within a stable institutional framework, such as adopting stakeholder participation as a new management tool.

The case study addressed the challenges faced when integrating collective frames. This included accommodating stakeholders' alternative value orientations, which may diverge from or even conflict with the prevailing institutional values. Aligning these varied perspectives was not straightforward and often required a nuanced approach to ensure that various stakeholders' views and interests were effectively represented and considered in project planning and implementation.

3.5.3. Framing effects

Exploring the interplay between factors that influenced stakeholder participation across projects emphasises the importance of costs and benefits, both actual and perceived, in shaping institutional and individual frames, including the uptake of different types of participatory processes. The following factors are crucial for understanding the operational space of the projects:

1. *Managerial autonomy versus organisational mandates*: Project managers frequently incorporated stakeholder participation as a requirement for funding, demonstrating a tension between managerial autonomy and organisational mandates. Furthermore, most organisations did not strictly control how participation was implemented. This aligned with an isomorphic response,[9] where institutions mimic the actions of others for legitimacy (George et al., 2006).
2. *Discrepancy in organisational actions*: There was a discrepancy between the actions imposed by the organisation and those implemented by the project managers. This gap stresses a core framing effect: institutions meet normative expectations by involving stakeholders;

however, they did not enforce how participation should be implemented. This conforms with the street-level bureaucracy perspective (Gofen et al., 2019, Lipsky, 1980), which argues that frontline workers have significant influence (or discretion) over how policies are implemented daily.

3. *Marginalisation of stakeholder participation*: While organisations may adapt superficially to changes in their operating environment, such as for legitimacy or funding, they do not necessarily change their established methods of operation. Participation was, as such, often relegated to the sidelines. This was especially true if the perceived risks, such as loss of power or disruption of the status quo, outweighed the perceived benefits of integrating new practices.

The results from this case study highlight the dominance of institutional frames over policy directives. The balance between the importance of a policy and the values the organisation holds is thus critical. In other words, a change in organisational behaviour is only observable when a policy is prioritised sufficiently. The implications for the landscape are twofold. At the meso level, institutional frames affect how policies are implemented in the landscape. At the micro level, individual frames imposed by managers influence operational conditions, including target setting for the landscape. This demonstrates the layered complexity of policy implementation and framing at the institutional and individual level.

3.6. Macro case study: longitudinal framing of the ecosystem concept in policy and science

The macro case study focused on a longitudinal analysis of the ecosystem concept, tracing the historical evolution of the concept back to its introduction by Arthur Tansley in 1935 (Tansley, 1935). The focus was on demonstrating how the concept had been framed and re-framed over time, focusing on scientific literature and policy documents. The aim was to develop a typology of frames and to identify the value frames that have shaped ecosystem perceptions. The documents that were analysed are, as such, considered not just static texts; they are seen as the products of a social and dynamic process. This involves negotiating and establishing dominant frames, which are then expressed at the macro level over an extended period. The analysis sought to uncover these underlying frames and understand how they evolved. Like the meso case, the implications of this study in terms of landscape management might be indirect; however, the case offers insights into how groups or organisations may value landscapes, which subsequently influence policymaking and landscape management. In addition, understanding the historical framing of ecosystems makes it possible to better understand current perspectives and attitudes towards landscape and ecosystem management.

3.6.1. Background

The study zooms in on value frames influencing scientific research and policymaking, focusing on the ecosystem concept. It demonstrates how the ecosystem concept transforms from a purely scientific term in biology and ecology to a widely adopted and normative concept used across scientific and policy domains and discourses. The analysis is based on an extensive literature review covering 501 scientific articles and 340 policy documents. This substantial body of content does not provide insights into the interactions between individuals or institutions; instead, it offers an overview of how value frames are communicated at the macro level over time. For example, the study categorises and describes the values associated with the ecosystem concept, exploring how they are connected to individual, collective, and policy frames. It also considers whether these value frames have evolved, offering a historical perspective on shifting paradigms associated with the ecosystem concept.

The widespread application of the ecosystem concept in landscape planning and management makes this case example relevant, even if the links are indirect. For instance, the results allow for inferences to be made about the impact of science and policy frames on the landscape. On both a conceptual and practical level, the study outlined how scientific and policy perspectives on ecosystems may have shaped landscape planning and management approaches.

3.6.2. Interactional act

The interactions between different value frames are indirect at the macro level; nevertheless, the analysis assumes some key interactions, particularly during policy development. Policy frames are understood as the product of a collective negotiation, which is shaped by institutional and individual perspectives:

- Institutional frames play a role in this process, such as those advocated by lobbying organisations. These frames represent the interests and perspectives of different groups or sectors trying to influence policymaking. Their impact can be seen in how policies are formulated, with institutional frames often shaping the priorities and focus of policy decisions.
- Individual frames also influence policy development, such as those of influential politicians or policymakers. These frames are shaped by personal beliefs, experiences, and expertise, often bringing a unique perspective to policy discussions. The influence of these individual frames is apparent in cases where policymakers champion specific causes or perspectives.

The case furthermore explores the relationship between academic literature and policy. For instance, it is assumed that individual frames, initially characterised more prominently in academic literature, gradually found their way into the policy discourse as the ecosystem concept gained traction as a scientific term. This migration of frames from academia to policy reflects the evolving nature of the scientific understanding of ecosystems and its impact on policy formulation. Conversely, value frames originating from policy also found their way into scientific discussions, demonstrating a two-way flow of perspectives. This two-way movement of frames between policy and science highlights that these realms are interconnected (e.g., policy considerations and academic developments inform and reshape each other over time).

3.6.3. Framing effects

The results reveal that both scientific literature and policy documents are characterised by stable value frames, which have remained relatively constant over 80 years. This stability persists despite the emergence of global challenges such as the biodiversity crisis and climate change.

1. Research on ecosystems commonly includes implicit value judgments and preferences, indicating a degree of subjectivity within what is typically framed as objective scientific inquiry. In contrast, policy documents are marked by explicit value statements. Policies are generally driven by management goals and reveal an anthropocentric approach.
2. The study identified six overarching collective frames within these documents, each with some internal variations: Humans First, Dual Systems, Eco-Science, Eco-Holism, Animals First, and Multicentric (see Aggestam (2015) for more details).

3. Most policy documents and scientific publications prioritise human interests, aligning with an anthropocentric viewpoint. Ecocentric or biocentric frames, which emphasise ecological and biological considerations, are rarely applied, especially in binding regulations that directly affect landscape management.
4. Certain global crises (like climate change) and the cross-disciplinary concept of ecosystem services have introduced new value perspectives into public discourse. These shifts have led to the modernisation of the collective frames, resulting in a typology of 14 sub-frames.

The identified frames shape how landscapes are perceived and managed, dependent on the aspect of the ecosystem concept that scientists and policymakers are prioritising. However, the actual impact of these frames on the landscape is inferred rather than directly observable. The dominant anthropocentric frame, exemplified by the 'Humans First' approach, sets out the system properties for land-use and landscape management. This would suggest that a significant shift in policy framing would be required to change landscape management practices.

3.7. Ontological and epistemological concerns

The three case studies address wetland restoration, stakeholder participation, and the ecosystem concept and operate at varying levels in the landscape, from the micro to macro level. Taken together, they demonstrate the role of frames and framing in shaping landscape management, directly and indirectly. These cases offer insights into horizontal and vertical framing effects, particularly in implementing new practices, such as ecosystem service valuation or innovative participatory processes. For example, macro-level value frames in scientific literature and policy documents, including guidelines and legislation, establish overarching norms and standards. These frames are subsequently interpreted and adapted by institutions and individuals at the meso level. For instance, a project manager might perceive new practices as a threat, influencing their implementation strategy. This can lead to new practices being relegated to the sidelines with minimal effects. At the micro level, individuals directly affected by these new practices may engage in participatory processes or negotiations to influence implementation. These varied relationships illustrate a cascading effect where each level adapts to operational conditions while simultaneously trying to influence or pass on its frames, downstream or upstream.

Table 9.2 underscores that a single theoretical perspective cannot fully capture the framing process. This is particularly evident when considering how frames are formed and influenced. For example, the case studies support the notion that frames are, in part, socially constructed. This means that they are shaped by the social environment, including cultural norms, societal values, and public discourse. In this context, frames emerged as a product of collective interactions, reflecting how individuals and groups perceived and interpreted information. At the same time, the results also suggest that frames are affected by cognitive predispositions. This points to the psychological aspect of framing, where mental models and biases, such as pre-existing attitudes, beliefs, and experiences, play a role in influencing individuals, groups and institutions. In other words, these predispositions shaped how information was perceived and understood. This interplay between social construction and cognition indicates a complex process that requires complementary theoretical perspectives. For example, the micro case demonstrates how cognitive factors such as values, beliefs, and knowledge influenced perspectives on wetlands. However, the negotiation process was also shaped by social construction, including power relations and purpose-driven frames from the political administration. This dual perspective is essential to understanding how frames were developed, how they influenced individuals and groups, and how they affected the landscape.

Table 9.2 Cognitive, contextual, and purpose-driven framing across the micro, meso, and macro level.

<table>
<tr><th></th><th>Cognitive frames</th><th>Contextual frames</th><th>Purpose-driven frames</th></tr>
<tr><td rowspan="2">Micro case: Wetland restoration</td><td>Educational and occupational backgrounds influence individual perceptions of wetlands, such as biodiversity, which was valued differently based on the individual's education and profession.</td><td>Strategic one-on-one interactions by the negotiator were crucial in persuading landowners. Engaging with people in their home environments facilitated stakeholder buy-in.</td><td>The negotiator used specific and tailored narratives to persuade landowners, often using locally relevant examples, to facilitate the provision of land for the restoration process.</td></tr>
<tr><td>Experiential factors, such as past experiences and emotions, affected conditions imposed on the restoration process, such as restoring the landscape to make it look more like it used to or, for agricultural purposes, as a water reservoir.</td><td>Power dynamics affected the negotiation process, whereby the interests of landowners and fishing associations were prioritised over environmental organisations.</td><td>Motivational factors were aligned with stakeholder interests during negotiations, regularly focusing on economic compensation or landscape restoration as the main driving factors.</td></tr>
<tr><td rowspan="2">Meso case: Stakeholder participation</td><td>Stakeholders' value-based predispositions towards participation, often related to perceived benefits, shaped the level of engagement. Stakeholders' input was typically only acted upon when there was a positive belief in the participatory process.</td><td>An institutional lack of control and support allowed managers to tailor the participatory process based on their individual framing. This commonly resulted in participation being limited to the end of the project cycle.</td><td>Institutional framing of stakeholder participation focused primarily on securing or maintaining project funding. The implementation and format of participation were often a secondary concern.</td></tr>
<tr><td>Perceptions of power influenced how stakeholders were involved; for example, managers who thought that participatory processes reduced their power would often limit stakeholder involvement.</td><td colspan="2">Reflecting both the context and purpose, participatory approaches depended on underlying managerial intent. Neither the institutional constellation nor relevant policy frameworks prevented managers from operationalising individual value frames when designing and implementing projects.</td></tr>
<tr><td rowspan="2">Macro case: The Ecosystem Concept</td><td>Science and policy definitions of the ecosystem concept are normative in terms of reflecting individual and collective values and preferences. This implies that specific value assertions characterise the ecosystem concept.</td><td>Collaborative writing for scientific publications and negotiation processes in policymaking are presumed to be influenced by contextual factors, such as power relationships.</td><td>Scientific literature and policy documents exhibit stable value structures associated with the ecosystem concept. This had implications for its practical use and operationalisation.</td></tr>
<tr><td>Core frames are linked to stable value systems, with anthropocentric perspectives dominating scientific writing and policy.</td><td></td><td>Re-framing implies changing sub-frames, not core frames (or value systems). Using arguments that do not depart too far from the core frame may allow for reconfiguring a person's perspectives on any given concept, such as ecosystems.</td></tr>
</table>

4. Unpacking framing effects across levels and disciplines

This chapter wanted to establish a connection between frames, framing processes, and their effects in the real world, using the landscape concept as a way to introduce an integrative analysis. The chapter uses three case studies to demonstrate framing effects at various operational levels, from micro to meso and macro, including their relation to the landscape. The first case study, focusing on wetland restoration, illustrates quantifiable framing effects at the micro level. It demonstrates how specific framing strategies can directly influence a restoration process. The second and third case studies considered less tangible framing effects at the meso and macro levels. These case studies explored stakeholder participation and the ecosystem concept, demonstrating how frames can shape perceptions, policies, and actions, including the relationship between framing and direct or indirect implications for landscape management. A novel aspect of this multi-level approach is the ability to distinguish the framing effects across different contexts, intending to showcase an integrated approach to framing research. By doing so, the chapter has aimed to demonstrate the added value of an integrative analysis. It highlights the varied framing effects and their significance in shaping policy decisions and the landscape. These types of analysis can help to improve our understanding of the role of frames in landscape management and policy.

However, the goal of this chapter extends beyond advocating for an integrated approach to framing. It has tried to link abstract macro-level concepts, like value frames in science and policy, with more tangible frames at the meso and micro levels. This ranges from how institutional dynamics at the meso level shape policy implementation to the direct impacts of individual frames at the micro level, such as those influencing wetland restoration. The aim is to advocate for a multi-level synthesis of the framing process, integrating theoretical perspectives and levels. The framing process was categorised into cognitive, contextual, and purpose-driven framing to facilitate this integration and offer a clearer understanding beyond theoretical constraints. This layered analysis allows for an in-depth review of the framing process, revealing how different levels of framing interact and how they can collectively shape policy outcomes.

While the case studies provide valuable insights, it should also be acknowledged that there are methodological limitations. These constraints emphasise the need for caution in generalising the findings. Notably, the lack of direct links between the case studies presents a challenge in correlating framing effects across different levels. It is difficult to conclusively link policy frames at a macro level to individual frames at a micro level due to the many factors influencing decision-making, even more so when the cases are not directly connected. This complexity is exacerbated by the inherently qualitative nature of framing research, such as the risk that the researchers' subjectivities influence the analytical process. This is, for example, a common concern for text-based frame analysis (Erlingsson and Brysiewicz, 2017, Krippendorff, 2004). The case studies mainly illustrate potential interconnections and the application of frame analysis in complex environments.

The main proposition made in this chapter is to adopt a more inductive and grounded approach to framing, moving beyond a singular focus on theoretical explanations, such as power dynamics. For example, while a Foucauldian approach,[10] which emphasises the role of power (in terms of its operation, enactment, and resistance), is prevalent in framing research; however, this perspective might overshadow other important explanatory factors. Authors like Entman (2007, 2008) and Van Buuren et al. (2014) have explored power dynamics in contexts ranging from the media's influence on political systems to power relationships in climate change debates, underscoring its significance:

- In the micro case study, the interplay of power dynamics was evident, with the negotiator skilfully navigating these dynamics to facilitate stakeholder engagement, even though this negatively affected the environmental performance of the restoration programs. This scenario exemplifies how power can directly shape the landscape as both an influential factor and a strategic tool during negotiations. However, an exclusive focus on power dynamics can obscure other influential factors such as varying value orientations, historical perspectives, and the impact of communication skills and credibility in the framing process (Brewer and Gross, 2005, Buijs et al., 2009, Druckman, 2001).

- In the meso case, the project manager's power position significantly influenced the design of participatory processes. When stakeholder participation was perceived as threatening their control, managers would implement strategies to limit its influence, aligning their actions with the institutional frames. However, concentrating only on power dynamics might overlook the importance of cultural perspectives, especially in projects implemented in multicultural settings, and the varied backgrounds of project managers, many of whom come from a technocratic background (Buijs and Lawrence, 2013, Buijs et al., 2009, George et al., 2006).

These examples highlight the need for diverse perspectives in framing research. Taking a broader view, considering other factors like societal norms, historical backgrounds, and individual attributes and experiences, allows for a more balanced, multi-level analysis of the framing process. Acknowledging this complexity leads to two key insights: First, framing processes must be viewed as temporally and contextually specific, reflecting assertions that the process is dynamic and shaped by its specific environment (Gillan, 2008). Second, adopting a grounded and theoretically neutral approach can enrich framing research (Glaser and Strauss, 1967, Strauss and Corbin, 1990, Thomas and James, 2006). Moreover, the validity of results, regardless of philosophical differences, depends on the robustness and integrity of the methodological approach and data. Consequently, the proposed approach emphasises empirical evidence over theoretical preconceptions.

To further explain the limitations of relying only on theoretical frameworks and the risks of theoretical tunnel vision, consider the example of neo-institutional theory (see Chapter 3)[11] in the context of framing within an institutional setting. This theory, as outlined by scholars like George et al. (2006) and Schmidt (2008), focuses on how individuals conform to prevailing rules and norms within institutions and how the institutions themselves play a role in shaping power arrangements as well as in establishing interactional patterns (Fligstein and McAdam, 2011). At its core, institutional research can bridge macro-level structures, such as new organisational practices, with micro-level individual actions. The macro case in this chapter demonstrates one aspect of this connection, such as the interactions between individuals and institutions that contribute to a negotiated definition of the ecosystem concept. However, this approach does not consider the direct, real-time interactions between individuals; instead, it examined collective interpretations as articulated in scientific literature and policy documents. These interpretations serve as frames extending beyond the documents, influencing how individuals and institutions perceive and interact with ecosystems (Norgaard, 2010, Barnaud and Antona, 2014). For example, the adoption of the ecosystem services concept in natural sciences literature, such as arguments advocating for natural resource conservation, demonstrates the impact of macro-level framing. Such framing can change how the landscape is valued; however, it tends to be anthropocentric.

The key point of these examples are that focusing solely on the institutional dimension risks a top-down perspective that may neglect the substantive actions of individuals, including the contextual elements of each framing process. This oversight is evident in the meso case, where the project manager's behaviour reveals the impact individual actions can have in an institutional setting. Similarly, in the micro case, contextual factors influencing negotiations highlight the importance of considering the circumstances of each framing event. In conclusion, different theoretical perspectives, like neo-institutional theory, provide valuable insights into the framing process; however, there is a need for comprehensive approaches that integrate macro-structural influences and micro-level individual actions, while also considering contextual and purpose-driven factors around each framing situation.

While the cases look beyond the forest-based sector, the approach demonstrates potential real-world implications. For instance, insights from the micro-level case can inform the application of deliberative processes. Results from the meso-level case can guide institutions in embracing new practices more effectively. Additionally, the macro-level analysis could offer insight for refining re-framing techniques in policymaking. In line with Lakoff (2010), there is a gap in public discourse concerning frames that shape our relationship with landscapes. There is a tendency to view humans as separate from the landscape, a perspective rooted in individual cognition and challenging to alter (Spangler, 2003, Jerneck and Olsson, 2011, Buijs et al., 2011). This distinction aligns with early anthropocentric framing, influencing landscape management, scientific research, and policy formulation. However, it is becoming increasingly apparent that there is a need to shift away from these anthropocentric perspectives (Stern, 2011, Dewulf and Bouwen, 2012, Van Buuren et al., 2014). Such a shift would involve moving beyond self-reinforcing systems that limit how we manage the landscape and prevent the inclusion of alternative strategies (Norgaard, 2010). This shift is not only crucial for effective landscape management but also for addressing broader environmental challenges.

Finally, the pursuit of an integrative approach has its challenges. The case studies highlight an ongoing debate between discipline-specific and integrated approaches in framing research, acknowledging that finding a perfect balance between the two is beyond the scope of this book. Integrated approaches face challenges compared to established methodologies and approaches of monodisciplinary research. For example, funding agencies tend to focus on single-level approaches rather than interdisciplinary research. As Holm et al. [60] suggested, this would require a shift in the structural and procedural frameworks governing academic research, including increased interdisciplinary collaboration. However, when properly implemented, an integrative approach can lead to more informed, contextually relevant, and impactful research.

4.1. Final thoughts

Exploring an integrative policy analysis using the landscape concept underscores that integration while challenging, offers significant benefits. It provides a more complete overview of a policy problem by combining theoretical and methodological perspectives. This approach is particularly relevant for complex policy environments, like EU forest-relevant policy, which is characterised by limited regulatory capacities and competing policy domains. This is further evident in how the different levels of framing – micro, meso, and macro – can interact and influence one another, facilitating a better understanding of real-world implications associated with policy decisions. In other words, the chapter demonstrates the potential of integrating different levels of framing analysis, highlighting cognitive, contextual, and purpose-driven frames. For example, the shift towards SFM in Europe may require a blend of theoretical and methodological perspectives for a comprehensive overview of complex socio-economic and environmental interactions affecting forest management.

The heuristic blueprint is designed to guide the selection of analytical approaches in a non-prescriptive manner, emphasising adaptability and integration of different perspectives. However, it should be noted that the utility of an integrative approach is context-dependent and far from being a panacea. For certain policy problems, especially those with clear and quantifiable outcomes, a single methodology or a focused, non-integrative approach might be more practical. The scale and scope of the policy analysis should influence the choice of an approach. Moreover, the proposed blueprint for developing an integrative approach, though flexible and adaptable, is not without its challenges. It requires researchers to look beyond ontological and epistemological biases, demanding an understanding of different analytical tools and methodologies. Furthermore, integration is also normative and a matter of value judgements concerning research designs under specific environmental, historical, cultural, and socio-economic contexts. While providing valuable insights, the case studies highlight the inherent challenges in integrative policy analysis, such as the complexity of linking different levels of framing and the qualitative nature of framing research.

Want to find out more?

For those seeking more knowledge on integrative policy analysis, please see the references provided below or some of the many references throughout the text.

- Public Policy-Analysis: an integrated approach (Dunn, 2017).
- From stakeholders analysis to cognitive mapping and Multi-Attribute Value Theory: An integrated approach for policy support (Ferretti, 2016).
- Integrative Analysis Strategies for Mixed Data Sources (Bazeley, 2012).
- Integrative Frontiers in Environmental Policy Theory and Research (Niles and Lubell, 2012).

The set of questions listed below may also help guide further inquiry into specific areas of interest:

1. How would an integrative policy analysis differ from traditional forest policy analysis, particularly for SFM?
2. What are the key challenges and benefits of adopting an integrative approach for policy analysis?
3. How could an integrative policy analysis address the complexities surrounding regulatory competencies and competing policy domains in the EU forest-based sector context?
4. How do epistemological and ontological concerns impact the framing and interpretation of policy issues?
5. In what ways can the results of integrative policy analysis be practically applied in policymaking?

5. Assignment IX: Developing a research proposal

Description

The student will start learning how to develop a research proposal, in this case, focusing on a forest-related policy problem that interests the student. Using Chapter 9, the purpose of the exercise will be to reflect on how to implement an integrative research approach that combines multiple theoretical and methodological perspectives to address a complex forest policy problem.

Purpose

Focusing on multidisciplinary approaches in forest policy research, students will be asked to prepare a research proposal on an interesting forest policy problem. The assignment will require the student to develop a research question that would benefit from an integrative approach and outline the methodologies used. It should also explain how the research could contribute to the broader work on the chosen topic.

Outline of task

- *Selecting a topic*: Choose a relevant and complex forest-relevant policy topic that can benefit from an integrative policy analysis.
- *Literature review*: Conduct a state-of-the-art review of existing research on the chosen topic and identify gaps your research could address.
- *Research questions and objectives*: Formulate a statement of the policy problem and research question(s) and define your study's objective(s).
- *Methodology*: Describe the interdisciplinary methodologies you plan to use, including how you plan to integrate different theoretical perspectives and data sources.
- *Expected outcomes*: Consider and outline the potential contributions of your research to forest policy research.

Suggested additional reading

- How to write a research proposal (available on Scribbr).[12]
- van Kerkhoff, L. (2005) Integrated research: concepts of connection in environmental science and policy. *Environmental Science & Policy*, 8(5), 452–463.
- Tress, G., Tress, B. & Fry, G. (2005) Clarifying integrative research concepts in landscape ecology. *Landscape Ecol.*, 20, 479–493.
- Colloff, J.M., Martín-López, B., Lavorel, S. et al. (2017) An integrative research framework for enabling transformative adaptation. *Environmental Science & Policy*, 68, 87–96.

Deliverable

- Prepare a research proposal (1500–2000 words) with an introduction, a literature review (focusing on the state of the art), research questions, objectives, and methodology. The proposal should demonstrate the rationale behind the chosen approach and how it will be implemented.

6. Notes

1 Epistemology is about the theory of knowledge. It concerns the nature, scope, and origins of knowledge, addressing questions such as what knowledge is, how knowledge is acquired, and how true and false knowledge can be differentiated. It is, as such, concerned with how knowledge is understood and interpreted in different research fields.
2 SNA relates to internationally agreed recommendations on how to collect data on economic activities.
3 Ontology refers to the nature of being, existence, or reality. It relates to questions concerning what entities exist and how such entities can be grouped, related within a hierarchy, and divided according to similarities and differences. Ontological considerations determine the categories of things that are assumed to exist and the relationships between them.

4 Theoretical tunnel vision refers to the limitation caused by adopting a monotheoretical perspective or approach. This can lead to a biased perspective or prevent the application of more suitable analytical approaches to a policy problem.

5 Unlike other research methods that start with a hypothesis followed by data collection, a grounded approach starts with data collection without defining theories or hypotheses (Glaser and Strauss, 1967, Strauss and Corbin, 1990). The underlying idea is that, as the data is being analysed, patterns, themes, and concepts emerge naturally, allowing for the development of theories grounded in the empirical data. This approach is equally applicable in setting the scope for an analysis.

6 People might make decisions based more on emotional reactions than objective assessment or rational thinking. This means that how a message is framed influences decision-making through cognitive biases, memory recall, neurological effects, and emotional resonance, generating a framing effect (Raffaelli et al., 2019).

7 Anthropocentrism is a philosophical viewpoint that interprets the environment in terms of human values and experiences. This perspective often leads to viewing nature primarily in terms of its utility or value to human beings, commonly to the exclusion of considering the intrinsic value of other species or ecosystems.

8 Ecocentrism is a philosophical perspective that recognises the inherent value of all living organisms and their natural environments, regardless of their utility to human beings. It extends ethical consideration to all elements of nature.

9 Isomorphic responses, rooted in institutional theory, suggest that organisations tend to conform to the norms and standards of their operational environment (e.g., to gain legitimacy or access to resources). Isomorphism can occur due to coercive pressures (e.g., laws and regulations), normative influences (e.g., professional standards), or mimetic processes (e.g., copying successful models).

10 Foucault's theory suggests that power is pervasive in all social relationships and is not just concentrated in political or economic institutions. It operates through accepted norms, knowledge systems, and discourses, influencing how individuals think and behave.

11 Neo-institutional theory focuses on the role of institutions in shaping social behaviour and organisational practices. It emphasises that institutions, comprising norms, values, and established routines, significantly influence how individuals and organisations behave and interact.

12 See www.scribbr.com/research-process/research-proposal/.

7. References

AGGESTAM, F. 2014a. Effects of the manager's value orientation on stakeholder participation: at the front line of policy implementation. *Water Policy*, 16, 62–78.

AGGESTAM, F. 2014b. Wetland restoration and the involvement of stakeholders: An analysis based on value-perspectives. *Landscape research*, 39, 680–697.

AGGESTAM, F. 2015. Framing the ecosystem concept through a longitudinal study of developments in science and policy. *Conservation Biology*, 29, 1052–1064.

AGGESTAM, F. 2017. Integrated Framing: A micro to macro case for the landscape *Int. J. Environ. Sci. Nat. Res.*, 2, 555578.

AGGESTAM, F., MIEDZINSKI, M. & BLEISCHWITZ, R. 2023. The mirage of integration: Taking a street-level perspective on the nexus approach. *Environmental Innovation and Societal Transitions*, 46, 100700.

BARDACH, E. & PATASHNIK, E. M. 2019. *A Practical Guide for Policy Analysis: The Eightfold Path to More Effective Problem Solving*, 6th edition. Sage.

BARNAUD, C. & ANTONA, M. 2014. Deconstructing ecosystem services: Uncertainties and controversies around a socially constructed concept. *Geoforum*, 56, 113–123.

BARTELMUS, P. 2014. Environmental–economic accounting: Progress and digression in the SEEA revisions. *Review of Income and Wealth*, 60, 887–904.

BATESON, G. 1954. A theory of play and fantasy. *Psychiatric Research Reports*, 2, 39–51.

BAZELEY, P. 2012. Integrative analysis strategies for mixed data sources. *American Behavioral Scientist*, 56, 814–828.

BENFORD, R. D. & SNOW, D. A. 2000. Framing processes and social movements: an overview and assessment. *Annual Review of Sociology*, 26, 611–639.

BREWER, R. P. & GROSS, K. 2005. Values, framing, and citizens' thoughts about policy issues: Effects on content and quantity. *Political Psychology*, 26, 929–948.

BROWNE, J., COFFEY, B., COOK, K., MEIKLEJOHN, S. & PALERMO, C. 2019. A guide to policy analysis as a research method. *Health Promot Int*, 34, 1032–1044.

BUIJS, A. & LAWRENCE, A. 2013. Emotional conflicts in rational forestry: Towards a research agenda for understanding emotions in environmental conflicts. *Forest Policy Econ.*, 33, 104–111. https://doi.org/10.1016/j.forpol.2012.09.002.

BUIJS, A. E., ARTS, J. M. B., ELANDS, H. M. B. & LENGKEEK, J. 2011. Beyond environmental frames: The social representation and cultural resonance of nature in conflicts over a Dutch woodland. *Geoforum*, 42, 329–341.

BUIJS, A. E., ELANDS, B. H. M. & LANGERS, F. 2009. No wilderness for immigrants: Cultural differences in images of nature and landscape preferences. *Landscape and Urban Planning*, 91, 113–123.

CHAZDON, R. L., BRANCALION, P. H. S., LAESTADIUS, L., BENNETT-CURRY, A., BUCKINGHAM, K., KUMAR, C., MOLL-ROCEK, J., VIEIRA, I. C. G. & WILSON, S. J. 2016. When is a forest a forest? Forest concepts and definitions in the era of forest and landscape restoration. *Ambio*, 45, 538–550.

CHONG, D. & DRUCKMAN, J. N. 2007. Framing theory. *Annual Review of Political Science*, 10, 103–126.

CHRISTOPOULOS, S., HORVATH, B. & KULL, M. 2012. Advancing the governance of cross-sectoral policies for sustainable development: A metagovernance perspective. *Public Administration and Development*, 32, 305–323.

CONINX, I., DEWULF, A. R. P. J., STEINGROVER, E. G., WAL, V. D. M. F., OPDAM, P. & VOS, C. C. 2015. Framing ecosystem services: Affecting behaviour of actors in collaborative landscape planning? *Land Use Policy*, 46, 223–231.

CORNELISSEN, J. P. & WERNER, M. D. 2014. Putting framing in perspective: A review of framing and frame analysis across the management and organizational literature. *The Academy of Management Annals*, 8, 181–235.

DAVITER, F. 2007. Policy framing in the European Union. *Journal of European Public Policy*, 14, 654–666.

DAVITER, F. 2019. Policy analysis in the face of complexity: What kind of knowledge to tackle wicked problems? *Public Policy and Administration*, 34, 62–83.

DE GROOT, R. S., ALKEMADE, R., BRAAT, L., HEIN, L. & WILLEMEN, L. 2010. Challenges in integrating the concept of ecosystem services and values in landscape planning, management and decision-making. *Ecological complexity*, 7, 260–272.

DE VREESE, C. H. 2012. New avenues for framing research. *American Behavioral Scientist*, 56, 365–375.

DEVIGNEMONT, F. & SINGER, T. 2006. The empathic brain: how, when and why? *Trends in Cognitive Sciences*, 10, 435–441.

DEWULF, A. & BOUWEN, R. 2012. Issue framing in conversations for change: Discursive interaction strategies for 'doing differences'. *The Journal of Applied Behavioral Science*, 48, 168–193.

DEWULF, A., GRAY, B., PUTNAM, L., LEWICKI, R., AARTS, N., BOUWEN, R. & VAN WOERKUM, C. 2009. Disentangling approaches to framing in conflict and negotiation research: A meta-paradigmatic perspective. *Human Relations*, 62, 155.

DONAHUE, W. A., ROGAN, R. G. & KAUFMAN, S. 2011. *Framing Matters: Perspectives on Negotiation Research and Practice in Communication*. New York: Peter Lang.

DRUCKMAN, J. N. 2001. The implications of framing effects for citizen competence. *Political Behavior*, 23, 225–256.

DUNN, N. W. 2017. *Public Policy Analysis: An Integrated Approach*, 6th edition. New York: Routledge.

EDER, K. 1996. *The Social Construction of Nature: A Sociology of Ecological Enlightenment*. London: Sage.

ELLINGSEN, T., JOHANNESSON, M., MOLLERSTROM, J. & MUNKHAMMAR, S. 2012. Social framing effects: Preferences or beliefs? *Games and Economic Behavior*, 76(1), 117–130.

EMERY, R. 1998. *Taking Complexity Seriously: Policy Analysis, Triangulation and Sustainable Development*. New York: Springer.

ENTMAN, R. M. 2007. Framing bias: Media in the distribution of power. *Journal of Communication*, 57, 163–173.

ENTMAN, R. M. 2008. Media framing biases and political power: Explaining slant in news of Campaign 2008. *Journalism*, 11, 389–408.

ERLINGSSON, C. & BRYSIEWICZ, P. 2017. A hands-on guide to doing content analysis. *Afr J Emerg Med*, 7, 93–99.

FERRETTI, V. 2016. From stakeholders analysis to cognitive mapping and multi-attribute value theory: An integrated approach for policy support. *European Journal of Operational Research*, 253, 524–541.

FISCHER, A. P. 2018. Forest landscapes as social-ecological systems and implications for management. *Landscape and Urban Planning*, 177, 138–147.
FLIGSTEIN, N. & MCADAM, D. 2011. Toward a general theory of strategic action fields. *Sociological Theory*, 29, 1–26.
GEORGE, E., CHATTOPADHYAY, P., SITKIN, S. B. & BARDEN, J. 2006. Cognitive underpinnings of institutional persistence and change: A framing perspective. *The Academy of Management Review*, 31, 347–365.
GIGERENZER, G. & TODD, M. P. 1999. *Simple Heuristics that Make us Smart*. New York: Oxford University Press.
GILLAN, K. 2008. Understanding meaning in movements: A hermeneutic approach to frames and ideologies. *Social Movement Studies*, 7, 247–263.
GLASER, B. G. & STRAUSS, A. 1967. *The Discovery of Grounded Theory: Strategies of Qualitative Research*. New York: Aldine.
GOFEN, A., SELLA, S. & GASSNER, D. 2019. Levels of analysis in street-level bureaucracy research *In:* HUPE, P. (ed.) *Research Handbook on Street-Level Bureaucracy.* Cheltenham: Edward Elgar Publishing.
GOFFMAN, E. 1974. *Frame Analysis: An Essay on the Organisation of Experience*. New York: Harper & Row.
GOODMAN, J. R. 2006. Beneficial or biohazard? How the media frame biosolids. *Public Understanding of Science*, 15, 359–375.
GREENE, J. C. 2007. *Mixed Methods in Social Inquiry*. San Francisco, CA: Jossey-Bass.
GROSS, K. 2008. Framing persuasive appeals: Episodic and thematic framing, emotional response, and policy opinion. *Political Psychology*, 29, 169–192.
GROSS, K. & D'AMBROSIO, L. 2004. Framing emotional response. *Political Psychology*, 25, 1–29.
HANGGLI, R. & KRIESI, H. 2012. Frame construction and frame promotion (strategic framing choices). *American Behavioral Scientist*, 56, 260–278.
HARIPRIYA, G. S. 2001. Integrated environmental and economic accounting: An application to the forest resources in India. *Environmental and Resource Economics*, 19, 73–95.
HEAD, B. W. & ALFORD, J. 2013. Wicked problems: Implications for public policy and management. *Administration & Society*, 47, 711–739.
HERTWIG, R., HOFFRAGE, U. & GROUP., A. R. 2012. *Simple Heuristics in a Social World*, Oxford University Press.
HOWLEY, P., DONOGHUE, O. C. & HYNES, S. 2011. Exploring public preferences for traditional farming landscapes. *Landscape and Urban Planning*, 104(1), 66–74.
JACOBY, W. G. 2000. Issue framing and public opinion on government spending. *American Journal of Political Science*, 44, 750–767.
JAX, K., FURMAN, E., SAARIKOSKI, H., BARTON, D. N., DELBAERE, B., DICK, J., DUKE, G., GÖRG, C., GÓMEZ-BAGGETHUN, E., HARRISON, P. A., MAES, J., PÉREZ-SOBA, M., SAARELA, S.-R., TURKELBOOM, F., VAN DIJK, J. & WATT, A. D. 2018. Handling a messy world: Lessons learned when trying to make the ecosystem services concept operational. *Ecosystem Services*, 29, 415–427.
JERNECK, A. & OLSSON, L. 2011. Breaking out of sustainability impasses: How to apply frame analysis, reframing and transition theory to global health challenges. *Environmental Innovation and Societal Transitions*, 1, 255–271.
KAHNEMAN, D., SLOVIK, P. & TVERSKY, A. 1982. *Judgment under Uncertainty: Heuristics and Biases*. Cambridge: Cambridge University Press.
KAHNEMAN, D. & TVERSKY, A. 2000. *Choices, Values and Frames*. Cambridge: Cambridge University Press.
KALTENBORN, B. P. & BJERKE, T. 2002. Associations between environmental value orientations and landscape preferences. *Landscape and Urban Planning*, 59, 1–11.
KREUTER, J. 2021. Ideas and objects, meaning and causation – Frame analysis from a modernist social constructivism perspective. *In:* KREUTER, J. (ed.) *Climate Engineering as an Instance of Politicization: Talking Tomorrow's Technology – Framing Political Choice?* Cham: Springer International Publishing.
KRIPPENDORFF, K. 2004. *Content Analysis: An Introduction to its Methodology*, 2nd ed. Thousand Oaks, CA: Sage.
LA NOTTE, A. & RHODES, C. 2020. The theoretical frameworks behind integrated environmental, ecosystem, and economic accounting systems and their classifications. *Environmental Impact Assessment Review*, 80, 106317.

LAKOFF, G. 2010. Why it matters how we frame the environment. *Environmental Communication*, 4, 70–81.

LAZDINIS, M., ANGELSTAM, P. & PÜLZL, H. 2019. Towards sustainable forest management in the European Union through polycentric forest governance and an integrated landscape approach. *Landscape Ecology*, 34, 1737–1749.

LEVIN, K., CASHORE, B., BERNSTEIN, S. & AULD, G. 2012. Overcoming the tragedy of super wicked problems: constraining our future selves to ameliorate global climate change. *Policy Sciences*, 45, 123–152.

LINSER, S., WOLFSLEHNER, B., BRIDGE, R. J. S., GRITTEN, D., JOHNSON, S., PAYN, T., PRINS, K., RAŠI, R. & ROBERTSON, G. 2018. 25 years of criteria and indicators for sustainable forest management: How intergovernmental C&I processes have made a difference. *Forests*, 9.

LIPSKY, M. 1980. *Street-Level Bureaucracy: Dilemmas of the Individual in Public Services*. New York: Russell Sage Foundation.

MAJCHRZAK, A. & MARKUS, M. L. 2013. *Methods for Policy Research: Taking Socially Responsible Action*, 2nd edition. Sage.

MANFREDO, M. J., BRUSKOTTER, J. T., TEEL, T. L., FULTON, D., SCHWARTZ, S. H., ARLINGHAUS, R., OISHI, S., USKUL, A. K., REDFORD, K., KITAYAMA, S. & SULLIVAN, L. 2017. Why social values cannot be changed for the sake of conservation. *Conserv Biol*, 31, 772–780.

MARKANDAY, A. & GALARRAGA, I. 2021. The cognitive and experiential effects of flood risk framings and experience, and their influence on adaptation investment behaviour. *Climate Risk Management*, 34, 100359.

MATTHES, J. 2012a. Framing politics: An integrative approach. *American Behavioral Scientist*, 56, 247–259.

MATTHES, J. 2012b. Special Issue: Framing politics – An integrative approach. *American Behavioral Scientist*, 56.

MORGAN, E. A., CADMAN, T. & MACKEY, B. 2021. Integrating forest management across the landscape: a three pillar framework. *Journal of Environmental Planning and Management*, 64, 1735–1769.

MUSKAT, M., BLACKMAN, D. & MUSKAT, B. 2012. Mixed methods: Combining expert interviews, cross-impact analysis and scenario development. *Electronic Journal of Business Research Methods*, 10, 9–21.

NELSON, T. E., OXLEY, Z. M. & CLAWSON, R. A. 1997. Towards a psychology of framing effects. *Political Behavior*, 19, 221–246.

NEWTON, A. C., CAYUELA, L., ECHEVERRÍA, C., ARMESTO, J. J., DEL CASTILLO, R. F., GOLICHER, D., GENELETTI, D., GONZALEZ-ESPINOSA, M., HUTH, A., LÓPEZ-BARRERA, F., MALIZIA, L., MANSON, R., PREMOLI, A., RAMÍREZ-MARCIAL, N., BENAYAS, J.-M. R., RÜGER, N., SMITH-RAMÍREZ, C. & WILLIAMS-LINERA, G. 2009. Toward Integrated Analysis of Human Impacts on Forest Biodiversity Lessons from Latin America. *Ecology and Society*, 14.

NICKERSON, R. S. 1998. Confirmation bias: A ubiquitous phenomenon in many guises. *Review of General Psychology*, 2, 175–220.

NILES, M. T. & LUBELL, M. 2012. Integrative frontiers in environmental policy theory and research. *Policy Studies Journal*, 40, 41–64.

NORGAARD, R. B. 2010. Ecosystem services: From eye-opening metaphor to complexity blinder. *Ecological Economics*, 69, 1219–1227.

PATTON, V. C., SAWICKI, D. & CLARK, J. 2016. *Basic Methods of Policy Analysis and Planning* New York, Routledge.

PETERS, B. G. & FONTAINE, G. 2020. *Handbook of Research Methods and Applications in Comparative Policy Analysis*. Cheltenham: Edward Elgar.

RAFFAELLI, R., GLYNN, M.-A. & TUSHMAN, M. 2019. Frame flexibility: The role of cognitive and emotional framing in innovation adoption by incumbent firms. *Strategic Management Journal*, 40, 1013–1039.

SCHMIDT, V. A. 2008. Discursive institutionalism: The explanatory power of ideas and discourse. *Annual Review of Political Science*, 11, 303–326.

SCHMITHÜSEN, F. J. 2003. Understanding cross-sectoral policy impacts: policy and legal aspects. *Working papers. International series 2003(3)*. Rome: FAO.

SHEN, F. & EDWARDS, H. H. 2005. Economic individualism, humanitarianism, and welfare reform: A value-based account of framing effects. *Journal of Communication*, 55, 795–809.

SNOW, D. A. & BENFORD, R. D. 1988. Ideology, frame resonance, and participant mobilization. *In:* KLANDERMANS, B., KRIESI, H. & TARROW, S. (eds.) *From Structure to Action: Social Movement Participation Across Cultures*. Greenwich: JAI Press.

SNOW, D. A., ROCHFORD, E. B., WORDEN, S. K. & BENFORD, R. D. 1986. Frame alignment processes, micromobilization, and movement participation. *American Sociological Review*, 51, 464–481.
SPANGLER, B. 2003. Reframing. Beyond intractability. *In:* BURGESS, G. & BURGESS, H. (eds.) *Conflict Research Consortium*. Boulder: University of Colorado.
STERN, C. P. 2011. Contributions of psychology to limiting climate change. *American Psychologist*, 66, 303–314.
STRAUSS, A. & CORBIN, J. 1990. *Basics of Qualitative Research: Grounded Theory Procedures and Techniques*. Newbury Park, CA: Sage.
TANSLEY, A. G. 1935. The use and abuse of vegetational concepts and terms. *Ecology*, 16, 284–307.
THOMAS, G. & JAMES, D. 2006. Reinventing grounded theory: some questions about theory, ground and discovery. *British Educational Research Journal*, 32, 767–795.
UN, EU, FAO, IMF, OECD & WB. 2014. *System of Environmental-Economic Accounting 2012: Central Framework*. New York: United Nations.
VAN BUUREN, A., VINK, M. & WARNER, J. 2014. Constructing authoritative answers to a latent crisis? Strategies of puzzling, powering and framing in Dutch climate adaptation practices compared. *Journal of Comparative Policy Analysis: Research and Practice*, 1–18.
VAN HULST, M. & YANOW, D. 2014. From policy 'frames' to 'framing': Theorizing a more dynamic, political approach. *The American Review of Public Administration*.
VARDON, M., CASTANEDA, J.-P., NAGY, M. & SCHENAU, S. 2018. How the System of Environmental-Economic Accounting can improve environmental information systems and data quality for decision making. *Environmental Science & Policy*, 89, 83–92.
WOLFSLEHNER, B. & VACIK, H. 2008. Evaluating sustainable forest management strategies with the analytic network process in a pressure-state-response framework. *Journal of Environmental Management*, 88, 1–10.

10 Conclusions

1. Introduction

European forests, as the largest land-based natural resource on the continent, are deeply intertwined with multiple socio-economic and environmental factors, playing an important role in industry, conservation, public well-being and health. Forest products and services have become integral and interconnected components across sectors and cultural settings, highlighting the relevance of forest ecosystems across various levels. It is this multi-sectoral governance system and the diverse range of products and services that make forests exciting to study; whether you are interested in the management of natural resources, biodiversity, climate change, construction, fashion or tourism, some aspects intersect with forest management and the use of timber as a raw material and feedstock.

Ten reasons why forests are important

1. *Provisioning services*: Forest ecosystems provide essential raw materials, such as timber for construction, biomass for energy and fibres for paper products and textiles.
2. *Biodiversity*: Forests are dynamic and complex ecosystems, contributing significantly to ecological resilience. They provide essential habitats for a vast array of species.
3. *Employment*: Forests substantially contribute to the European and global economy, supporting professions ranging from logging to construction, ecotourism, and conservation.
4. *Subsistence*: A large portion of the global rural population lives near forests, relying on forests for subsistence.
5. *Food security*: Forests are crucial for food security, with a significant portion of the global rural population still using wood fuel for cooking and as a food source.
6. *Climate regulation*: Forests act as carbon sinks, sequestering significant quantities of carbon dioxide (CO_2) and regulating temperatures. Their role in climate mitigation is substantial.
7. *Air quality*: Forests enhance air quality by absorbing pollutants, which is important in mitigating air pollution-related health risks.
8. *Mitigation*: Forests stabilise soils and mitigate the impact of natural hazards, such as reducing landslide and flooding risks and acting as natural protective barriers.
9. *Water quality maintenance*: Forest ecosystems reduce soil erosion and water runoff while supporting sediment and pollutant filtration and refilling aquifers.
10. *Recreation and public health*: Forests are important recreational and culturally significant spaces.

DOI: 10.4324/9781003198987-10

Forest policy research is particularly relevant when considering the sustainable management and use of forest resources associated with different segments of the forest value chain (see Chapter 5). This research is key for improving our understanding of how forest ecosystem functions can be preserved while meeting societal needs and demands. In the face of numerous threats confronting forests today, such as deforestation, climate change, and habitat fragmentation, policy research becomes instrumental in identifying effective solutions and strategies that can mitigate or counter these threats. Forest policy research is fundamental for developing informed and adaptive approaches to sustainable forest management (SFM) and evidence-based policymaking.

Bearing in mind the diversity of topics within the field of forest policy research, this book has endeavoured to introduce a selection of research areas related to forests (Figure 10.1). It has covered a variety of subjects, including European Union (EU) forest governance, discussed in chapters 2 and 4, and methodologies in forest policy research, explored in Chapters 3 and 9.

Figure 10.1 Areas of research that feature forests.

Source: own figure

Additionally, the book considered the prospects of implementing a circular bioeconomy in Chapter 5, examined forest ownership dynamics in Chapter 6, analysed forest fire policy in Chapter 7, and investigated the multifunctional forest use concept in Chapter 8. While the compilation of examples does not provide an exhaustive overview of forest policy research, it has shed light on various approaches and themes currently being explored by researchers in the forest policy domain.

Emphasising a comprehensive exploration of different aspects and approaches in forest policy research, this book has gone beyond the traditional confines of the forest-based sector, adopting an integrative view of what constitutes forest-focused and forest-relevant policy (see Chapter 2). Understanding the complex and diverse nature of the forest-based sector, which covers a wide spectrum of policy perspectives and objectives, is a prerequisite for analysing interactions within forest ecosystems, including interactions with society and policymaking. The crux of the matter is the need for different research methods to analyse and understand the European forest governance framework, which is characterised by distinct and, at times, conflicting perceptions regarding the purpose of forests. This complexity is compounded by varying priorities and specificities of European countries and regions, each with its own environmental and socio-economic considerations regarding forest use and management. This makes it difficult to make simple comparisons.

This concluding chapter serves as a synthesis of the different topics explored throughout this book. It will offer an overview, emphasising key points and findings from each section. In doing so, the chapter will reiterate the core ideas and insights, shedding light on some of the themes covered by forest policy research. The following section offers an overview of EU forest governance and the theoretical frameworks shaping forest policy research. It then extends to the potential of a bio-based economy, as interlinked with the European Green Deal and the new EU Forest Strategy, and essential systemic changes needed for the circular bioeconomy within the EU's forest-based sector. Additionally, it addresses the complexities surrounding forest ownership and providing forest ecosystem service (FES), challenges associated with forest fires, resilience, climate change, and implications associated with multifunctional forest management. The section finishes by covering work on an integrative blueprint for forest policy analysis. The approaches and results introduced in this book are intended to be an informative contribution to forest policy research and help equip new researchers with a nuanced understanding of forest governance.

2. Synthesis of European forest policymaking and research

2.1. Forest governance: the EU perspective

Forest governance in the EU corresponds to an ever-evolving policy landscape. Although the EU lacks a common forest policy, it has developed forest-focused and forest-relevant policies and engaged in international efforts to address forests. The EU's approach to forest governance is mainly characterised by non-legally binding (soft) policy instruments due to its limited competency on forests under the existing Treaty on the Functioning of the European Union (TFEU). This means that the principle of subsidiarity applies to EU forests, which has resulted in a prevailing reliance on soft policy instruments despite the EU's involvement in policy domains like agriculture, trade, and environmental policy. This approach has led to policy incoherence and unresolved trade-offs between sectors affecting forest use and management (Edwards and Kleinschmit, 2013, Aggestam and Pülzl, 2020, Lier et al., 2021).

A cornerstone of EU forest-focused policy is the EU Forest Strategy. Since its inception in 1998, the Strategy has undergone several iterations, with the latest in 2021 focusing on protecting, restoring, and sustainably managing forests. The new Strategy aligns with EU policy instruments such as the European Green Deal, Renewable Energy Directive, European Climate Law, and the EU Biodiversity Strategy, aiming to support EU targets to reduce greenhouse gas (GHG) emissions, improve biodiversity conservation, and promote SFM. However, the new Strategy has faced criticism for not sufficiently addressing the incoherence in EU forest policy and failing to propose significant changes to improve governance and inter-service communication on forests within the Union (Aggestam and Giurca, 2021).

The selection of hard and soft EU policy instruments introduced in Chapter 2 indicates that a broad range of EU policies affects the forest-based sector:

- Legislative frameworks like the EU Timber Regulation and the Common Agricultural Policy (CAP) have been critical in shaping European forest governance. The EU regulation on deforestation-free products (replacing the EU Timber Regulation) targets illegal logging. At the same time, the CAP, particularly through the European Agricultural Fund for Rural Development (EAFRD), provides funding for forestry across EU Member States (Alliance Environnement et al., 2018).
- EU products and environmental protection regulations, such as the General Products and Safety Directive, REACH, and the Industrial Emission Directive, affect forest-based industries by setting standards and requirements for production, safety, and environmental compliance (Rivera León et al., 2016).
- The EU's sustainable finance framework, particularly the EU Taxonomy Regulation, guides investments towards ecologically responsible practices (Begemann et al., 2023).
- Employment and transport policies, such as EU directives on working hours, health and safety, and transport regulations, including those regulating marine fuel sulphur content and waste shipment, shape how the forest-based value chain can operate (Aggestam and Pülzl, 2018).

The EU has made progress in addressing the multifunctionality of forests and in integrating forest issues into a broader range of environmental and economic frameworks. However, challenges remain in achieving policy coherence and effective forest governance (Aggestam and Pülzl, 2018, 2020). Moreover, while the idea of a common EU forest policy may be advantageous from a policy coherence perspective, it is currently not a political priority (Aggestam and Giurca, 2021, Blattert et al., 2022, Gordeeva et al., 2022). Until this changes, the need for improved collaborative efforts and enhanced coordination between EU forest-relevant policy domains and national strategies remains. In addition, evolving policy developments in areas like the bio- and circular economy suggest the need for more integration across the forest-based value chain. Particularly as forests are not only about producing carbon-neutral biomass, timber, and non-wood forest products (NWFPs) but also about the approaches employed in their production and processing. Future efforts must reconcile competing interests and objectives, enhance inter-service communication, and adapt policies to the dynamic environmental and socio-economic policy landscapes affecting forests.

- The current form of EU forest governance is limited in terms of impacts and relevance, highlighting the need for new and innovative governance solutions, including improved dialogue and information exchange mechanisms that can help to identify synergies and resolve trade-offs.
- It is crucial to thoroughly examine all policy domains that impact forest-based value chains, particularly those reliant on wood as a raw material, and to develop a shared understanding of what constitutes relevant EU forest-based policy.

2.2. *Overview of theories and approaches used in forest policy research*

Forest policy research, which gained prominence in the latter half of the twentieth century, addresses the socio-economic and environmental impacts of forest management practices and policymaking. Initially focused on practical aspects, like sustainable timber production, the field of forest policy research has evolved to become more theory-driven. Researchers have turned to different theoretical frameworks and models to analyse and explain forest policymaking, including the development, implementation, and impact of forest-relevant policies (Glück, 2000, Arts, 2012). For instance, institutional policy analysis, policy network analysis, the Advocacy Coalition Framework, critical policy analysis, and rational policy analysis are some of the frameworks that have been used to analyse international, regional, and national forest policies and practices (Dunn, 2017, Browne et al., 2019, Daviter, 2019).

The application of the approaches introduced in Chapter 3 emphasises the importance of theory-driven policy analysis:

- Forest ecosystems are fundamentally complex, involving many biological, ecological, socio-economic, and political factors. Different theoretical approaches allow researchers to analyse forest policy from multiple perspectives (de Jong et al., 2012, Dunn, 2017).
- Forest policy decisions commonly involve many stakeholders, including governments, local communities, NGOs, and industry. Theoretical frameworks, such as policy network analysis and stakeholder theory, can help explain these different groups' motivations, influences, and power dynamics (Secco et al., 2016, Deuffic et al., 2018, Pülzl et al., 2018).
- Theories like the Advocacy Coalition and Multiple Streams Framework can provide insights into how forest policies are formulated, negotiated, and implemented. These frameworks help to analyse how different policy problems come to the forefront in the policy discourse, how coalitions form, and how policy windows open, which can lead to policy change (Sabatier, 1988, Winkel and Leipold, 2016, Zahariadis, 2018).
- The increasing impacts of climate change and other stressors emphasise that theories focusing on adaptive management and resilience are also relevant. They allow policy researchers to consider how forest policies can be designed to be flexible and adaptive to changing climatic and environmental conditions (Curtis and Cicchetti, 2003, Wilkinson, 2011, Chelleri and Baravikova, 2021).
- Social constructionist theories and frameworks that emphasise cultural and social values are important when considering the impact of societal norms and cultural values. These theories help to identify non-material benefits of forests, such as cultural identity, spiritual values, and recreational benefits, which tend to be overlooked in traditional policy analysis (Gamson and Modigliani, 1989, Van Gorp, 2007).
- Theories focusing on policy feedback and evaluation can be important when reviewing the effectiveness of forest policies. They offer methods and tools for monitoring, learning, and adapting policies based on their outcomes and impacts (Sewerin et al., 2020).

The forest policy research community has made significant strides over the last few decades, reflected in the increasing use of multiple theoretical lenses that provide valuable insights into forest governance as a complex socio-ecological system. Even more so, forest policy research often goes beyond the boundaries of traditional disciplines. This highlights the added value of integrating different research methods and data sources, from quantitative to qualitative analyses, to ensure that forest policy research is grounded in robust evidence. In other words, forest policy research benefits from using multiple theoretical and interdisciplinary approaches, which has brought fresh perspectives to a complex policy domain.

- Forest policy research has evolved from focusing on practical aspects of forest management to incorporating social science models and frameworks.
- The use of theory, such as institutional policy analysis, policy network analysis, and the advocacy coalition framework, has improved our understanding of forest policies' development, implementation, and impact.
- Embracing multidisciplinary approaches is crucial, enabling researchers to look beyond disciplinary boundaries to uncover synergies and trade-offs in forest management and policy.
- Theoretical frameworks in forest policy research offer practical applications in policymaking, providing insights for practitioners and policymakers, particularly in understanding and navigating the dynamics of forest resource management.

2.3. Prospects for a bio-based economy: the European Green Deal

A significant challenge for EU forest-focused and forest-related policy is their incoherence and fragmentation, failing to consider relevant policy interactions and trade-offs adequately. This has led to conflicting policy targets across EU policy domains, such as agriculture and rural development, energy, water, and climate. The absence of an effective steering instrument for EU forests has enhanced this effect, emphasising the need for a coherent EU policy mix to balance trade-offs, such as biodiversity conservation versus climate change targets.

As introduced by the European Green Deal, the new EU Forest Strategy faces the challenge of having to account for the wide range of environmental and socio-economic policy objectives while navigating a complex and fragmented EU policy landscape. Nevertheless, as set out in the European Green Deal, the ambition for Europe to become climate-neutral and to operationalise a circular (bio)economy underscores the important role of forests. Moreover, the Green Deal introduced several important policies and regulatory proposals directly addressing forests. This includes the EU Biodiversity Strategy for 2030, the EU taxonomy for sustainable activities, and the Climate Change Mitigation Strategy. These instruments underpin the Green Deal's environmental objectives. Additionally, the new EU Forest Strategy, a non-binding (soft) policy instrument, and the Green Deal set out a vision for European forests that depend on the active participation of Member States.

Chapter 4 demonstrates that effective, adaptive, and integrative strategies are essential for EU forest-related policy:

- The European Green Deal affects EU forest policy directly. For example, the Green Deal introduced the new EU Forest Strategy, developed alongside instruments like the EU Biodiversity Strategy for 2030 and the EU taxonomy for sustainable activities. These advances reflect the interconnectedness of environmental policy objectives (Pianta and Lucchese, 2020, Siddi, 2020, Aggestam and Giurca, 2021).
- Perspectives from representative organisations and scientific studies provide valuable insights into the EU Forest Strategy's prospective implementation and alignment with the Green Deal's objectives (CEPF, 2019, CEPF et al., 2019, FERN, 2019, IUCN, 2019).
- Integrating objectives under the Green Deal, such as biodiversity protection, sustainable land use, and carbon sequestration, poses challenges, particularly for implementation. Policy pathways that could shape the Green Deal and the EU Forest Strategy emphasise the need for improved policy coordination and adaptable strategies (Aggestam and Giurca, 2021).

- The incoherence in EU forest-focused and forest-related policies often results from inadequate consideration of policy interactions. This highlights the need for a coherent policy mix to balance trade-offs (Aggestam and Pülzl, 2018, 2020).

The complexity and conflicting objectives inherent in EU forest policy present a significant governance challenge, requiring a revision of the tools and mechanisms used to implement the new EU Forest Strategy. This challenge is exacerbated by the complex interplay between Member States' priorities and EU political directives. For instance, while actions outlined in the Green Deal, such as those related to biodiversity, trade, and climate, fall under the purview of the European Commission, the approach to other policy domains remains less clear. Moreover, the Commission's limited enforcement capacity casts doubt on its ability to effect changes in the forest-based sector. Consequently, there is a need for the new EU Forest Strategy to innovate in terms of implementation mechanisms, enforcement, and processes that can coordinate EU forest-relevant policies.

- The new EU Forest Strategy has emerged in a dynamic and evolving EU policy landscape, heavily influenced by the Green Deal's focus on sustainability, biodiversity conservation, and measures concerned with carbon sequestration.
- Significant incoherence and fragmentation in EU forest-relevant policy highlight the need for improved policy coordination.
- Implementing the Green Deal's forest-related objectives requires an adaptable EU Forest Strategy to address and integrate conflicting policy objectives.

2.4. Systemic change for the forest-based sector: implementing the circular bioeconomy

The forest-based sector is an interesting case study for exploring the circular bioeconomy. This chapter focused on two value chains, covering five sectors: sawnwood, bioenergy, construction, and pulp and paper manufacturing, including cellulose production. Unlike other materials, like metals or glass, which can be recycled indefinitely, wood as a renewable raw material has its limitations in terms of recyclability. Woody biomass is more versatile than other materials, providing the main ingredient for textiles, bioplastics, paper, and bioenergy. Forests contribute to services beyond material goods, including public health and carbon sequestration. In the context of different conceptual frameworks for circularity, including the 3R model, the more expansive 9R model, and the cascading use principle for wood, the question is whether circularity principles can be effectively integrated across the different segments of the forest-based value chain. With this in mind, this chapter examines sector-specific challenges and opportunities in transitioning to a circular bioeconomy through a value chain analysis. The goal was to review the unique requirements of each value chain and sector.

Chapter 5 demonstrates that the forest-based sector has the potential to become circular, identifying sector-specific challenges and opportunities:

- The circular bioeconomy narratives cover varied conceptual perspectives, from operationalising circularity within current economic systems to advocating for a transformation of the socio-economic order. The 9R model is one approach regarding resource sustainability strategies that emphasise system change. The 9R components, each representing a different approach towards resource use, can be applied variably within the forest-based sector (Khaw-ngern et al., 2021).

- Insights into circular efficiency indicate significant variability in the solutions needed to achieve circularity. The analysis underscores the diversity and complexity of the forest-based industry ecosystem, illustrating that the transition to circularity is a complex process that requires tailored approaches that are adjusted to the specificities of each value chain (Aggestam et al., 2021, Aggestam and Giurca, 2022).
- Implementing circularity across the respective forest-based sectors will require new technologies, business models, and increased awareness and understanding of circularity principles. For instance, systematic issues, such as promoting a circular lifestyle, could be tackled through cross-cutting factors like integration, education, certification, and innovation (Hetemäki et al., 2017, D'Amato et al., 2020, Toppinen et al., 2020).
- Circularity aims to optimise resource use and minimise waste; however, this does not equate to sustainable practices, especially when considering negative externalities, such as increased emissions generated by transportation and other potential spillovers (Lazaridou et al., 2021).

The two value chains highlight the need for customised strategies across different sectors to address the varied use of resources, energy, and waste management practices. It can also be noted that forest-based industries need to overcome significant barriers in transitioning towards a circular bioeconomy, particularly due to lock-in (or path dependency), where existing processes make it more difficult to shift towards more circular practices. For example, sawmill residues are commonly underutilised, often relegated to low-value products or energy production rather than fully utilised. Additionally, there is a prevailing misconception that the renewable nature of wood inherently means the sector is circular. This means that the sector often overlooks creative and innovative solutions to enhance the acceptance and application of circular models and address recyclability barriers.

- Circularity aims to extend the lifecycle of wood products and increase resource use efficiency. The concept emphasises the reuse, recycling, and cascading use of wood, not just as material for timber and paper but also for other by-products and bioenergy.
- The forest-based sector must consider the limitations of wood as a raw material and ensure that sustainability criteria are integrated into the design of circular systems.
- Applying value retention loops has been challenging from a research perspective, indicating a need for a more systematic approach in future studies.

2.5. Forest ownership and the provision of forest ecosystem services

The provision of forest ecosystem services (FES) is intrinsically interlinked with forest ownership structures across Europe and internationally. Different forest ownership types, whether public or private, influence how forests are managed, affecting the quantity and quality of FES provision. This includes provisioning services like timber and non-wood forest products (NWFP), regulating services like climate control and water purification, cultural services, and supporting services like nutrient cycling. Chapter 6 outlines different forest ownership categories, such as state, municipal, common, private, and open access properties, where each category has distinct implications regarding the rights and responsibilities associated with forest ownership. This includes different access, withdrawal, management, exclusion, and alienation rights associated with legal and social frameworks. These variations across European countries stress the importance of understanding the legal and socio-economic contexts in which forest management operates. These factors ultimately shape how forest resources can be used.

Chapter 6 highlights the role of different forest ownership types in influencing FES provision:

- There are substantial variations in forest ownership structures across European countries and regions, with noteworthy differences in public and private ownership influenced by historical, cultural, and legal factors. These national and regional variations affect forest management strategies and objectives (Weiss et al., 2019, Aggestam et al., 2020b).
- Legal structures and institutional rules play a significant role in shaping forest ownership and use, emphasising the crucial role of government in setting forest management guidelines. The sustainability of forest management practices is, in effect, mediated by the type of forest ownership and the regulatory environment (ECE, 2020).
- There are significant challenges in gathering comprehensive data on forest ownership and its relationship with FES. These data gaps and the lack of information on forest ownership stress the need for more harmonised data collection and categorisation methods to improve our understanding of the relationship between forest ownership frameworks and FES provision (Živojinović et al., 2015).

The analysis of European forest ownership structures reveals significant interlinkages between ownership and forest management choices. The diverse range of ownership types, each representing a distinct set of rights and responsibilities, is shaped by institutional, legal, and socio-cultural factors. This diversity leads to a complex mosaic of management objectives and strategies, influencing forest use and conservation efforts. As a result, forest management in Europe is characterised by many approaches, reflecting the varied motivations and constraints of different forest owners. For example, different ownership structures present context-specific challenges and opportunities for SFM, climate change mitigation, and adaptation strategies. Private owners might prioritise economic returns, while public and communal owners may focus on biodiversity conservation and climate resilience. The role of policies and legislative frameworks may shape how these forest owners can incentivise SFM practices, integrate climate change considerations into forest management plans, and balance the interests of different ownership types. Conversely, large-scale owners typically have resources for intensive forest management, whereas smallholders might adopt less intensive practices due to resource constraints. Addressing these variations would involve creating context-specific frameworks that encourage sustainable practices and facilitate access to resources and information for forest owners.

- Forest ownership involves more than just holding the deed for land; it includes a broad range of rights, obligations, and management practices under different national legislative frameworks and societal norms.
- Institutional frameworks, property rights, land use policies, and forest management regulations are all interlinked and important for understanding forest ownership.

2.6. Forest fires, resilience, and climate change

EU forest fire policy reveals some of the complexity and variability of forest fire challenges across Europe. The chapter stresses the importance of understanding how forest fires are framed within national, regional, and EU policymaking and the need for interdisciplinary approaches to forest fire management and research. By analysing EU policy documents, the chapter identified six collective frames that have shaped forest fire narratives in the EU over time: (1) climate adaptation and resilience, (2) sustainable energy transformation, (3) technocratic perspectives on forest fires, (4) ecological diversity and stewardship, (5) agriculture and rural development, and (6) risk mitigation and protective governance. These collective frames have been instrumental in shaping the EU strategies and approaches towards forest fire management and monitoring, demonstrating a strategic use of narratives that promote specific policy objectives when addressing fires, such as climate change and resilience. Although the analysis focused on the adopted policy document, it offers a nuanced picture of EU forest fire policy, stressing the importance of an integrated approach to policy formulation and implementation.

Chapter 7 demonstrates that policies on forest fires have been contextualised within broader environmental trends and changing EU competencies:

- The six collective frames have used 'forest fires' to advance specific narratives. These narratives range from land use planning to public health and safety to socio-economic concerns adapted to the policy domains' priorities and objectives (Aggestam, 2024).
- Four frames were identified as particularly dominant: (1) climate adaptation and resilience, (2) risk mitigation and protective governance, (3) agriculture and rural development, and (4) technocratic perspectives on forest fires. These frames have developed in isolation within their respective policy domains, which suggests a gap in cross-sectoral communication and collaboration (Aggestam, 2024).
- The evolving nature of forest fire policy is evident in how the EU approach has changed focus and scope over time. The EU's approach to forest governance has broadened and become more integrative, moving away from protection-oriented strategies to cover a wider range of societal concerns, including climate change adaptation, biodiversity conservation and SFM. However, certain dominant frames have remained stable, especially regarding forest utility (EC, 2018, Mourao and Martinho, 2019).
- Risks associated with forest fire have been consistently used as a narrative device, highlighting a negative framing where inaction is strategically equated with negative consequences, such as damage to infrastructure, loss of human life, and impacts on biodiversity (Seijo, 2009, Castelló and Montagut, 2019).
- Different frames emphasise different forest fire impacts. These differences indicate that the framing of forest fires is closely linked with the concerns and expertise of the epistemic communities that have drafted the respective policy documents (Bosomworth, 2015, Nilsson and Enander, 2020).

Despite the expanding scope of EU forest fire policy, the relationship between forest fires, policy framing, and EU competencies highlights how framing research can add value when addressing policy challenges posed by forest fires. For instance, it reveals how narratives within the forest fire policy area have developed, influenced by content and context, across different policy domains. The analysis also demonstrates the prevailing fragmentation across EU policy domains while suggesting coherence within individual domains. This points towards a compartmentalised approach. Moreover, the consistent use of collective forest fire frames suggests coordinated and domain-specific strategies regarding forest fires.

- EU forest fire policy reveals variations in framing across sectors and policy instruments, with each sector adjusting the forest fire narrative to specific mandates and objectives.
- Forest fire frames influence the direction and content of policies and affect how policymakers and the public perceive and understand forest fires.

- The influence of epistemic communities at various levels raises concerns about the inclusiveness of the policymaking process.

2.7. Nature protection, productivity, and multifunctional forest use

Understanding integrative approaches is relevant when considering the future of forest management, particularly as it may offer a more holistic approach to governing forest resources. More importantly, integrated forest management can also help reconcile fundamental trade-offs and minimise negative environmental and societal impacts. With this in mind, the focus in Chapter 8 was set on one complex aspect of integrated forest management: how to harmonise objectives associated with nature conservation vis-a-vis wood production. Using the STEEP approach, it considered mechanisms influencing the adoption of integrative practices. It identified 37 factors affecting forest management, from dynamic elements like policy change to static ones like forest ownership structures. Notably, these factors are significant in themselves and how they interact with each other, such as laws and regulations affecting forest ownership patterns. This web of interactions can determine the success or failure of an integrative approach.

Chapter 8 confirms the need for adaptive strategies when considering integrated forest management, recognising the complexity of intertwining social-cultural, technological, economic, environmental and political (STEEP) factors:

- Individual and family values, local customs, and social networks are significant determinants in terms of shaping forest use, often being the defining factor for the uptake of integrative approaches (Borrass et al., 2017, Sotirov et al., 2017, Konczal et al., 2023).
- The competitiveness of the forest-based sector emerged as an important economic driver, essential for its long-term viability and the effective implementation of integrative practices (Ustaoglu and Collier, 2018, Aggestam et al., 2020a).
- Local environmental conditions, including climate, soil composition, and biodiversity, are critical factors in tailoring integrated forest management strategies that effectively fit specific countries or regions (Aggestam et al., 2020a, Augusto and Boča, 2022).
- Effective leadership is essential at every stage, from the formulation and development of policies to the uptake and practical implementation of integrated forest management practices (Aggestam et al., 2020a, Geels, 2021).
- Empowering stakeholders, including forest owners, forestry professionals, and conservationists, with the necessary knowledge to effectively balance nature conservation with timber production (Reed et al., 2019, Aggestam et al., 2020a).
- There is a need for detailed and context-specific insights that can provide forest managers and owners with guidance towards more informed and effective implementation strategies (Konczal et al., 2023).

The findings of this research underscore the central role that commitment and proactive actions by key individuals at the operational level play in adopting integrated forest management. These individuals often include early adopters, innovative entrepreneurs, and influential community leaders who can demonstrate the practical benefits of integrative practices. Such front-runners do more than endorse new forest management approaches; they become instrumental in bridging the gap between policy formulation and practical implementation. In other words, any forest management approach's initial efficacy and performance, including integrated forest management, depend on these pioneering individuals. At the policymaking level, the role of leadership is equally important. Leaders with a clear and forward-looking vision for the future are essential for developing and implementing policies that acknowledge and leverage the multifunctional role of forests.

- Integrated forest management considers the intricate relationship between environmental and socio-economic dimensions of forest ecosystems, highlighting the importance of context-specific insights for informed implementation of forest management strategies.
- Forest policy instruments must be tailored to account for specific contextual factors relevant to integrating biodiversity considerations into the practices of integrated forest management.
- Governance approaches for integrated forest management must be further evaluated, such as determining whether forest owners should be mandated or incentivised and identifying who should hold decision-making authority regarding conservation objectives.

2.8. A framework for integrative forest policy analysis

As considered in Chapter 3, traditional forest policy analysis often operates under the assumption that policy systems are well-defined and manageable, implying a certain level of controllability. However, this perspective overlooks inherent complexities and uncertainties associated with policy problems, which are often characterised by unpredictable future outcomes and are commonly managed by policymakers who possess varying, and sometimes limited, legal and technical capacities. In other words, traditional methods used in policy analysis can fall short when addressing complex systems and policy problems. This is particularly evident in policy systems undergoing organisational and structural transformations, which are unpredictable and typically do not allow for straightforward control. Given these dynamics, there is a growing recognition that new analytical approaches are needed, approaches that can capture complex and cross-sectoral policy frameworks and governance structures affecting forests.

Using the landscape concept to demonstrate an integrated approach to framing research, Chapter 9 explored the connection between framing processes, frames, and their impacts in real-world case studies. The chapter presented three case studies, each focusing on framing effects at different operational levels – micro, meso, and macro – and how these levels related to the landscape concept:

- Framing processes should be understood as temporally and contextually specific, reflecting the dynamic nature of these processes and how they are shaped by their specific environments (Gillan, 2008).
- The case studies provided insights into horizontal and vertical framing effects, especially in implementing new practices like ecosystem service valuation and innovative participatory processes (Aggestam, 2014a, 2014b, 2015).
- Insights from the micro-level case study can inform and enhance the application of deliberative processes. Results from the meso-level case study provide valuable guidance for institutions looking to adopt new practices more effectively, and the macro-level analysis offers insights that could be used to refine re-framing techniques (Aggestam, 2014a, 2014b, 2015).

- The layered analysis facilitated a review of different framing processes, uncovering the interplay between various levels of framing and how, collectively, they can influence and shape policy outcomes (Aggestam, 2017).

The main purpose of Chapter 9 was to introduce a blueprint for an integrative policy analysis, emphasising the need for multidisciplinary approaches. The blueprint provides a practical, structured, and adaptable framework designed to help students and researchers when selecting an analytical approach. It stresses the importance of understanding the policy environment, carefully reviewing and selecting suitable methodologies and integrating different perspectives before applying analytical approaches, methods, and tools. The proposed framework comprises seven steps (see Chapter 9) and was developed to provide flexibility when tailoring approaches to meet a research project's specific demands and objectives. Intentionally structured as a non-prescriptive toolkit, it offers guidance while aiming to encourage a reflexive process in research. This adaptability is further extended to cover different analysis levels, ensuring that the blueprint can be applied across different policy domains and disciplines.

- Merging theoretical approaches with unique ontological and epistemological backgrounds can yield a more comprehensive and integrative understanding of a framing process.
- The efficacy of an integrative approach depends on the research's specific context and objectives. There are instances where an integrative approach may not be necessary, particularly in scenarios where policy issues can be more effectively tackled using a single methodology.
- Integration in research design is normative, involving value judgments dependent on specific environmental, historical, and socio-economic contexts.

3. Final thoughts

As a multifunctional and renewable natural resource, forests provide a unique case study of conflicting and synergistic priorities in natural resource management. Unlike other resources, such as metals, forests offer many ecosystem products and services, from carbon sequestration and climate change mitigation to biodiversity conservation, recreation, and timber. This diversity in FES has resulted in a complex web of competing priorities and interests between sectors and stakeholders, reflecting varied societal demands on forest resources across Europe and globally. The complexity behind multi-sectoral forest governance systems (or the human-forest nexus) highlights the need for robust and appropriate tools, methodologies, and interdisciplinary insights to analyse forest policy instruments and interests, including their effects. Moreover, it is evident that researchers in this field are not only tasked with understanding how forest ecosystems work – as a natural environment – but are commonly required to integrate socio-economic and political considerations when carrying out research. This highlights the increasing relevance of interdisciplinary knowledge and approaches in understanding the socio-ecological systems of forests.

Whether integrative or mono-disciplinary, forest policy research offers a pathway for new and innovative perspectives on forests. With this concept at its core, one objective in this book has been to demonstrate some of the theoretical and practical approaches to forest policy analysis, in this case, also emphasising and demonstrating the need for more integrative research. For example, the book focuses extensively on the wide range of forest-focused and forest-related policies and programs in Europe (Aggestam et al., 2017, Pülzl et al., 2018, Wolfslehner et al., 2020), guiding readers towards a broader understanding of forest resource use and management. A fundamental premise of this work has been the perceived need to broaden our perspectives on forest policy to address the interconnected challenges that forests currently face effectively (Kleinschmit et al., 2018, Pülzl et al., 2018, Winkel et al., 2022). In other words, this book highlights the diverse academic approaches in forest policy research and demonstrates that we need to move beyond siloed and traditional perspectives on the forest-based sector. It is argued that we should broaden our perspectives to encompass different sectors while adopting different research tools and methodologies. This can lead to more policy-relevant knowledge and a more comprehensive understanding of forest ecosystems and their role in environmental sustainability and socio-economic development.

Additionally, the rapidly changing environmental and socio-economic landscape calls for adaptive approaches in forest policy research, especially considering the complex interplay of policy instruments affecting forest-based value chains. Recent advancements in the bioeconomy and circular economy sectors underscore the need for perspectives that can account for the entire forest-based value chain, from sustainable production and processing practices to the end-use of forest products (Toppinen et al., 2020, Jarre et al., 2020). Therefore, innovation and adaptive strategies in forest policy research are crucial as they would allow for the evaluation and updating of policies in response to developments affecting the forest-based sector.

Evaluating the suitability and efficiency of tools and methodologies in generating relevant scientific insights is another relevant aspect of forest policy research. An appropriate approach should be selected based on its capacity to meet a given project's specific objectives and constraints effectively. This necessitates researchers to explore different theoretical frameworks and assess their potential to offer distinct perspectives on a particular forest policy problem. This is important as choosing a theory or approach can significantly influence the researcher's perception and understanding of forest-related issues. It is imperative, therefore, to select an analytical framework that aligns closely with the particularities of the case being studied (de Jong et al., 2012, Dunn, 2017). While the importance of selecting the most suitable framework might seem self-evident, researchers often choose approaches or theories based on personal preferences or prevailing trends rather than their applicability to specific contexts (Arts, 2012, Nagasaka et al., 2016). This tendency can lead to biased or incomplete analysis, underscoring the need for more objective and case-specific approaches in forest policy research. In other words, adopting methodologically diverse and context-specific approaches is vital to increase the practical relevance and effectiveness of the research process.

In conclusion, this book started with the idea that we must broaden our analytical scope to encompass different sectors while maintaining an open-minded approach towards the extensive range of research tools and methodologies available. Such an approach is academically rewarding and important in fostering a deeper and more cross-cutting understanding of forest ecosystems and their role in our global environment. Emphasising tailored and objective methodologies is crucial in preventing biased perspectives and ensuring that research outcomes are scientifically sound and practically valuable. As the forest policy research community addresses the continuously evolving challenges inherent to forest management and conservation, this work may guide innovative and adaptable research strategies. I express my sincere gratitude to the reader and hope this book will encourage further inquiry and progress in forest policy research.

4. References

AGGESTAM, F. 2014a. Effects of the manager's value orientation on stakeholder participation: at the front line of policy implementation. *Water Policy*, 16, 62–78.

AGGESTAM, F. 2014b. Wetland restoration and the involvement of stakeholders: An analysis based on value-perspectives. *Landscape Research*, 39, 680–697.

AGGESTAM, F. 2015. Framing the ecosystem concept through a longitudinal study of developments in science and policy. *Conservation Biology*, 29, 1052–1064.

AGGESTAM, F. 2017. Integrated framing: A micro to macro case for the landscape *Int. J. Environ. Sci. Nat. Res.*, 2, 555578.

AGGESTAM, F. 2024. Through smoke to policy: Framing the EU forest fire policy landscape. *Land*, 13(9), 1450, DOI: https://doi.org/10.3390/land13091450.

AGGESTAM, F. & GIURCA, A. 2021. The art of the 'green' deal: Policy pathways for the EU Forest Strategy. *Forest Policy and Economics*, 128, 102456.

AGGESTAM, F. & GIURCA, A. 2022. Implementing circular-bioeconomy principles across two value chains of the wood-based sector: A conceptual approach. *Land* [Online], 11.

AGGESTAM, F., KACPRZAK, A. & WOLLNIK, R. 2021. Circularity concepts in forest-based industries. Geneva: United Nations Economic Commission for Europe.

AGGESTAM, F., KONCZAL, A., SOTIROV, M., WALLIN, I., PAILLET, Y., SPINELLI, R., LINDNER, M., DERKS, J., HANEWINKEL, M. & WINKEL, G. 2020a. Can nature conservation and wood production be reconciled in managed forests? A review of driving factors for integrated forest management in Europe. *Journal of Environmental Management*, 268, 110670.

AGGESTAM, F. & PÜLZL, H. 2018. Coordinating the uncoordinated: the EU Forest Strategy. *Forests*, 9, 125, DOI: https://doi.org/10.3390/f9030125.

AGGESTAM, F. & PÜLZL, H. 2020. Downloading Europe: A regional comparison in the uptake of the EU Forest Action Plan. *Sustainability*, 12.

AGGESTAM, F., QUIROGA, S., SUÁREZ, C., MARTÍNEZ, P. & SARVASOVA, Z. 2020b. Forest ownership and the provision of wood and other forest ecosystem services. *In:* UNECE (ed.) *Who Owns Our Forests? Forest Ownership in the ECE Region.* Geneva: United Nations Economic Commission for Europe.

AGGESTAM, F., WINKEL, G., PÜLZL, H. & SOTIROV, M. 2017. The EU policy framework. *In:* WINKEL, G. (ed.) *Towards a Sustainable European Forest Based Bioeconomy – Assessment and the Way Forward. What Science Can Tell Us 8.* European Forest Institute.

ALLIANCE ENVIRONNEMENT, ORÉADE-BRÈCHE SARL, IEEP & EFI. 2018. Evaluation study of the forestry measures under Rural Development. Luxembourg: Publications Office of the European Union.

ARTS, B. 2012. Forests policy analysis and theory use: Overview and trends. *Forest Policy and Economics*, 16, 7–13.

AUGUSTO, L. & BOČA, A. 2022. Tree functional traits, forest biomass, and tree species diversity interact with site properties to drive forest soil carbon. *Nature Communications*, 13, 1097.

BEGEMANN, A., DOLRIIS, C. & WINKEL, G. 2023. Rich forests, rich people? Sustainable finance and its links to forests. *Journal of Environmental Management*, 326, 116808.

BLATTERT, C., EYVINDSON, K., HARTIKAINEN, M., BURGAS, D., POTTERF, M., LUKKARINEN, J., SNÄLL, T., TORAÑO-CAICOYA, A. & MÖNKKÖNEN, M. 2022. Sectoral policies cause incoherence in forest management and ecosystem service provisioning. *Forest Policy and Economics*, 136, 102689.

BORRASS, L., KLEINSCHMIT, D. & WINKEL, G. 2017. The 'German model' of integrative multifunctional forest management – Analysing the emergence and political evolution of a forest management concept. *Forest Policy Econ.*, 77, 16–23. https://doi.org/10.1016/j.forpol.2016.06.028.

BOSOMWORTH, K. 2015. Climate change adaptation in public policy: frames, fire management, and frame reflection. *Environment and Planning C: Government and Policy*, 33, 1450–1466.

BROWNE, J., COFFEY, B., COOK, K., MEIKLEJOHN, S. & PALERMO, C. 2019. A guide to policy analysis as a research method. *Health Promot Int*, 34, 1032–1044.

CASTELLÓ, E. & MONTAGUT, M. 2019. Framing forest fires and environmental activism: A storytelling contest about human intervention in nature. *Communication and Society*, 32, 291–306.

CEPF. 2019. *EU forests in the European Green Deal* [Online]. www.cepf-eu.org/news/eu-forests-european-green-deal.

CEPF, EUSTAFOR, COPA COGECA, CEI-BOIS, CEPI, EOS, ELO, FECOF, CEETTAR, UEF & EUROPE., B. 2019. Multifunctional European forests are crucial to reach the objectives of a European Green Deal. Brussels: http://www.cepf-eu.org/sites/default/files/Multifunctional%20European%20forests%20are%20crucial%20to%20reach%20the%20objectives%20of%20a%20European%20Green%20Deal_20191010.pdf (accessed 16 June 2020).
CHELLERI, L. & BARAVIKOVA, A. 2021. Understandings of urban resilience meanings and principles across Europe. *Cities*, 108, 102985.
CURTIS, W. J. & CICCHETTI, D. 2003. Moving research on resilience into the 21st century: theoretical and methodological considerations in examining the biological contributors to resilience. *Dev Psychopathol*, 15, 773–810.
D'AMATO, D., VEIJONAHO, S. & TOPPINEN, A. 2020. Towards sustainability? Forest-based circular bioeconomy business models in Finnish SMEs. *Forest Policy and Economics*, 110, 101848.
DAVITER, F. 2019. Policy analysis in the face of complexity: What kind of knowledge to tackle wicked problems? *Public Policy and Administration*, 34, 62–83.
DE JONG, W., ARTS, B. & KROTT, M. 2012. Political theory in forest policy science. *Forest Policy and Economics*, 16, 1–6.
DEUFFIC, P., SOTIROV, M. & ARTS, B. 2018. 'Your policy, my rationale'. How individual and structural drivers influence European forest owners' decisions. *Land Use Policy*, 79, 1024–1038. https://doi.org/10.1016/j.landusepol.2016.09.021.
DUNN, N. W. 2017. *Public Policy Analysis: An Integrated Approach*, 6th edition. New York: Routledge.
EC 2018. Forest fires: Sparking firesmart policies in the EU. *In:* FAIVRE, N. (ed.) *Research & Innovation: Projects for Policy.* Luxembourg: Publications Office of the European Union.
ECE 2020. Who owns our forests? Forest ownership in the ECE region. Geneva, Switzerland: United Nations Economic Commission for Europe.
EDWARDS, P. & KLEINSCHMIT, D. 2013. Towards a European forest policy: conflicting goals. *Forest Policy and Economics*, 33, 87–91.
FERN 2019. European Green Deal offers hope for the future of forests. www.fern.org/news-resources/european-green-deal-offers-hope-for-the-future-of-forests-2063/: FERN.
GAMSON, W. A. & MODIGLIANI, A. 1989. Media discourse and public opinion on nuclear power: A constructionist approach. *American Journal of Sociology*, 95, 1–37.
GEELS, F. W. 2021. From leadership to followership: A suggestion for interdisciplinary theorising of mainstream actor reorientation in sustainability transitions. *Environmental Innovation and Societal Transitions*, 41, 45–48.
GILLAN, K. 2008. Understanding meaning in movements: A hermeneutic approach to frames and ideologies. *Social Movement Studies*, 7, 247–263.
GLÜCK, P. 2000. Theoretical perspectives for enhancing biological diversity in forest ecosystems in Europe. *Forest Policy and Economics*, 1, 195–207.
GORDEEVA, E., WEBER, N. & WOLFSLEHNER, B. 2022. The new EU Forest Strategy for 2030: An analysis of major interests. *Forests*, 13, 1503.
HETEMÄKI, L., HANEWINKEL, M., MUYS, B., OLLIKAINEN, M., PALAHÍ, M. & TRASOBARES, A. 2017. Leading the way to a European circular bioeconomy strategy. *From Science to Policy 5.* Joensuu: European Forest Institute.
IUCN 2019. Will the European Green Deal meet its ambitious expectations? www.iucn.org/news/europe/201912/will-european-green-deal-meet-its-ambitious-expectations: International Union for Conservation of Nature and Natural Resources.
JARRE, M., PETIT-BOIX, A., PRIEFER, C., MEYER, R. & LEIPOLD, S. 2020. Transforming the bio-based sector towards a circular economy – What can we learn from wood cascading? *Forest Policy and Economics*, 110, 101872.
KHAW-NGERN, K., PEUCHTHONGLANG, P., KLOMKUL, L. & KHAW-NGERN, C. 2021. The 9Rs strategies for the circular economy 3.0. *Psychology and Education*, 58, 1440–1446.
KLEINSCHMIT, D., PÜLZL, H., SECCO, L., SERGENT, A. & WALLIN, I. 2018. Orchestration in political processes: Involvement of experts, citizens, and participatory professionals in forest policy making. *Forest Policy Econ.*, 89, 4–15. https://doi.org/10.1016/j.forpol.2017.12.011.
KONCZAL, A. A., DERKS, J., DE KONING, J. H. C. & WINKEL, G. 2023. Integrating nature conservation measures in european forest management – An exploratory study of barriers and drivers in 9 european countries. *Journal of Environmental Management*, 325, 116619.

LAZARIDOU, D. C., MICHAILIDIS, A. & TRIGKAS, M. 2021. Exploring environmental and economic costs and benefits of a forest-based circular economy: A literature review. *Forests*, 12, 436.
LIER, M., KÖHL, M., KORHONEN, K. T., LINSER, S. & PRINS, K. 2021. Forest relevant targets in EU policy instruments – can progress be measured by the pan-European criteria and indicators for sustainable forest management? *Forest Policy and Economics*, 128, 102481.
MOURAO, P. R. & MARTINHO, V. D. 2019. Forest fire legislation: Reactive or proactive? *Ecological Indicators*, 104, 137–144.
NAGASAKA, K., BÖCHER, M. & KROTT, M. 2016. Are forest researchers only scientists? Case studies on the roles of researchers in Japanese and Swedish forest policy processes. *Forest Policy and Economics*, 70, 147–154.
NILSSON, S. & ENANDER, A. 2020. 'Damned if you do, damned if you don't': Media frames of responsibility and accountability in handling a wildfire. *Journal of Contingencies and Crisis Management*, 28, 69–82.
PIANTA, M. & LUCCHESE, M. 2020. Rethinking the European Green Deal: An industrial policy for a just transition in Europe. *Review of Radical Political Economics*, 52, 633–641.
PÜLZL, H., WYDRA, D. & HOGL, K. 2018. Piecemeal integration: Explaining and understanding 60 years of European Union forest policymaking. *Forests*, 9, 719.
REED, J., BARLOW, J., CARMENTA, R., VAN VIANEN, J. & SUNDERLAND, T. 2019. Engaging multiple stakeholders to reconcile climate, conservation and development objectives in tropical landscapes. *Biological Conservation*, 238, 108229.
RIVERA LEÓN, L., BOUGAS, K., AGGESTAM, F., PÜLZL, H., ZOBOLI, E., RAVET, J., GRINIECE, E., VERMEER, J., MAROULIS, N., ETTWEIN, F., VAN BRUSSELEN J. & GREEN, T. 2016. An assessment of the cumulative cost impact of specified EU legislation and policies on the EU forest-based industries. Brussels: DG GROW.
SABATIER, P. A. 1988. An advocacy coalition framework of policy change and the role of policy-oriented learning therein. *Policy Sciences*, 21, 129–168.
SECCO, L., WALLIN, I., SERGENT, A., PÜLZL, H. & KLEINSCHMIT, D. 2016. Orchestrating forest policymaking: involvement of scientists & stakeholders in political processes in Europe *Natural Resources and Bioeconomy Studies*, 41, 72–74.
SEIJO, F. 2009. Who framed the forest fire? State framing and peasant counter-framing of anthropogenic forest fires in Spain since 1940. *Journal of Environmental Policy & Planning*, 11, 103–128.
SEWERIN, S., BÉLAND, D. & CASHORE, B. 2020. Designing policy for the long term: agency, policy feedback and policy change. *Policy Sciences*, 53, 243–252.
SIDDI, M. 2020. The European Green Deal: Assessing its current state and future implementation. *FIIA Working Paper 114*. Helsinki: Finnish Institute of International Affairs.
SOTIROV, M., BLUM, M., STORCH, S., SELTER, A. & SCHRAML, U. 2017. Do forest policy actors learn through forward-thinking? Conflict and cooperation relating to the past, present and futures of sustainable forest management in Germany. *Forest Policy Econ.*, 85, 256–268. https://doi.org/10.1016/j.forpol.2016.11.011.
TOPPINEN, A., D'AMATO, D. & STERN, T. 2020. Forest-based circular bioeconomy: matching sustainability challenges and novel business opportunities? *Forest Policy and Economics*, 110, 102041.
USTAOGLU, E. & COLLIER, M. J. 2018. Farmland abandonment in Europe: an overview of drivers, consequences, and assessment of the sustainability implications. *Environmental Reviews*, 26, 396–416.
VAN GORP, B. 2007. The constructionist approach to framing: Bringing culture back in. *Journal of Communication*, 57, 60–78.
WEISS, G., LAWRENCE, A., HUJALA, T., LIDESTAV, G., NICHIFOREL, L., NYBAKK, E., QUIROGA, S., SARVAŠOVÁ, Z., SUAREZ, C. & ŽIVOJINOVIĆ, I. 2019. Forest ownership changes in Europe: State of knowledge and conceptual foundations. *Forest Policy Econ.*, 99, 9–20. https://doi.org/10.1016/j.forpol.2018.03.003.
WILKINSON, C. 2011. Social-ecological resilience: Insights and issues for planning theory. *Planning Theory*, 11, 148–169.
WINKEL, G. & LEIPOLD, S. 2016. Demolishing dikes: Multiple streams and policy discourse analysis. *Policy Studies Journal*, 44, 108–129.
WINKEL, G., LOVRIĆ, M., MUYS, B., KATILA, P., LUNDHEDE, T., PECURUL, M., PETTENELLA, D., PIPART, N., PLIENINGER, T., PROKOFIEVA, I., PARRA, C., PÜLZL, H., ROITSCH, D., ROUX, J.-L., THORSEN, B. J., TYRVÄINEN, L., TORRALBA, M., VACIK, H., WEISS, G. & WUNDER,

S. 2022. Governing Europe's forests for multiple ecosystem services: Opportunities, challenges, and policy options. *Forest Policy and Economics*, 145, 102849.

WOLFSLEHNER, B., PÜLZL, H., KLEINSCHMIT, D., AGGESTAM, F., WINKEL, G., CANDEL, J., ECKERBERG, K., FEINDT, P., MCDERMOTT, C., SECCO, L., SOTIROV, M., LACKNER, M. & LAZYA ROUX, J. 2020. European forest governance post-2020. *From Science to Policy 10.* Joensuu, Finland: European Forest Institute. https://doi.org/10.36333/fs10

ZAHARIADIS, N. 2018. The multiple streams framework: Structure, limitations, prospects. *In:* WEIBLE, M. C. & SABATIER, P. A. (eds.), *Theories of the Policy Process*, 4th edition. New York: Routledge.

ŽIVOJINOVIĆ, I., WEISS, G., LIDESTAV, G., FELICIANO, D., HUJALA, T., DOBŠINSKÁ, Z., LAWRENCE, A., NYBAKK, E., QUIROGA, S. & SCHRAML, U. 2015. Forest land ownership change in Europe. COST Action FP1201 FACESMAP Country Reports, Joint Volume. *EFICEEC-EFISEE Research Report.* Vienna: University of Natural Resources and Life Sciences.

Index

FSC
www.fsc.org
MIX
Papier | Fördert gute Waldnutzung
FSC® C083411

Kolibri 360 GmbH
Ferdinand-Jühlke-Straße 7
99095 Erfurt, Deutschland
produktsicherheit@kolibri360.de